ERGEBNISSE DER
ANGEWANDTEN MATHEMATIK

UNTER MITWIRKUNG DER SCHRIFTLEITUNG DES
„ZENTRALBLATT FÜR MATHEMATIK“

HERAUSGEGEBEN VON
F. L. BAUER · L. COLLATZ · F. LÖSCH · C. TRUESDELL

8

ANTIPLANE ELASTIC SYSTEMS

BY

L. M. MILNE-THOMSON

WITH 53 FIGURES

1962

NEW YORK
ACADEMIC PRESS INC. PUBLISHERS

BERLIN · GÖTTINGEN · HEIDELBERG
SPRINGER-VERLAG

SPRINGER-VERLAG
BERLIN · GÖTTINGEN · HEIDELBERG

Published in USA and Canada by

ACADEMIC PRESS INC., PUBLISHERS

111 Fifth Avenue, New York 3, New York

ISBN 978-3-540-02805-5 ISBN 978-3-642-85627-3 (eBook)
DOI 10.1007/978-3-642-85627-3

Library of Congress Catalog Card Number 62-14 930

SOFTCOVER REPRINT OF THE HARDCOVER 1ST EDITION 1962

Preface

The term antiplane was introduced by L. N. G. Filon to describe such problems as tension, push, bending by couples, torsion, and flexure by a transverse load. Looked at physically these problems differ from those of plane elasticity already treated* in that certain shearing stresses no longer vanish.

This book is concerned with antiplane elastic systems in equilibrium or in steady motion within the framework of the linear theory, and is based upon lectures given at the Royal Naval College, Greenwich, to officers of the Royal Corps of Naval Constructors, and on technical reports recently published at the Mathematics Research Center, United States Army.

My aim has been to tackle each problem, as far as possible, by direct rather than inverse or guessing methods. Here the complex variable again assumes an important role by simplifying equations and by introducing order into much of the treatment of anisotropic material.

The work begins with an introduction to tensors by an intrinsic method which starts from a new and simple definition. This enables elastic properties to be stated with conciseness and physical clarity. This course in no way commits the reader to the exclusive use of tensor calculus, for the structure so built up merges into a more familiar form. Nevertheless it is believed that the tensor methods outlined here will prove useful also in other branches of applied mathematics.

It is found to be just as simple, and certainly more instructive, to develop results for general anisotropic materials in Chapters I and II, and then to apply them to isotropic bodies in Chapters III, IV and V. Chapters VI and VII deal with anisotropic bodies and it is thought that by simple general applications the reader will be led to a clear pathway through what has often been regarded as a jungle of unrelated particular results.

Each chapter is followed by a set of examples, there are 169 in all, which in many cases amplify the text.

Although the treatment takes a theoretical form, it is directly adapted to practical physical and engineering applications.

The references to literature are mainly directed to elucidating the text. No attempt whatever has been made to achieve even the semblance of completeness or to settle questions of priority.

---

* L. M. Milne-Thomson: "Plane Elastic Systems" Springer (1960).

The sections of the book are numbered in decimal notation; the integer preceding the decimal point gives the number of the chapter. Thus the section numbered **7.41** belongs to Chapter VII and occurs later than section **7.4** but earlier than section **7.42**. Reference to the diagrams is greatly facilitated by giving them the number of the section to which they belong.

It is my privilege to express my lively thanks to my friend, Dr W. M. Shepherd, Professor of Theoretical Mechanics in the University of Bristol for again undertaking the arduous task of proof reading. To him I owe a great debt of gratitude for this friendly act and for the valuable suggestions for improvement which he has made. My colleague Dr D. Trifan has performed the same kindly office and to him also I extend my warmest thanks.

L. M. Milne-Thomson

Mathematics Department

The University of Arizona
Tucson, Arizona

1 October, 1961

Contents

Chapter I

The law of elasticity

In this introduction we shall be concerned with a simple approach to tensor algebra, and the tensor formulation of stress, deformation and energy, leading to HOOKE's law, the properties of anisotropy, and considerations of elastic symmetry.

1.1. Continued dyadic products

Consider a scalar λ, vectors $\boldsymbol{a}, \boldsymbol{b}, \boldsymbol{c}, \ldots$, and an operation, at present undefined, denoted by a semicolon (;) called *dyadic multiplication*. We can then write down the sequence of terms

$$\lambda, \quad \boldsymbol{a}, \quad \boldsymbol{a}; \boldsymbol{b}, \quad \boldsymbol{a}; \boldsymbol{b}; \boldsymbol{c}, \quad \boldsymbol{a}; \boldsymbol{b}; \boldsymbol{c}; \boldsymbol{d}, \ldots \tag{1}$$

which we shall call *continued dyadic products* of ranks 0, 1, 2, 3, 4, ... respectively. Observe that the rank corresponds with the number of vectors in the product.

The particular product of two vectors, $\boldsymbol{a}; \boldsymbol{b}$, we shall call a *dyad*.

We proceed to assign meaning to the operation (;) by setting down four laws of operation which must be obeyed. No interpretations will be admitted except those deducible from the use of these laws.

I. *The associative law*. A continued dyadic product can be bracketed in any manner without change of meaning.

Thus for example,

$$\boldsymbol{a}; \boldsymbol{b}; \boldsymbol{c} = (\boldsymbol{a}; \boldsymbol{b}); \boldsymbol{c} = \boldsymbol{a}; (\boldsymbol{b}; \boldsymbol{c}), \tag{2}$$

$$\boldsymbol{a}; \boldsymbol{b}; \boldsymbol{c}; \boldsymbol{d} = (\boldsymbol{a}; \boldsymbol{b}); (\boldsymbol{c}; \boldsymbol{d}) = \boldsymbol{a}; (\boldsymbol{b}; \boldsymbol{c}); \boldsymbol{d}. \tag{3}$$

II. *The scalar law*. If in a continued dyadic product we replace a vector by a scalar, one semicolon adjacent to the scalar must be suppressed.

For example replacing $\boldsymbol{b}$ in (2) by the scalar λ we get

$$\boldsymbol{a}; \lambda; \boldsymbol{c} = \boldsymbol{a}\lambda; \boldsymbol{c} = \boldsymbol{a}; \lambda\boldsymbol{c}. \tag{4}$$

Similarly from $(\boldsymbol{b}; \boldsymbol{a}); \boldsymbol{c} = \boldsymbol{b}; (\boldsymbol{a}; \boldsymbol{c})$ we get

$$\lambda\boldsymbol{a}; \boldsymbol{c} = \lambda(\boldsymbol{a}; \boldsymbol{c}). \tag{5}$$

III. *The contraction law*. In an equality of continued dyadic products we can replace dyadic multiplication (;) by scalar multiplication (·) in one and the same position on both sides of the equality.

Thus from (2), choosing the last semicolon, we get

$$(\boldsymbol{a};\boldsymbol{b})\cdot\boldsymbol{c}=\boldsymbol{a};(\boldsymbol{b}\cdot\boldsymbol{c})=\boldsymbol{a}(\boldsymbol{b}\cdot\boldsymbol{c}) \tag{6}$$

on use of the scalar law. Thus the scalar product of a dyad and a vector is a vector.

This process, called *contraction*, lowers the rank of the original product by two units.

Contracting (3) with respect to the middle semicolon we get

$$(\boldsymbol{a};\boldsymbol{b})\cdot(\boldsymbol{c};\boldsymbol{d})=\boldsymbol{a};(\boldsymbol{b}\cdot\boldsymbol{c});\boldsymbol{d}=(\boldsymbol{a};\boldsymbol{d})\,(\boldsymbol{b}\cdot\boldsymbol{c}) \tag{7}$$

on using the scalar law. Thus the scalar product of two dyads is a dyad.

Again consider the identity

$$(\boldsymbol{a};\boldsymbol{b};\boldsymbol{c};\boldsymbol{d});(\boldsymbol{p};\boldsymbol{q};\boldsymbol{r})=\boldsymbol{a};\boldsymbol{b};\boldsymbol{c};(\boldsymbol{d};\boldsymbol{p});\boldsymbol{q};\boldsymbol{r}\,. \tag{8}$$

Contracting with respect to the 4th semicolon and using the scalar law we get

$$(\boldsymbol{a};\boldsymbol{b};\boldsymbol{c};\boldsymbol{d})\cdot(\boldsymbol{p};\boldsymbol{q};\boldsymbol{r})=(\boldsymbol{a};\boldsymbol{b};\boldsymbol{c};\boldsymbol{q};\boldsymbol{r})\,(\boldsymbol{d}\cdot\boldsymbol{p})\,. \tag{9}$$

A second contraction of the right-hand sides of (7) and (9) with respect to the semicolon which follows $\boldsymbol{a}$ in the right-hand side of (7) and with respect to the semicolon which follows $\boldsymbol{c}$ in the right-hand side of (9) gives the *twice contracted* or double *scalar products* of these factors written

$$(\boldsymbol{a};\boldsymbol{b})\cdot\cdot(\boldsymbol{c};\boldsymbol{d})=(\boldsymbol{a}\cdot\boldsymbol{d})\,(\boldsymbol{b}\cdot\boldsymbol{c})\,, \tag{10}$$

$$(\boldsymbol{a};\boldsymbol{b};\boldsymbol{c};\boldsymbol{d})\cdot\cdot(\boldsymbol{p};\boldsymbol{q};\boldsymbol{r})=(\boldsymbol{a};\boldsymbol{b};\boldsymbol{r})\,(\boldsymbol{c}\cdot\boldsymbol{q})\,(\boldsymbol{d}\cdot\boldsymbol{p})\,. \tag{11}$$

Thus double contraction lowers the rank of the original continued dyadic product by 4 units.

The double scalar product (10) of two dyads is particularly important. Since the scalar product of two vectors is independent of the order, we have

$$(\boldsymbol{a}\cdot\boldsymbol{d})\,(\boldsymbol{b}\cdot\boldsymbol{c})=(\boldsymbol{c}\cdot\boldsymbol{b})\,(\boldsymbol{a}\cdot\boldsymbol{d})=(\boldsymbol{b}\cdot\boldsymbol{c})\,(\boldsymbol{a}\cdot\boldsymbol{d})=(\boldsymbol{d}\cdot\boldsymbol{a})\,(\boldsymbol{c}\cdot\boldsymbol{b}) \tag{12}$$

Therefore from (10)

$$(\boldsymbol{a};\boldsymbol{b})\cdot\cdot(\boldsymbol{c};\boldsymbol{d})=(\boldsymbol{c};\boldsymbol{d})\cdot\cdot(\boldsymbol{a};\boldsymbol{b})=(\boldsymbol{b};\boldsymbol{a})\cdot\cdot(\boldsymbol{d};\boldsymbol{c})=(\boldsymbol{d};\boldsymbol{c})\cdot\cdot(\boldsymbol{b};\boldsymbol{a})\,. \tag{13}$$

If in a dyad we reverse the order of the vectors which compose it, we obtain the *conjugate* dyad, denoted by suffix c. Thus

$$(\boldsymbol{a};\boldsymbol{b})_c=(\boldsymbol{b};\boldsymbol{a})\,,\qquad (\boldsymbol{b};\boldsymbol{a})_c=(\boldsymbol{a};\boldsymbol{b})\,, \tag{14}$$

so that the conjugate of the conjugate is the original dyad. We then see from (13) that

$$(\boldsymbol{a};\boldsymbol{b})\cdot\cdot(\boldsymbol{c};\boldsymbol{d})=(\boldsymbol{c};\boldsymbol{d})\cdot\cdot(\boldsymbol{a};\boldsymbol{b})=(\boldsymbol{a};\boldsymbol{b})_c\cdot\cdot(\boldsymbol{c};\boldsymbol{d})_c=(\boldsymbol{c};\boldsymbol{d})_c\cdot\cdot(\boldsymbol{a};\boldsymbol{b})_c \tag{15}$$

that is to say double scalar multiplication of two dyads is completely commutative whether we use both the originals or both their conjugates.

IV. *The distributive law.* Let B and C denote continued dyadic products of the same rank, and let A be any continued dyadic product. Then

$$\mathsf{A};(\mathsf{B}+\mathsf{C}) = \mathsf{A};\mathsf{B} + \mathsf{A};\mathsf{C}, \qquad (\mathsf{B}+\mathsf{C});\mathsf{A} = \mathsf{B};\mathsf{A} + \mathsf{C};\mathsf{A}. \quad (16)$$

From the distributive law it follows that any continued dyadic product A is a *linear vector operator* in the sense that, $\boldsymbol{x}$ and $\boldsymbol{y}$ being arbitrary vectors,

$$\mathsf{A}\cdot(\boldsymbol{x}+\boldsymbol{y}) = \mathsf{A}\cdot\boldsymbol{x} + \mathsf{A}\cdot\boldsymbol{y}, \qquad (\boldsymbol{x}+\boldsymbol{y})\cdot\mathsf{A} = \boldsymbol{x}\cdot\mathsf{A} + \boldsymbol{y}\cdot\mathsf{A}. \quad (17)$$

1.12. Definition of a tensor

A linear vector operator $\mathsf{T}^{(r)}$ is called a tensor of rank r if $\mathsf{T}^{(0)}$ is a scalar and if, for every positive integer $r \geqq 1$ and for an arbitrary vector $\boldsymbol{x}$, $\mathsf{T}^{(r)}\boldsymbol{x}$ is a tensor of rank $r-1$. [MILNE-THOMSON (2)].

Thus a given vector $\boldsymbol{a}$ is a tensor of rank 1, for $\boldsymbol{a}\cdot\boldsymbol{x}$ is a scalar i.e. a tensor of rank zero.

A dyad $\boldsymbol{a};\boldsymbol{b}$ is a tensor of rank 2, for $(\boldsymbol{a};\boldsymbol{b})\cdot\boldsymbol{x} = \boldsymbol{a}(\boldsymbol{b}\cdot\boldsymbol{x})$ which is a vector i.e. a tensor of rank 1.

In this way we show that a continued dyadic product of rank r is a tensor, so that tensors of all ranks exist.

For brevity we call a tensor of rank r an r-tensor. It should be observed that while we have proved that a dyad $\boldsymbol{a};\boldsymbol{b}$ is a 2-tensor, the converse that every 2-tensor can be put in the form $\boldsymbol{a};\boldsymbol{b}$ is false.

A particularly important 2-tensor is *the unit* 2 *tensor* or *idemfactor* I which has the property that for an arbitrary vector $\boldsymbol{x}$,

$$\boldsymbol{x} = \mathsf{I}\cdot\boldsymbol{x} = \boldsymbol{x}\cdot\mathsf{I}. \quad (1)$$

That the idemfactor actually exists is most simply seen from the representation of it in terms of a triply orthogonal system of unit vectors $\boldsymbol{i}_1$, $\boldsymbol{i}_2$, $\boldsymbol{i}_3$, in the case of 3-space, namely

$$\mathsf{I} = \boldsymbol{i}_1;\boldsymbol{i}_1 + \boldsymbol{i}_2;\boldsymbol{i}_2 + \boldsymbol{i}_3;\boldsymbol{i}_3. \quad (2)$$

Since we can write any vector $\boldsymbol{x}$ in the form

$$\boldsymbol{x} = x_1\boldsymbol{i}_1 + x_2\boldsymbol{i}_2 + x_3\boldsymbol{i}_3 \quad (3)$$

(1) follows at once.

1.14. Properties of tensors

We now prove that every 2-tensor can be expressed as the sum of at most N dyads where N is the dimension-number of the space. For simplicity consider Euclidean 3-space.

Proof. From 1.12 since a 2-tensor $\mathsf{T}^{(2)}$ is a linear vector operator

$$\begin{aligned}\mathsf{T}^{(2)}\boldsymbol{x} &= \mathsf{T}^{(2)}x_1\boldsymbol{i}_1 + \mathsf{T}^{(2)}x_2\boldsymbol{i}_2 + \mathsf{T}^{(2)}x_3\boldsymbol{i}_3\\ &= \mathsf{T}^{(2)}\boldsymbol{i}_1(\boldsymbol{i}_1\cdot\boldsymbol{x}) + \mathsf{T}^{(2)}\boldsymbol{i}_2(\boldsymbol{i}_2\cdot\boldsymbol{x}) + \mathsf{T}^{(2)}\boldsymbol{i}_3(\boldsymbol{i}_3\cdot\boldsymbol{x}),\end{aligned}$$

where by definition of a tensor $\mathsf{T}^{(2)}\boldsymbol{i}_1$, $\mathsf{T}^{(2)}\boldsymbol{i}_2$, $\mathsf{T}^{(2)}\boldsymbol{i}_3$ are vectors. Therefore

$$\mathsf{T}^{(2)}\boldsymbol{x} = [(\mathsf{T}^{(2)}\boldsymbol{i}_1)\,;\,\boldsymbol{i}_1 + (\mathsf{T}^{(2)}\boldsymbol{i}_2)\,;\,\boldsymbol{i}_2 + (\mathsf{T}^{(2)}\boldsymbol{i}_3)\,;\,\boldsymbol{i}_3]\cdot\boldsymbol{x}\,.$$

But $\boldsymbol{x}$ is an arbitrary vector. Therefore

$$\mathsf{T}^{(2)} = (\mathsf{T}^{(2)}\boldsymbol{i}_1)\,;\,\boldsymbol{i}_1 + (\mathsf{T}^{(2)}\boldsymbol{i}_2)\,;\,\boldsymbol{i}_2 + (\mathsf{T}^{(2)}\boldsymbol{i}_3)\,;\,\boldsymbol{i}_3 \tag{1}$$

Q. E. D.

We can rewrite the above proof to show that

$$\mathsf{T}^{(2)} = \boldsymbol{i}_1;(\boldsymbol{i}_1\mathsf{T}^{(2)}) + \boldsymbol{i}_2;(\boldsymbol{i}_2\mathsf{T}^{(2)}) + \boldsymbol{i}_3;(\boldsymbol{i}_3\mathsf{T}^{(2)})\,, \tag{2}$$

and since the proof also shows that $\mathsf{T}^{(2)}$ operates by scalar multiplication, we can put (1), (2) together in the form

$$\mathsf{T}^{(2)}\cdot\mathsf{I} = \mathsf{I}\cdot\mathsf{T}^{(2)} = \mathsf{T}^{(2)}. \tag{3}$$

More generally the above proof can be adapted to show that in N-space a tensor of rank r can be expressed as the sum of at most N^{r-1} continued dyadic products each of rank r, and that

$$\mathsf{T}^{(r)}\cdot\mathsf{I} = \mathsf{I}\cdot\mathsf{T}^{(r)} = \mathsf{T}^{(r)} \tag{4}$$

so that the idemfactor indeed behaves like a unit in regard to scalar multiplication with a tensor.

Since tensors can be expressed as sums of continued dyadic products, by distributing the products we can use 1.1 (11) to define the double scalar product $\mathsf{T}^{(4)}\cdot\cdot\,\mathsf{T}^{(3)}$ of two tensors of the 4th and 3rd ranks or more generally $\mathsf{T}^{(r)}\cdot\cdot\,\mathsf{T}^{(s)}$, $(r, s \geqq 2)$ for an r-tensor and an s-tensor. In the case of a pair of 2-tensors T, T' the double scalar product is a scalar and the multiplication is commutative

$$\mathsf{T}\cdot\cdot\,\mathsf{T}' = \mathsf{T}'\cdot\cdot\,\mathsf{T} = \mathsf{T}_c\cdot\cdot\,\mathsf{T}'_c = \mathsf{T}'_c\cdot\cdot\,\mathsf{T}_c\,, \tag{5}$$

where the *conjugate tensor* T_c is such that

$$\mathsf{T}\cdot\boldsymbol{x} = \boldsymbol{x}\cdot\mathsf{T}_c\,,$$

and could for example be obtained from an expression

$$\mathsf{T} = (\boldsymbol{a};\boldsymbol{b}) + (\boldsymbol{c};\boldsymbol{d}) + (\boldsymbol{e};\boldsymbol{f}) \tag{6}$$

by reversing the vectors to give, cf. 1.1 (14),

$$\mathsf{T}_c = (\boldsymbol{b};\boldsymbol{a}) + (\boldsymbol{d};\boldsymbol{c}) + (\boldsymbol{f};\boldsymbol{e})\,. \tag{7}$$

A 2-tensor T is said to be *symmetric* if

$$\mathsf{T} = \mathsf{T}_c\,, \tag{8}$$

and *antisymmetric* or *skew* if

$$\mathsf{T} = -\,\mathsf{T}_c\,. \tag{9}$$

The identity

$$\mathsf{T} = \frac{1}{2}(\mathsf{T} + \mathsf{T}_c) + \frac{1}{2}(\mathsf{T} - \mathsf{T}_c) \tag{10}$$

shows that an arbitrary 2-tensor can be expressed as the sum of a symmetric tensor $\frac{1}{2}(\mathsf{T}+\mathsf{T}_c)$ and an antisymmetric tensor $\frac{1}{2}(\mathsf{T}-\mathsf{T}_c)$.

Let S be a symmetric and A an antisymmetric 2-tensor. Then from (5)

$$\mathsf{S}\cdot\cdot\,\mathsf{A}=\mathsf{S}_c\cdot\cdot\,\mathsf{A}_c=\mathsf{S}\cdot\cdot\,(-\mathsf{A})=-(\mathsf{S}\cdot\cdot\,\mathsf{A}),$$

and therefore $\mathsf{S}\cdot\cdot\,\mathsf{A}=0$. Thus *the double scalar product of a symmetric and an antisymmetric 2-tensor is zero.*

It follows that, if S is a symmetric and T is a general 2-tensor,

$$\mathsf{S}\cdot\cdot\,\mathsf{T}=\mathsf{S}\cdot\cdot\,\frac{1}{2}(\mathsf{T}+\mathsf{T}_c), \tag{11}$$

where we have used (10).

The double scalar product of a 2-tensor T with the idemfactor I is called the *first scalar invariant* of T, written

$$\mathsf{T}_I=\mathsf{T}\cdot\cdot\,\mathsf{I}=\mathsf{T}_c\cdot\cdot\,\mathsf{I}. \tag{12}$$

We can therefore obtain T_I from (6) in the form

$$\mathsf{T}_I=\boldsymbol{a}\cdot\boldsymbol{b}+\boldsymbol{c}\cdot\boldsymbol{d}+\boldsymbol{e}\cdot\boldsymbol{f}. \tag{13}$$

1.2. The stress tensor

Consider a tetrahedron, centroid P, of continuous material, isolated in thought from the material which surrounds it. Let the lengths of the edges be infinitesimal of order l.

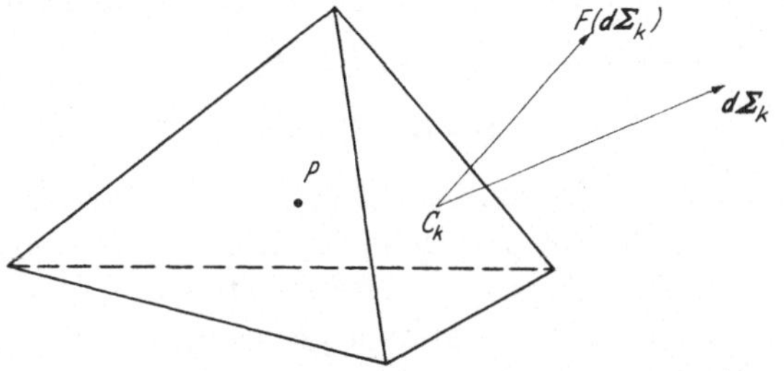

Fig. 1.2

Let the vector areas of the four faces directed along the outward normals be $d\boldsymbol{\Sigma}_k$, $k=1, 2, 3, 4$.

Then by projection on an arbitrary plane we have

$$d\boldsymbol{\Sigma}_1+d\boldsymbol{\Sigma}_2+d\boldsymbol{\Sigma}_3+d\boldsymbol{\Sigma}_4=0. \tag{1}$$

Hypothesis. The force exerted on any face k by the material on that side of the face into which $d\boldsymbol{\Sigma}_k$ points is a vector $F(d\boldsymbol{\Sigma}_k)$ which is a function of the vector area $d\boldsymbol{\Sigma}_k$ of the face, and acts at the centroid C_k of the face.

From this hypothesis it follows by the law of action and reaction that

$$F(-\,d\boldsymbol{\Sigma}_k) = -\,F(d\boldsymbol{\Sigma}_k)\,, \tag{2}$$

so that F is an odd function.

We can neglect the inertial forces which are proportional to the volume (of order l^3) in comparison with the surface forces which are proportional to the area (of order l^2). Therefore the surface forces form a system in equilibrium, that is to say

$$F(d\boldsymbol{\Sigma}_1) + F(d\boldsymbol{\Sigma}_2) + F(d\boldsymbol{\Sigma}_3) + F(d\boldsymbol{\Sigma}_4) = 0\,. \tag{3}$$

Therefore

$$\begin{aligned} F(d\boldsymbol{\Sigma}_1) + F(d\boldsymbol{\Sigma}_2) + F(d\boldsymbol{\Sigma}_3) &= -\,F(d\boldsymbol{\Sigma}_4) = F(-\,d\boldsymbol{\Sigma}_4) \\ &= F(d\boldsymbol{\Sigma}_1 + d\boldsymbol{\Sigma}_2 + d\boldsymbol{\Sigma}_3)\,, \end{aligned}$$

on using (2) and then (1).

Thus the vector $F(d\boldsymbol{\Sigma})$ is a *linear function* of $d\boldsymbol{\Sigma}$ so that we can write, the dot denoting the scalar product,

$$F(d\boldsymbol{\Sigma}) = d\boldsymbol{\Sigma}\cdot \mathsf{S}\,,$$

where S is a tensor of the second rank, called the *stress tensor* at P.

The vector $d\boldsymbol{\Sigma}\cdot\mathsf{S}$ is the force exerted on the directed element of area $d\boldsymbol{\Sigma}$. Let $d\boldsymbol{\Sigma} = \boldsymbol{n}\,d\Sigma$, where $\boldsymbol{n}$ is the unit normal vector. The force per unit area $\boldsymbol{n}\cdot\mathsf{S}$ is called the *stress vector* for an infinitesimal area whose unit normal is $\boldsymbol{n}$.

Let $\boldsymbol{i}_h$ be a unit vector in the direction h. The component of the stress vector in this direction is denoted by $\widehat{nh}$, where

$$\widehat{nh} = (\boldsymbol{n}\cdot\mathsf{S})\cdot\boldsymbol{i}_h\,,$$

so that in words; $\widehat{nh}$ is the component of stress in the h-direction across an element of area which is perpendicular to the n-direction.

Thus, for example, the components of stress referred to cartesian axes x, y, z across an element of area perpendicular to the x-axis will be $\widehat{xx}$, $\widehat{xy}$, $\widehat{xz}$ in the x-, y-, z-directions respectively.

1.22. Directions of principal stress

Let S be the stress tensor at the point P whose position vector is $\boldsymbol{r}$ and let any point Q near to P have the position vector $\boldsymbol{r}+\boldsymbol{\eta}$. Then $\boldsymbol{\eta}$ is the position vector of Q referred to P as origin.

The locus

$$\boldsymbol{\eta}\cdot\mathsf{S}\cdot\boldsymbol{\eta} = K\,, \tag{1}$$

where K is a constant is a surface of the second degree. It is indeed a central quadric with P as centre, since if $\boldsymbol{\eta}$ is a point on it, so is $-\,\boldsymbol{\eta}$.

The quadric (1) is called a *stress quadric* at P. The stress quadrics obtained by attributing to K different constant values of the same sign are homothetic.

If we put

$$\boldsymbol{\eta} = \xi_\alpha \boldsymbol{i}_\alpha, \qquad \mathsf{S} = \widehat{x_\beta x_\gamma}(\boldsymbol{i}_\beta;\boldsymbol{i}_\gamma), \tag{2}$$

where $\boldsymbol{i}_1$, $\boldsymbol{i}_2$, $\boldsymbol{i}_3$ are mutually orthogonal unit vectors corresponding to cartesian coordinates (ξ_1, ξ_2, ξ_3) at P, the stress quadric becomes

$$\xi_\alpha \xi_\beta \cdot \widehat{x_\alpha x_\beta} = K, \tag{3}$$

so that the coefficient of $\xi_\alpha \xi_\beta$ is the corresponding stress component $\widehat{x_\alpha x_\beta}$ at P. From the definition (1) this property is clearly invariant for rotation of the axes of reference about P.

Referred to principal axes of the quadric its equation will therefore be of the form

$$\eta_1^2 \widehat{XX} + \eta_2^2 \widehat{YY} + \eta_3^2 \widehat{ZZ} = K \tag{4}$$

in cartesian coordinates (η_1, η_2, η_3).

This shows that, at any point P of the material, there are, in general, three mutually orthogonal planes, namely the principal planes of the stress quadratic at P, such that the stress across each of them is purely normal. These planes are called the *principal planes of stress* at P, and their normals, namely the principal axes of the stress quadric, are called the *directions of principal stress* at P, while the corresponding stress components $\widehat{XX}$, $\widehat{YY}$, $\widehat{ZZ}$ are the *principal stresses* at P.

A line such that its tangent at every point is in a direction of principal stress at that point is called a *line of principal stress.*

If the principal stresses at P all have the same sign, the stress quadric is an ellipsoid and the state of stress at P is tension or compression according as the sign is positive or negative.

If one principal stress differs in sign from the other two, the stress quadric is a hyperboloid of one or two sheets according to the sign pattern. Thus for example if $\widehat{XX}$ alone is negative, the stress quadric is

$$-\eta_1^2 |\widehat{XX}| + \eta_2^2 \widehat{YY} + \eta_3^2 \widehat{ZZ} = K, \tag{5}$$

which is a hyperboloid of one sheet when K is positive and a hyperboloid of two sheets when K is negative.

The asymptotic cone

$$-\eta_1^2 |\widehat{XX}| + \eta_2^2 \widehat{YY} + \eta_3^2 \widehat{ZZ} = 0 \tag{6}$$

separates the material near P into two regions such that the stress across a plane normal to a ray through P is tension or compression according to the region in which the ray lies.

A point at which the principal stresses are equal is called an *isotropic point.*

At an isotropic point the stress quadric is a sphere and near P there is a state of hydrostatic stress namely all round tension or all round compression.

1.3. The deformation tensor

Consider the particles of deformable material which lie on a line C_0 at time t_0, the first state. At time t the same particles will occupy a line C, the second state.

Let P_0, P_0' be two neighbouring particles on C_0 at time t_0; P, P' the *same* particles on C at time t. We shall denote the position vectors of P_0, P by $\boldsymbol{P}_0$, $\boldsymbol{P}$ respectively and write

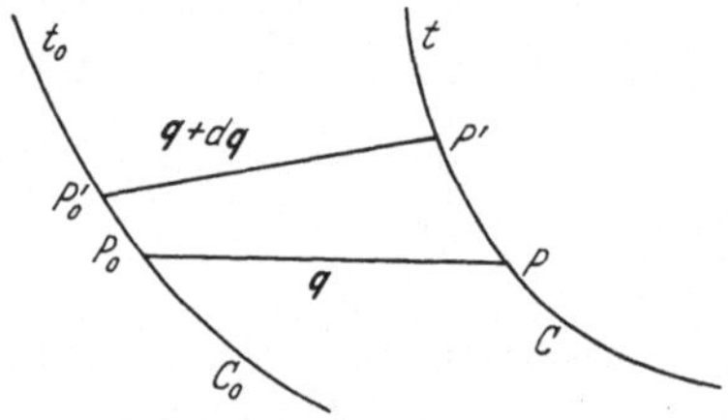

Fig. 1.3

$$\overrightarrow{P_0P_0'} = d\boldsymbol{P}_0, \quad \overrightarrow{PP'} = d\boldsymbol{P}, \tag{1}$$

where $d\boldsymbol{P}$ will be supposed infinitesimal.

Consider the scalar quantity

$$d\boldsymbol{P}\cdot d\boldsymbol{P} - d\boldsymbol{P}_0\cdot d\boldsymbol{P}_0. \tag{2}$$

If the material is rigid, the distance between the same two particles is invariant in time and (2) vanishes.

If the material is deformable, we take (2) as the measure of the deformation. Then

$$d\boldsymbol{P}_0 = d\boldsymbol{P}\cdot\frac{\partial;\boldsymbol{P}_0}{\partial\boldsymbol{P}} =: \frac{\boldsymbol{P}_0;\partial}{\partial\boldsymbol{P}}\cdot d\boldsymbol{P},$$

where we write $\boldsymbol{P}_0;\partial/\partial\boldsymbol{P}$ for the tensor conjugate to $\partial;\boldsymbol{P}_0/\partial\boldsymbol{P}$. For a cartesian interpretation see (12), (13) below.

For brevity write

$$\mathsf{E} = \frac{\partial;\boldsymbol{P}_0}{\partial\boldsymbol{P}}, \qquad \mathsf{E}_c = \frac{\boldsymbol{P}_0;\partial}{\partial\boldsymbol{P}} \tag{3}$$

$$\mathsf{L} = \frac{\partial;\boldsymbol{P}}{\partial\boldsymbol{P}_0}, \qquad \mathsf{L}_c = \frac{\boldsymbol{P};\partial}{\partial\boldsymbol{P}_0}. \tag{4}$$

Then

$$d\boldsymbol{P}_0 = d\boldsymbol{P}\cdot\mathsf{E} = \mathsf{E}_c\cdot d\boldsymbol{P}, \qquad d\boldsymbol{P} = d\boldsymbol{P}_0\cdot\mathsf{L} = \mathsf{L}_c\cdot d\boldsymbol{P}_0, \tag{5}$$

and therefore

$$\begin{aligned} d\boldsymbol{P}\cdot d\boldsymbol{P} - d\boldsymbol{P}_0\cdot d\boldsymbol{P}_0 &= d\boldsymbol{P}\cdot d\boldsymbol{P} - d\boldsymbol{P}\cdot\mathsf{E}\cdot\mathsf{E}_c\cdot d\boldsymbol{P} \\ &= d\boldsymbol{P}\cdot(\mathsf{I} - \mathsf{E}\cdot\mathsf{E}_c)\cdot d\boldsymbol{P} \\ &= d\boldsymbol{P}_0\cdot(\mathsf{L}\cdot\mathsf{L}_c - \mathsf{I})\cdot d\boldsymbol{P}_0 \end{aligned}$$

by a similar argument, where I is the idemfactor or unit tensor of the second rank.

Therefore

$$d\boldsymbol{P}\cdot d\boldsymbol{P} - d\boldsymbol{P}_0\cdot d\boldsymbol{P}_0 = d\boldsymbol{P}\cdot 2\,\mathsf{D}^E\cdot d\boldsymbol{P} = d\boldsymbol{P}_0\cdot 2\,\mathsf{D}^L\cdot d\boldsymbol{P}_0 \tag{6}$$

where

$$2\,\mathsf{D}^E = \mathsf{I} - \mathsf{E}\cdot\mathsf{E}_c, \qquad 2\,\mathsf{D}^L = \mathsf{L}\cdot\mathsf{L}_c - \mathsf{I}. \tag{7}$$

The tensor D^E we call the *Eulerian deformation tensor*, since the present position $\boldsymbol{P}$ at time t is taken as independent variable. The tensor D^L we call the *Lagrangian deformation tensor*, since the initial position $\boldsymbol{P}_0$ at time t_0 is taken as independent variable.

The factor 2 is conventional being introduced to constrain agreement with established notation for the components deduced from these tensors.

From (5)

$$d\boldsymbol{P}_0 = (d\boldsymbol{P}_0\cdot\mathsf{L})\cdot\mathsf{E} = d\boldsymbol{P}_0\cdot(\mathsf{L}\cdot\mathsf{E}),$$
$$d\boldsymbol{P} = (d\boldsymbol{P}\cdot\mathsf{E})\cdot\mathsf{L} = d\boldsymbol{P}\cdot(\mathsf{E}\cdot\mathsf{L}),$$

and therefore

$$\mathsf{L}\cdot\mathsf{E} = \mathsf{E}\cdot\mathsf{L} = \mathsf{I} = \mathsf{E}_c\cdot\mathsf{L}_c = \mathsf{L}_c\cdot\mathsf{E}_c \tag{8}$$

similarly.

Therefore

$$\mathsf{D}^L = \mathsf{L}\cdot\mathsf{D}^E\cdot\mathsf{L}_c, \qquad \mathsf{D}^E = \mathsf{E}\cdot\mathsf{D}^L\cdot\mathsf{E}_c. \tag{9}$$

Observe that D^L and D^E are *symmetric* tensors in the sense that

$$\mathsf{D}^L = \mathsf{D}^L_c, \qquad \mathsf{D}^E = \mathsf{D}^E_c \tag{10}$$

so that for an arbitrary vector $\boldsymbol{x}$

$$\mathsf{D}^L\cdot\boldsymbol{x} = \boldsymbol{x}\cdot\mathsf{D}^L, \qquad \mathsf{D}^E\cdot\boldsymbol{x} = \boldsymbol{x}\cdot\mathsf{D}^E. \tag{11}$$

Notes: (i) If we use orthogonal cartesian coordinates (x_1, x_2, x_3), if $\boldsymbol{i}_1, \boldsymbol{i}_2, \boldsymbol{i}_3$ are unit vectors along the axes and if

$$\boldsymbol{P} = \boldsymbol{i}_1 x_1 + \boldsymbol{i}_2 x_2 + \boldsymbol{i}_3 x_3, \qquad \boldsymbol{P}_0 = \boldsymbol{i}_1 x_1^{(0)} + \boldsymbol{i}_2 x_2^{(0)} + \boldsymbol{i}_3 x_3^{(0)},$$

then we can write

$$\frac{\partial}{\partial\boldsymbol{P}} = \boldsymbol{i}_1\frac{\partial}{\partial x_1} + \boldsymbol{i}_2\frac{\partial}{\partial x_2} + \boldsymbol{i}_3\frac{\partial}{\partial x_3} = \nabla \tag{12}$$

$$\frac{\partial}{\partial\boldsymbol{P}_0} = \boldsymbol{i}_1\frac{\partial}{\partial x_1^{(0)}} + \boldsymbol{i}_2\frac{\partial}{\partial x_2^{(0)}} + \boldsymbol{i}_3\frac{\partial}{\partial x_3^{(0)}} = \nabla_0 \tag{13}$$

in terms of HAMILTON's operator ∇ *(nabla)*.

(ii) From (6) we have $d\boldsymbol{P}_0\cdot d\boldsymbol{P}_0 = d\boldsymbol{P}\cdot(\mathsf{I} - 2\mathsf{D}^E)\cdot d\boldsymbol{P}$ and this expression for $(ds_0)^2 = d\boldsymbol{P}_0\cdot d\boldsymbol{P}_0$ must give the metric of a Euclidean space, assuming that only that type of space concerns us. It follows that the Riemann-Christoffel tensor [MCCONNELL] for this metric must vanish, and this gives the *condition of compatibility* which must be satisfied by a given symmetric 2-tensor in order that it may be capable of representing an actual deformation. We shall not carry the matter further for the general case than this indication.

1.32. Deformation tensor in terms of displacement

Introduce the displacement

$$\boldsymbol{q} = \boldsymbol{P} - \boldsymbol{P}_0 . \tag{1}$$

Then 1.3 (4) gives

$$\mathsf{L} = \frac{\partial;\boldsymbol{P}}{\partial \boldsymbol{P}_0} = \mathsf{I} + \frac{\partial;\boldsymbol{q}}{\partial \boldsymbol{P}_0}, \qquad \mathsf{L}_c = \mathsf{I} + \frac{\boldsymbol{q};\partial}{\partial \boldsymbol{P}_0} \tag{2}$$

and therefore by substitution in 1.3 (7)

$$\mathsf{D}^L = \frac{1}{2}\left(\frac{\partial;\boldsymbol{q}}{\partial \boldsymbol{P}_0} + \frac{\boldsymbol{q};\partial}{\partial \boldsymbol{P}_0}\right) + \frac{1}{2}\frac{\partial;\boldsymbol{q}}{\partial \boldsymbol{P}_0} \cdot \frac{\boldsymbol{q};\partial}{\partial \boldsymbol{P}_0} . \tag{3}$$

Similarly we show that

$$\mathsf{D}^E = \frac{1}{2}\left(\frac{\partial;\boldsymbol{q}}{\partial \boldsymbol{P}} + \frac{\boldsymbol{q};\partial}{\partial \boldsymbol{P}}\right) - \frac{1}{2}\frac{\partial;\boldsymbol{q}}{\partial \boldsymbol{P}} \cdot \frac{\boldsymbol{q};\partial}{\partial \boldsymbol{P}} . \tag{4}$$

A specially important case arises when the displacement and its gradient* are alike infinitesimal. To this case we give the name *infinitesimal deformation*. Then, for infinitesimal deformation, we can write

$$\mathsf{D}^E = \mathsf{D}^L = \mathsf{D} = \frac{1}{2}\left(\frac{\partial;\boldsymbol{q}}{\partial \boldsymbol{P}} + \frac{\boldsymbol{q};\partial}{\partial \boldsymbol{P}}\right). \tag{5}$$

Using ∇ for $\partial/\partial \boldsymbol{P}$ we can write

$$\mathsf{D} = \frac{1}{2}(\nabla;\boldsymbol{q} + \boldsymbol{q};\nabla) \tag{6}$$

whence taking the curl

$$\nabla_\wedge \mathsf{D} = \frac{1}{2}\nabla_\wedge(\boldsymbol{q};\nabla)$$

since $\nabla_\wedge(\nabla;\boldsymbol{q}) \equiv 0$.

Repeating the operation

$$\nabla_\wedge \mathsf{D}_\wedge \nabla \equiv 0 \tag{7}$$

and this is the *compatibility equation* which must be satisfied by a 2-tensor D if it is to be capable of representing the deformation tensor of an infinitesimal deformation [MILNE-THOMSON (1)].

1.34. Virtual displacements

Starting from the second state let us imagine the material to undergo further displacement in such a way that the particle at $\boldsymbol{P}$ whose displacement is already $\boldsymbol{q}$ has its displacement changed to $\boldsymbol{q} + \delta\boldsymbol{q}$ where $\delta\boldsymbol{q}$ is an infinitesimal vector, while the particle at $\boldsymbol{P}'$ has its displacement changed from $\boldsymbol{q} + d\boldsymbol{q}$ to $\boldsymbol{q} + d\boldsymbol{q} + \delta(\boldsymbol{q} + d\boldsymbol{q})$ so that

$$\delta(\boldsymbol{q} + d\boldsymbol{q}) = \delta\boldsymbol{q} + \delta(d\boldsymbol{q}) . \tag{1}$$

* If we denote the *norm* of an arbitrary 2-tensor A by $||\mathsf{A}|| = [\mathsf{A} \cdot\cdot\, \mathsf{A}_c]^{1/2}$, we can say that A is infinitesimal of the first order when its norm is infinitesimal of the first order.

In the *virtual* (or imagined) displacements just described the particle at $\boldsymbol{P}_0$ in the first state is necessarily unaffected and since $\boldsymbol{P} = \boldsymbol{P}_0 + \boldsymbol{q}$ we have

$$d\boldsymbol{P} = d\boldsymbol{P}_0 + d\boldsymbol{q}, \qquad \delta\boldsymbol{P}_0 = 0, \delta\boldsymbol{P} = \delta\boldsymbol{q}. \tag{2}$$

Also since $\boldsymbol{P}$ is at $\boldsymbol{P} + \delta\boldsymbol{q}$ after virtual displacement $\boldsymbol{P}'$ is at $\boldsymbol{P} + \delta\boldsymbol{q} + d(\boldsymbol{P} + \delta\boldsymbol{q})$ and comparison with (1) shows that

$$\delta(d\boldsymbol{q}) = d(\delta\boldsymbol{q}) \tag{3}$$

i.e. the operators d and δ are commutative.

From 1.32 (5) the deformation tensor for a virtual displacement $\delta\boldsymbol{q}$ is

$$\frac{1}{2}\left(\frac{\partial;\delta\boldsymbol{q}}{\partial\boldsymbol{P}} + \frac{\delta\boldsymbol{q};\partial}{\partial\boldsymbol{P}}\right) \tag{4}$$

and for a *rigid* virtual displacement, namely one which leaves unaltered the distance of each pair of particles, (4) vanishes.

1.36. Variation of the Lagrangian deformation tensor

We have from 1.3 (7)

$$2\delta\mathsf{D}^L = \delta\mathsf{L}\cdot\mathsf{L}_c + \mathsf{L}\cdot\delta\mathsf{L}_c, \qquad \mathsf{L} = \partial;\boldsymbol{P}/\partial\boldsymbol{P}_0.$$

But

$$\delta\mathsf{L} = \frac{\partial;\delta\boldsymbol{P}}{\partial\boldsymbol{P}_0} = \frac{\partial;\boldsymbol{P}}{\partial\boldsymbol{P}_0}\cdot\frac{\partial;\delta\boldsymbol{P}}{\partial\boldsymbol{P}} = \mathsf{L}\cdot\frac{\partial;\delta\boldsymbol{q}}{\partial\boldsymbol{P}}$$

with a similar result for $\delta\mathsf{L}_c$. Therefore the variation of D^L is

$$\delta\mathsf{D}^L = \mathsf{L}\cdot\frac{1}{2}\left(\frac{\partial;\delta\boldsymbol{q}}{\partial\boldsymbol{P}} + \frac{\delta\boldsymbol{q};\partial}{\partial\boldsymbol{P}}\right)\cdot\mathsf{L}_c. \tag{1}$$

It follows that for a rigid virtual displacement $\delta\mathsf{D}^L = 0$, i.e. variation of the Lagrangian deformation tensor vanishes.

1.4. The equation of motion

Consider the material which occupies the entire region τ within a closed convex surface Σ. If $\boldsymbol{n}$ is the unit vector drawn normally outwards at the area element $d\Sigma$ of the surface, the force exerted upon the material in the region τ by the material outside Σ is

$$\int_{\Sigma} \boldsymbol{n}\cdot\mathsf{S}\,d\Sigma.$$

Let ϱ be the density, $\boldsymbol{g}$ the body-force per unit mass, $\boldsymbol{v}$ the velocity and $\boldsymbol{a}$ the acceleration at any point of the material. Then NEWTON's second law of motion gives

$$\int_{\Sigma} \boldsymbol{n}\cdot\mathsf{S}\,d\Sigma + \int_{\tau}\boldsymbol{g}\varrho\,d\tau = \frac{d}{dt}\int_{\tau}\boldsymbol{v}\varrho\,d\tau = \int_{\tau}\boldsymbol{a}\varrho\,d\tau, \tag{1}$$

since conservation of mass demands that $d(\varrho\, d\tau)/dt = 0$. To the first integral on the left of (1) we apply Gauss's theorem, 5.46, and so obtain the equation of motion in the intrinsic form

$$\nabla \cdot \mathsf{S} = \varrho(\boldsymbol{a} - \boldsymbol{g})\,, \tag{2}$$

where ∇ is HAMILTON's vector differentiation operator nabla, 1.3.

The vector

$$\boldsymbol{b} = \varrho(\boldsymbol{a} - \boldsymbol{g}) \tag{3}$$

will be called the *body-vector*. It includes mass acceleration and reversed body force.

We shall assume the stress tensor to be symmetric. In the absence of volume couples and certain types of discontinuity of stress, the symmetry follows by equating to zero the sum of the moments about the centroid of the stress forces which act upon the faces of an infinitesimal rectangular parallelepiped of the material.

The condition of symmetry is expressed by

$$\boldsymbol{x} \cdot \mathsf{S} = \mathsf{S} \cdot \boldsymbol{x} \tag{4}$$

where $\boldsymbol{x}$ is an arbitrary vector.

Take the orthogonal cartesian axes* Ox, Oy, OR with corresponding unit vectors $\boldsymbol{i}, \boldsymbol{j}, \boldsymbol{k}$. The stress tensor can then be written in the form

$$\begin{aligned} \mathsf{S} = {} & \widehat{xx}(\boldsymbol{i};\boldsymbol{i}) + \widehat{xy}(\boldsymbol{i};\boldsymbol{j}) + \widehat{xz}(\boldsymbol{i};\boldsymbol{k}) \\ & + \widehat{yx}(\boldsymbol{j};\boldsymbol{i}) + \widehat{yy}(\boldsymbol{j};\boldsymbol{j}) + \widehat{yz}(\boldsymbol{j};\boldsymbol{k}) \\ & + \widehat{zx}(\boldsymbol{k};\boldsymbol{i}) + \widehat{zy}(\boldsymbol{k};\boldsymbol{j}) + \widehat{zz}(\boldsymbol{k};\boldsymbol{k})\,, \end{aligned} \tag{5}$$

where the semi-colon denotes dyadic multiplication, 1.1.

The symmetry is expressed by the equalities

$$\widehat{xy} = \widehat{yx}, \quad \widehat{yz} = \widehat{zy}, \quad \widehat{zx} = \widehat{xz} \tag{6}$$

the first pair of which follow by writing $\boldsymbol{j}$ for $\boldsymbol{x}$ in (4).

The equation of motion (2) is then equivalent to the three scalar equations

$$\begin{aligned} \frac{\partial \widehat{xx}}{\partial x} + \frac{\partial \widehat{yx}}{\partial y} + \frac{\partial \widehat{zx}}{\partial R} &= b_1 \\ \frac{\partial \widehat{xy}}{\partial x} + \frac{\partial \widehat{yy}}{\partial y} + \frac{\partial \widehat{zy}}{\partial R} &= b_2 \\ \frac{\partial \widehat{xz}}{\partial x} + \frac{\partial \widehat{yz}}{\partial y} + \frac{\partial \widehat{zz}}{\partial R} &= b_3 \end{aligned} \tag{7}$$

where b_1, b_2, b_3 are components of the body-vector $\boldsymbol{b}$.

* The reason for calling the third axis OR instead of Oz is that in the sequel we shall require z for the complex variable $x + iy$, and differentiations with respect to this variable will occur. On the other hand there need be no confusion in using $\widehat{xz}$, $\widehat{zz}$ etc. for stress components.

We have preferred the above notation since in the antiplane systems with which we shall deal, the *applicate* R plays an asymmetric part with respect to the *abscissa* x and the *ordinate* y.

The general tensor notation in terms of cartesian coordinates x_1, x_2, x_3 with corresponding unit vectors $\boldsymbol{i}_1$, $\boldsymbol{i}_2$, $\boldsymbol{i}_3$ would replace (5) by*

$$\mathsf{S} = \widehat{x_\alpha x_\beta}(\boldsymbol{i}_\alpha; \boldsymbol{i}_\beta) . \tag{8}$$

The corresponding representation of the Lagrangian deformation tensor in terms of displacement of a particle from (x_1, x_2, x_3) to $(x_1+q_1, x_2+q_2, x_3+q_3)$ is

$$\mathsf{D}^L = e_{\alpha\beta}(\boldsymbol{i}_\alpha; \boldsymbol{i}_\beta) , \tag{9}$$

$$e_{\alpha\beta} = \frac{1}{2}\left(\frac{\partial q_\alpha}{\partial x_\beta} + \frac{\partial q_\beta}{\partial x_\alpha}\right) + \frac{1}{2}\frac{\partial q_\gamma}{\partial x_\alpha}\frac{\partial q_\gamma}{\partial x_\beta} . \tag{10}$$

The deformation tensor D^L used here is always symmetric.

1.5. Internal energy

We consider a system which undergoes a series of transformations which bring the system from a first to a second state. During these processes the forces do work W_e and a quantity H of heat is supplied to the system, say, for example, by conduction through the boundary or by radiation, while the kinetic energy increases from K_1 to K_2.

Definition: The excess of the energy supplied over and above the increase in kinetic energy is the increase of *internal energy* of the system.

In our case the increase of internal energy is

$$W_e + H - (K_2 - K_1) .$$

Kelvin's hypothesis. The increase of internal energy depends solely on the first and second states and not on the manner in which these states are attained.

This implies the existence of a function E, determinate save for an arbitrary added constant, such that

$$E_2 - E_1 = W_e + H - (K_2 - K_1) ,$$

where E_1 is the value of E for the first state and E_2 is the value of E for the second state.

The function E is called the *internal energy* of the system, and is in general a function of all the variables which define the *state of the system*, its temperature, density, stress, and so forth.

---

* Note that here we are using the summation convention, whereby for every repeated index we substitute in turn the numbers 1, 2, 3 and then add the results. Thus the expression on the right-hand side of (8) stands for

$$\sum_{\alpha=1}^{3}\sum_{\beta=1}^{3} \widehat{x_\alpha x_\beta}(\boldsymbol{i}_\alpha; \boldsymbol{i}_\beta) .$$

We now make two further hypotheses, namely

(i) The internal energy is the sum of the internal energies of each of the parts into which we can suppose the system to be subdivided.

(ii) The material which composes the system is continuously distributed, that is to say, is capable of unlimited subdivision.

Clearly these hypotheses define an ideal system whose consideration is not incommoded by limitations imposed by subdivision into parts whose dimensions might be comparable with intermolecular distances.

It then follows that when the material of the system has density ϱ and occupies a volume τ we can *define a density of internal energy* e_i *per unit mass* such that

$$E = \int_\tau \varrho e_i d\tau .$$

Like E, the density e_i of internal energy is independent of the manner in which the state is attained and depends solely on the variables which define the state. Calling these variables $q_1, q_2, \ldots$, we have

$$d e_i = \frac{\partial e_i}{\partial q_1} d q_1 + \frac{\partial e_i}{\partial q_2} d q_2 + \cdots .$$

1.54. Energy of deformation

Consider deformable material which has passed from a first to a second state in which the material occupies the region τ bounded by a closed surface Σ. Let $\boldsymbol{q}$ be the displacement of a particle in the second state as measured from its position in the first. Suppose that in the second state this particle has temperature T, mass $\varrho d\tau$, entropy $\sigma\varrho d\tau$ and internal energy $\varrho e_i d\tau$, so that σ is the entropy and e_i the internal energy per unit mass.

Let us now undertake a virtual displacement of the whole system in which the particle of displacement $\boldsymbol{q}$ undergoes an additional infinitesimal displacement $\delta\boldsymbol{q}$.

Since the work done by the forces plus the heat supplied is equal to the increase of energy of the material within Σ, we have

$$\begin{aligned} \int_\Sigma d\boldsymbol{\Sigma}\cdot \mathsf{S}\cdot \delta\boldsymbol{q} + \int_\tau \varrho \boldsymbol{g}\cdot \delta\boldsymbol{q} d\tau + \int_\tau T\delta(\sigma\varrho d\tau) = \int_\tau \delta\left(\frac{1}{2}\boldsymbol{v}\cdot\boldsymbol{v}\varrho d\tau\right) \\ + \int_\tau \delta(e_i\varrho d\tau) , \end{aligned} \tag{1}$$

where $d\boldsymbol{\Sigma}$ is an outwardly directed element of area of the surface Σ, $\boldsymbol{g}$ is the body force per unit mass and $\boldsymbol{v} = d\boldsymbol{q}/dt$ is the velocity.

Since the mass of a particle does not change during its motion, $\delta(\varrho\, d\tau) = 0$ and therefore

$$\delta(\sigma\varrho\, d\tau) = (\delta\sigma)\,\varrho\, d\tau\,, \qquad \delta(e_i\varrho\, d\tau) = (\delta e_i)\,\varrho\, d\tau \tag{2}$$

$$\delta\left(\frac{1}{2}\boldsymbol{v}\cdot\boldsymbol{v}\varrho\, d\tau\right) = (\boldsymbol{v}\cdot\delta\boldsymbol{v})\,\varrho\, d\tau = (\boldsymbol{a}\cdot\delta\boldsymbol{q})\,\varrho\, d\tau$$

where $\boldsymbol{a}$ is the acceleration. Also by Gauss's theorem, 5.46,

$$\int_{\Sigma} d\boldsymbol{\Sigma}\cdot\mathsf{S}\cdot\delta\boldsymbol{q} = \int_{\tau}\frac{\partial}{\partial\boldsymbol{P}}\cdot(\mathsf{S}\cdot\delta\boldsymbol{q})\,d\tau\,, \tag{3}$$

while

$$\frac{\partial}{\partial\boldsymbol{P}}\cdot(\mathsf{S}\cdot\delta\boldsymbol{q}) = \left(\frac{\partial}{\partial\boldsymbol{P}}\cdot\mathsf{S}\right)\cdot\delta\boldsymbol{q} + \left(\mathsf{S}\cdot\frac{\partial}{\partial\boldsymbol{P}}\right)\cdot\delta\boldsymbol{q}\,.$$

Combining (1), (2) and (3) we get

$$\int_{\tau}\left\{\left(\frac{\partial}{\partial\boldsymbol{P}}\cdot\mathsf{S} + \boldsymbol{g}\varrho\right)\cdot\delta\boldsymbol{q} + \left(\mathsf{S}\cdot\frac{\partial}{\partial\boldsymbol{P}}\right)\cdot\delta\boldsymbol{q} + T\varrho\,\delta\sigma\right\}d\tau$$

$$= \int_{\tau}(\boldsymbol{a}\varrho\cdot\delta\boldsymbol{q} + \varrho\,\delta e_i)\,d\tau\,.$$

But observing that $\nabla = \partial/\partial\boldsymbol{P}$ we have from the equation of motion 1.4 (2)

$$\frac{\partial}{\partial\boldsymbol{P}}\cdot\mathsf{S} + \boldsymbol{g}\varrho = \boldsymbol{a}\varrho\,.$$

Therefore

$$\int_{\tau}\left\{\left(\mathsf{S}\cdot\frac{\partial}{\partial\boldsymbol{P}}\right)\cdot\delta\boldsymbol{q} + T\varrho\,\delta\sigma - \varrho\,\delta e_i\right\}d\tau = 0\,.$$

Since the volume of integration is arbitrary we have

$$\left(\mathsf{S}\cdot\frac{\partial}{\partial\boldsymbol{P}}\right)\cdot\delta\boldsymbol{q} + T\varrho\,\delta\sigma - \varrho\,\delta e_i = 0 \tag{4}$$

for an arbitrary virtual infinitesimal displacement $\delta\boldsymbol{q}$.

We shall call $đv_s$, where

$$\varrho\, đv_s = \left(\mathsf{S}\cdot\frac{\partial}{\partial\boldsymbol{P}}\right)\cdot\delta\boldsymbol{q}\,, \tag{5}$$

the *virtual work per unit mass of the stress*, due to the virtual displacement $\delta\boldsymbol{q}$. The notation $đv_s$ instead of Dv_s is intended to remove any implication that we are dealing with an exact differential. Now

$$\left(\mathsf{S}\cdot\frac{\partial}{\partial\boldsymbol{P}}\right)\cdot\delta\boldsymbol{q} = \mathsf{S}\cdot\cdot\frac{\partial;\delta\boldsymbol{q}}{\partial\boldsymbol{P}} = \mathsf{S}\cdot\cdot\frac{1}{2}\left(\frac{\partial;\delta\boldsymbol{q}}{\partial\boldsymbol{P}} + \frac{\delta\boldsymbol{q};\partial}{\partial\boldsymbol{P}}\right), \tag{6}$$

since the double scalar product of a symmetric and an antisymmetric tensor is zero, 1.14 (11). From (5), (6) and 1.36 it follows that the virtual work of the stress vanishes for a rigid virtual displacement.

With the above notation (4) becomes

$$\mathfrak{a} v_s = \delta e_i - T\,\delta\sigma\,. \tag{7}$$

Since δe_i and $\delta\sigma$ are always exact differentials it follows that, in general, $\delta e_i - T\,\delta\sigma$ is not an exact differential. Among exceptions to this statement two cases are of special importance.

Case i. Isothermal deformation. If the deformation takes place at constant temperature T we have

$$\mathfrak{a} v_s = \delta(e_i - T\sigma) = \delta u\,,$$

where

$$u = e_i - T\sigma + \text{constant} \tag{8}$$

and u is the *density of energy of deformation per unit mass (or strain-energy density) for isothermal deformation.*

Isothermal deformation is characteristic of phenomena which take place infinitely slowly.

Case ii. Adiabatic deformation. For adiabatic deformation the particle neither gains nor loses heat. Therefore $\delta\sigma = 0$ and (7) becomes

$$\mathfrak{a} v_s = \delta e_i = \delta u\,,$$

where

$$u = e_i + \text{constant} \tag{9}$$

and u is the *density of energy of deformation per unit mass (or strain-energy density) for adiabatic deformation.*

Adiabatic deformation is characteristic of phenomena which take place infinitely fast.

Clearly isothermal and adiabatic deformations are idealizations. Nevertheless they are approximated by phenomena which take place sufficiently slowly or sufficiently fast.

1.6. Elastic deformation

Definition. When material is deformed in conditions in which a strain-energy density exists, which is a one-valued function of the *Lagrangian* deformation tensor, the deformation is said to be *elastic.*

Thus if u is the strain-energy density,

$$u = u(\mathsf{D}^L)\,. \tag{1}$$

Energy of deformation is energy stored by the deformation and can be recovered if the material goes through the reverse series of transformations.

Observe that elastic response is a property of the circumstances of deformation. Not all materials are capable of elastic response, for example cast metals. Elastic materials are those capable of elastic response, but this does not imply that an arbitrary method of deformation will lead to elastic response.

1.62. The law of elasticity

In an elastic deformation

$$u = u(\mathsf{D}^L) \tag{1}$$

wherein u is a scalar and D^L is a symmetric 2-tensor. In a virtual displacement u will become $u + \delta u$ and D^L will become $\mathsf{D}^L + \delta\mathsf{D}^L$ and so

$$\delta u = u(\mathsf{D}^L + \delta\mathsf{D}^L) - u(\mathsf{D}^L) \tag{2}$$

and $\delta u \to 0$ when $\delta\mathsf{D}^L \to 0$.

This motivates the following considerations. Just as for a scalar function $f(x)$ of a scalar variable x we can write

$$\delta f(x) = \frac{df(x)}{dx}\,\delta x$$

so we introduce the derivative $\partial u/\partial\mathsf{D}^L$ of the *scalar* $u(\mathsf{D}^L)$ *with respect to the tensor* D^L by

$$\delta u = \frac{\partial u}{\partial\mathsf{D}^L}\cdot\cdot\,\delta\mathsf{D}^L \tag{3}$$

on the understanding that the derivative $\partial u/\partial\mathsf{D}^L$ is to be a *symmetric* tensor of the second rank. Since $\delta\mathsf{D}^L$ is itself a symmetric tensor and since, 1.14, the double scalar product of a symmetric and an antisymmetric tensor is zero, (3) furnishes a unique derivative $\partial u/\partial\mathsf{D}^L$ [MILNE-THOMSON (3)]. Then from 1.36 it follows that

$$\begin{aligned}\delta u &= \frac{\partial u}{\partial\mathsf{D}^L}\cdot\cdot\,\mathsf{L}\cdot\frac{1}{2}\left(\frac{\partial;\delta q}{\partial P} + \frac{\delta q;\partial}{\partial P}\right)\cdot\mathsf{L}_c\,.\\ &= \mathsf{L}_c\cdot\frac{\partial u}{\partial\mathsf{D}^L}\cdot\mathsf{L}\cdot\cdot\,\frac{1}{2}\left(\frac{\partial;\delta q}{\partial P} + \frac{\delta q;\partial}{\partial P}\right)\end{aligned} \tag{4}$$

But from 1.54 (5), (6) the virtual work of the stress is

$$\varrho\,\delta u = \mathsf{S}\cdot\cdot\,\frac{1}{2}\left(\frac{\partial;\delta q}{\partial P} + \frac{\delta q;\partial}{\partial P}\right). \tag{5}$$

Comparing (4) and (5) we have

$$\mathsf{S} = \varrho\,\mathsf{L}_c\cdot\frac{\partial u}{\partial\mathsf{D}^L}\cdot\mathsf{L}, \qquad \mathsf{L} = \frac{\partial;P}{\partial P_0} = \mathsf{I} + \frac{\partial;q}{\partial P_0} \tag{6}$$

and this expresses the stress tensor in terms of the derivative of the strain-energy density, the tensor L and its conjugate. We may call (6) the *law of elasticity*. It applies to finite or infinitesimal deformation and makes no hypothesis on the nature of the material beyond capacity for elastic response. Let ϱ_0 be the density in the first state. We write $\varrho = \varrho_0(1+s)$ and the law of elasticity assumes the form

$$\mathsf{S} = (1+s)\left(\mathsf{I} + \frac{q;\partial}{\partial P_0}\right) \cdot \frac{\partial(\varrho_0 u)}{\partial \mathsf{D}^L} \cdot \left(\mathsf{I} + \frac{\partial;q}{\partial P_0}\right). \tag{7}$$

In the case of infinitesimal deformation, 1.32, when the displacement and its gradient are alike infinitesimal, s also will be infinitesimal, and $\mathsf{D}^L = \mathsf{D}$. Thus, for this case, the law of elasticity simplifies to Castigliano's first theorem

$$\mathsf{S} = \frac{\partial U}{\partial \mathsf{D}}, \quad U = \varrho_0 u, \quad \mathsf{D} = \frac{1}{2}\left(\frac{\partial;q}{\partial P} + \frac{q;\partial}{\partial P}\right). \tag{8}$$

We shall call U the *strain-energy function*. It is obtained from the strain-energy density on multiplication by ϱ_0 the density in the first state. Since the displacement and its gradient are infinitesimal $\varrho - \varrho_0 = \varrho_0 s$ is likewise infinitesimal and we could write $U = \varrho u$. Thus we may also regard U as the strain-energy per unit *volume*.

1.64. Form of the strain-energy density for elastic deformation

When the Lagrangian deformation tensor D^L is small, the derivative $\partial u/\partial \mathsf{D}^L$ will be sensibly linear in D^L so that we can write

$$\frac{\partial u}{\partial \mathsf{D}_L} = \mathsf{H}^{(2)} + \mathsf{H}^{(4)} \cdot\cdot\, \mathsf{D}^L, \qquad \mathsf{D}^L \text{ small}, \tag{1}$$

where $\mathsf{H}^{(n)}$ denotes a tensor of rank n independent of D^L. This suggests considering an expansion in "powers" of D^L of the type

$$\begin{aligned}\frac{\partial u}{\partial \mathsf{D}^L} = {} & \mathsf{H}^{(2)} + \mathsf{H}^{(4)} \cdot\cdot\, \mathsf{D}^L + (\mathsf{H}^{(6)} \cdot\cdot\, \mathsf{D}^L) \cdot\cdot\, \mathsf{D}^L \\ & + [(\mathsf{H}^{(8)} \cdot\cdot\, \mathsf{D}^L) \cdot\cdot\, \mathsf{D}^L] \cdot\cdot\, \mathsf{D}^L + \cdots.\end{aligned} \tag{2}$$

Should such an expansion exist, we should have for the strain-energy density u

$$\begin{aligned}u = {} & u_0 + (\mathsf{H}^{(2)} \cdot\cdot\, \mathsf{D}^L) + \frac{1}{2}[(\mathsf{H}^{(4)} \cdot\cdot\, \mathsf{D}^L) \cdot\cdot\, \mathsf{D}^L] \\ & + \frac{1}{3}[(\{\mathsf{H}^{(6)} \cdot\cdot\, \mathsf{D}^L\} \cdot\cdot\, \mathsf{D}^L) \cdot\cdot\, \mathsf{D}^L] + \cdots.\end{aligned} \tag{3}$$

In order that (3) may lead to the definition of $\partial u/\partial \mathsf{D}^L$ given in 1.62 (3) it is necessary that certain identities should exist of the types

$$(\mathsf{H}^{(4)} \cdot\cdot\, \delta \mathsf{D}^L) \cdot\cdot\, \mathsf{D}^L = (\mathsf{H}^{(4)} \cdot\cdot\, \mathsf{D}^L) \cdot\cdot\, \delta \mathsf{D}^L \tag{4}$$

$$\begin{aligned}[(\mathsf{H}^{(6)} \cdot\cdot\, \delta \mathsf{D}^L) \cdot\cdot\, \mathsf{D}^L] \cdot\cdot\, \mathsf{D}^L &= [(\mathsf{H}^{(6)} \cdot\cdot\, \mathsf{D}^L) \cdot\cdot\, \delta \mathsf{D}^L] \cdot\cdot\, \mathsf{D}^L \\ &= [(\mathsf{H}^{(6)} \cdot\cdot\, \mathsf{D}^L) \cdot\cdot\, \mathsf{D}^L] \cdot\cdot\, \delta \mathsf{D}^L\end{aligned} \tag{5}$$

and so on. Such identities imply certain symmetry properties for the tensors $\mathsf{H}^{(n)}$.

Substitution in 1.62 (6) leads to the following form for the law of elasticity

$$S^{(2)} = \varrho L_c \cdot \{H^{(2)} + (H^{(4)} \cdot\cdot D^L) + (H^{(6)} \cdot\cdot D^L) \cdot\cdot D^L + \cdots\} \cdot L\,, \quad (6)$$

where $S^{(2)}$ denotes the stress tensor in the second state.

By introducing the tensor

$$S^{(1)} = \varrho L_c \cdot H^{(2)} \cdot L\,, \quad (7)$$

which corresponds to the stress *induced by initial stress* that is by stress which already exists in the first state, we have finally [MILNE-THOMSON (3)]

$$S^{(2)} - S^{(1)} = S = \varrho L_c \cdot \{(H^{(4)} \cdot\cdot D^L) + (H^{(6)} \cdot\cdot D^L) \cdot\cdot D^L + \cdots\} \cdot L\,. \quad (8)$$

Here S denotes a stress tensor which is independent of the stress tensor induced by the initial stress. When the initial stress is known (7) and (8) suffice to describe the stress at any subsequent time. When the initial stress is unknown, we have (8) alone which describes the stress to be superimposed on that determined by the (unknown) initial stress. This situation must be accepted and to avoid circumlocution we make the *hypothesis of the unstressed state,* namely that there exists a state, which we take to be the first state, for which the material is unstressed.

Observe that this amounts to taking the tensor $H^{(2)}$ to vanish identically. With this hypothesis we have idealized the problem.

1.66. The tensor $H^{(4)}$

Let us take rectangular cartesian coordinates x_1, x_2, x_3 with unit vectors $\boldsymbol{i}_1$, $\boldsymbol{i}_2$, $\boldsymbol{i}_3$ parallel to the axes. Then the tensor $H^{(4)}$ of the 4th rank can be represented in the form

$$H^{(4)} = \{pq\ rs\}\,(\boldsymbol{i}_p;\boldsymbol{i}_q;\boldsymbol{i}_r;\boldsymbol{i}_s)\,, \quad (1)$$

where we are using the summation convention that the terms due to any repeated index must be summed over the values 1, 2, 3 of the index. Expressed in this form the number of components $\{pq\ rs\}$ is $3^4 = 81$.

The deformation tensor and its variation we represent in the forms

$$D^L = e_{\alpha\beta}(\boldsymbol{i}_\alpha;\boldsymbol{i}_\beta)\,, \qquad \delta D^L = \delta e_{\sigma\tau}(\boldsymbol{i}_\sigma;\boldsymbol{i}_\tau)\,. \quad (2)$$

Now from 1.64 (4) and the commutativity of the double scalar product for 2-tensors we have

$$D^L \cdot\cdot (H^{(4)} \cdot\cdot \delta D^L) = \delta D^L \cdot\cdot (H^{(4)} \cdot\cdot D^L)$$

so that

$$e_{\alpha\beta}\{\beta\alpha\ \tau\sigma\}\delta e_{\sigma\tau} = \delta e_{\sigma\tau}\{\tau\sigma\ \beta\alpha\}e_{\alpha\beta} \quad (3)$$

and therefore

$$\{\beta\alpha\ \tau\sigma\} = \{\tau\sigma\ \beta\alpha\}\,. \quad (4)$$

Moreover the deformation tensor is symmetric so that

$$e_{\alpha\beta}=e_{\beta\alpha}, \qquad \delta e_{\sigma\tau}=\delta e_{\tau\sigma}.$$

Therefore in (3) we can interchange α and β, and or σ and τ. Thus finally

$$\begin{aligned}\{\alpha\beta\ \ \sigma\tau\}&=\{\beta\alpha\ \ \sigma\tau\}=\{\beta\alpha\ \ \tau\sigma\}=\{\alpha\beta\ \ \tau\sigma\}=\{\sigma\tau\ \ \beta\alpha\}=\{\tau\sigma\ \ \beta\alpha\}\\&=\{\tau\sigma\ \ \alpha\beta\}=\{\sigma\tau\ \ \alpha\beta\}.\end{aligned}$$

Thus the component $\{\alpha\beta\ \ \sigma\tau\}$ depends only on the pairs $\alpha\beta$, $\sigma\tau$ and not on the order of the pairs or on the order of the numbers in the pairs. Since there are six distinct pairs (11), (22), (33), (23), (31), (12), and since there are six components for which $(\alpha\beta)$ is the same as $(\sigma\tau)$, of the 36 couples of pairs only $36-6=30$ give rise to distinct components and of these 30, 15 alone are different on account of (4). Thus there are at most $15+6=21$ components whose values are different.

We exhibit them in the following tableau

{11 11}	{11 22}	{11 33}	{11 23}	{11 31}	{11 12}
{22 11}	{22 22}	{22 33}	{22 23}	{22 31}	{22 12}
{33 11}	{33 22}	{33 33}	{33 23}	{33 31}	{33 12}
{23 11}	{23 22}	{23 33}	{23 23}	{23 31}	{23 12}
{31 11}	{31 22}	{31 33}	{31 23}	{31 31}	{31 12}
{12 11}	{12 22}	{12 33}	{12 23}	{12 31}	{12 12}

which is symmetrical about its leading diagonal.

1.7. Hooke's law

In the case of infinitesimal deformation, 1.32, the Lagrangian and Eulerian deformation tensors coincide with the infinitesimal deformation tensor

$$\mathsf{D}=\frac{1}{2}\left\{\frac{\partial;q}{\partial P}+\frac{q;\partial}{\partial P}\right\}. \tag{1}$$

Therefore in the law of elasticity in the form 1.64 (8) we can neglect all powers of D beyond the first so that, with $\varrho=\varrho_0(1+s)$, we can write

$$\mathsf{S}=(1+s)\,\varrho_0\left(\mathsf{I}+\frac{q;\partial}{\partial P}\right)\cdot\{\mathsf{H}^{(4)}\cdot\cdot\,\mathsf{D}\}\cdot\left(\mathsf{I}+\frac{\partial;q}{\partial P}\right),$$

which with the hypothesis of infinitesimal deformation reduces to

$$\mathsf{S}=\varrho_0(\mathsf{H}^{(4)}\cdot\cdot\,\mathsf{D}). \tag{2}$$

If we define the tensor $\mathsf{H}_{(4)}$ of the fourth rank by

$$\mathsf{H}_{(4)}=\varrho_0\,\mathsf{H}^{(4)} \tag{3}$$

we have

$$\mathsf{S}=\mathsf{H}_{(4)}\cdot\cdot\,\mathsf{D}, \tag{4}$$

which is the generalized form of *Hooke's law**, stating that the stress tensor is a homogeneous linear function of the deformation tensor, when we make the hypothesis of the unstressed state and when the displacement and its gradient are alike infinitesimal. We shall call the 4-tensor $\mathsf{H}_{(4)}$ *Hooke's tensor.* In the above form Hooke's law gives the stress tensor in terms of the deformation tensor. We can invert this to give

$$\mathsf{D} = \mathsf{K}_{(4)} \cdot\cdot \mathsf{S}\,, \tag{5}$$

where $\mathsf{K}_{(4)}$, *the inverse Hooke's tensor,* is a tensor of the fourth rank such that

$$\mathsf{H}_{(4)} \cdot\cdot \mathsf{K}_{(4)} = \mathsf{J}_{(4)}\,, \tag{6}$$

where $\mathsf{J}_{(4)}$ is the *fourth rank idemfactor for double scalar multiplication,* with the cartesian representation

$$\mathsf{J}_{(4)} = \boldsymbol{i}_\alpha; \boldsymbol{i}_\beta; \boldsymbol{i}_\beta; \boldsymbol{i}_\alpha\,. \tag{7}$$

It is easy to verify that for an arbitrary tensor $\mathsf{T}_{(n)}$ of rank $n \geqq 2$,

$$\mathsf{T}_{(n)} \cdot\cdot \mathsf{J}_{(4)} = \mathsf{J}_{(4)} \cdot\cdot \mathsf{T}_{(n)} = \mathsf{T}_{(n)}\,. \tag{8}$$

Let

$$\mathsf{H}_{(4)} = H_{pq\,rs}(\boldsymbol{i}_p; \boldsymbol{i}_q; \boldsymbol{i}_r; \boldsymbol{i}_s) \tag{9}$$

be a cartesian representation (cf. 1.66 (1)) of $\mathsf{H}_{(4)}$. If we change to other cartesian axes with the same origin, the representation will be say

$$\mathsf{H}_{(4)} = H'_{pq\,rs}(\boldsymbol{i}'_p; \boldsymbol{i}'_q; \boldsymbol{i}'_r; \boldsymbol{i}'_s) \tag{10}$$

and for $\mathsf{J}_{(4)}$

$$\mathsf{J}_{(4)} = \boldsymbol{i}'_\alpha; \boldsymbol{i}'_\beta; \boldsymbol{i}'_\beta; \boldsymbol{i}'_\alpha\,. \tag{11}$$

We have the identity

$$\mathsf{H}_{(4)} = \mathsf{J}_{(4)} \cdot\cdot \mathsf{H}_{(4)} \cdot\cdot \mathsf{J}_{(4)} \tag{12}$$

and therefore using the representations (10), (11), (9)

$$\begin{aligned} &H'_{\alpha\beta\gamma\delta}(\boldsymbol{i}'_\alpha; \boldsymbol{i}'_\beta; \boldsymbol{i}'_\gamma; \boldsymbol{i}'_\delta) \\ &= (\boldsymbol{i}'_\alpha; \boldsymbol{i}'_\beta; \boldsymbol{i}'_\beta; \boldsymbol{i}'_\alpha) \cdot\cdot H_{pq\,rs}(\boldsymbol{i}_p; \boldsymbol{i}_q; \boldsymbol{i}_r; \boldsymbol{i}_s) \cdot\cdot (\boldsymbol{i}'_\delta; \boldsymbol{i}'_\gamma; \boldsymbol{i}'_\gamma; \boldsymbol{i}'_\delta) \\ &= H_{pq\,rs} l_{\alpha' p} l_{\beta' q} l_{\gamma' r} l_{\delta' s}(\boldsymbol{i}'_\alpha; \boldsymbol{i}'_\beta; \boldsymbol{i}'_\gamma; \boldsymbol{i}'_\delta)\,, \end{aligned}$$

where

$$l_{\alpha' p} = \boldsymbol{i}'_\alpha \cdot \boldsymbol{i}_p\,, \quad l_{\beta' q} = \boldsymbol{i}'_\beta \cdot \boldsymbol{i}_q\,, \quad l_{\gamma' r} = \boldsymbol{i}'_\gamma \cdot \boldsymbol{i}_r\,, \quad l_{\delta' s} = \boldsymbol{i}'_\delta \cdot \boldsymbol{i}_s\,. \tag{13}$$

Therefore

$$H'_{\alpha\beta\gamma\delta} = l_{\alpha' p} l_{\beta' q} l_{\gamma' r} l_{\delta' s} H_{pq\,rs} \tag{14}$$

which permits us to express the components of $\mathsf{H}_{(4)}$ referred to a set of axes of reference in terms of the components referred to a second set.

The components of $\mathsf{K}_{(4)}$ obey the same rule of transformation.

* Robert Hooke, Professor of Geometry in Gresham College and Surveyor to the City of London, stated (1678) that for springy material the force is proportional to the extension it produces. On this statement all subsequent generalizations have been based.

On account of (3) the problem of determining a representation

$$\mathsf{K}_{(4)} = K_{pq\ rs}(\boldsymbol{i}_p;\boldsymbol{i}_q;\boldsymbol{i}_r;\boldsymbol{i}_s) \tag{15}$$

is essentially that of inverting the matrix of the tableau of 1.66. Let $||H^{(0)}||$ be the determinant of this matrix and let $h_{pq\ rs}$ be the cofactor of $\{pq\ rs\}$ in this determinant. Then

$$K_{pq\ rs} = \frac{h_{pq\ rs}}{\varrho_0 ||H^{(0)}||}. \tag{16}$$

1.72. Matrix expression of Hooke's law

Hooke's law may be conveniently expressed in matrix form as follows

$$\begin{bmatrix} \widehat{xx} \\ \widehat{yy} \\ \widehat{zz} \\ \widehat{yz} \\ \widehat{zx} \\ \widehat{xy} \end{bmatrix} = \begin{bmatrix} h_{11} & h_{12} & h_{13} & h_{14} & h_{15} & h_{16} \\ h_{21} & h_{22} & h_{23} & h_{24} & h_{25} & h_{26} \\ h_{31} & h_{32} & h_{33} & h_{34} & h_{35} & h_{36} \\ h_{41} & h_{42} & h_{43} & h_{44} & h_{45} & h_{46} \\ h_{51} & h_{52} & h_{53} & h_{54} & h_{55} & h_{56} \\ h_{61} & h_{62} & h_{63} & h_{64} & h_{65} & h_{66} \end{bmatrix} \begin{bmatrix} e_{xx} \\ e_{yy} \\ e_{zz} \\ 2e_{yz} \\ 2e_{zx} \\ 2e_{xy} \end{bmatrix} \tag{1}$$

$$\begin{bmatrix} e_{xx} \\ e_{yy} \\ e_{zz} \\ 2e_{yz} \\ 2e_{zx} \\ 2e_{xy} \end{bmatrix} = \begin{bmatrix} k_{11} & k_{12} & k_{13} & k_{14} & k_{15} & k_{16} \\ k_{21} & k_{22} & k_{23} & k_{24} & k_{25} & k_{26} \\ k_{31} & k_{32} & k_{33} & k_{34} & k_{35} & k_{36} \\ k_{41} & k_{42} & k_{43} & k_{44} & k_{45} & k_{46} \\ k_{51} & k_{52} & k_{53} & k_{54} & k_{55} & k_{56} \\ k_{61} & k_{62} & k_{63} & k_{64} & k_{65} & k_{66} \end{bmatrix} \begin{bmatrix} \widehat{xx} \\ \widehat{yy} \\ \widehat{zz} \\ \widehat{yz} \\ \widehat{zx} \\ \widehat{xy} \end{bmatrix} \tag{2}$$

The second formulation is obtained by inverting the square matrix of the first.

If H denotes the determinant of the square matrix of (1) and H_{rs} denotes the cofactor of h_{rs} in this determinant, then

$$k_{rs} = \frac{H_{rs}}{H}. \tag{3}$$

The symmetry is expressed by

$$h_{rs} = h_{sr}, \qquad k_{rs} = k_{sr}. \tag{4}$$

The coefficients h_{rs} can be related to the components of Hooke's tensor $\mathsf{H}_{(4)} = H_{\alpha\beta\ \gamma\delta}(\boldsymbol{i}_\alpha;\boldsymbol{i}_\beta;\boldsymbol{i}_\gamma;\boldsymbol{i}_\delta)$ by numbering the pairs $\alpha\beta$, $\gamma\delta$ as follows

Pair	11	22	33	23	31	12
Number	1	2	3	4	5	6

We then have for example

$$H_{11\,11} = h_{11}, \qquad H_{23\,31} = h_{45}, \qquad H_{11\,12} = h_{16} \tag{5}$$

and so on.

We note that the coefficients h_{rs} are of the physical dimensions of stress $[ML^{-1}T^{-2}]$. We shall call them *Hooke's moduli of elasticity* or briefly *moduli*. The relations of the k_{rs} to the components of the inverse Hooke's tensor $K_{(4)}$ of 1.7 are easily shown to be of the types

$$\left.\begin{array}{l} k_{11} = K_{11,\,11}, \quad k_{12} = K_{11,\,22}, \quad k_{13} = K_{11,\,33}, \\ k_{14} = 2K_{11,\,23}, \; k_{15} = 2K_{11,\,31}, \; k_{16} = 2K_{11,\,12}, \\ k_{44} = 4K_{23,\,23}, \; k_{45} = 4K_{23,\,31}, \; k_{46} = 4K_{23,\,12}. \end{array}\right\} \qquad (6)$$

It follows that for change of axes of reference the coefficients h_{rs}, k_{rs} will be found by first finding $H_{pq\,rs}$ and $K_{pq\,rs}$ from 1.7 (14), (16) and then using (5) and (6).

Owing to the inverse relation between the h_{rs} and k_{rs} the latter are of the inverse dimensions of stress $[M^{-1}LT^{2}]$. We shall therefore call* the k_{rs} *inverse elastic moduli* or *inverse moduli.*

1.8. Anisotropy

It appears from the formulation in 1.72, that for a given system of axes of reference, the determination of the components of Hooke's tensor, and Hooke's moduli of elasticity namely the elements of the matrix 1.72 (1) are equivalent problems. It also appears from 1.7 (14) that the moduli depend, in general, on the directions of the axes of reference.

Material for which the moduli depend upon the directions chosen for the axes of reference are termed *anisotropic* or *aeolotropic.*

It appears from the tableau in 1.66 that at a given point of the material and for given directions there of the axes of reference, there are not more than 21 distinct moduli.

A material for which the number is 21 is said to possess the highest degree of anisotropy, or *general anisotropy.*

Anisotropic material is said to be *elastically homogeneous* when Hooke's tensor is independent of position. Such material is said to possess *linear anisotropy.*

As a consequence of this all identical and identically directed elements in the form of a rectangular parallelepiped are identical in respect of their elastic properties.

Another type of anisotropy, of which linear anisotropy is a particular case, may be described as follows.

Suppose there exists in the material a system of triply orthogonal surfaces, thereby defining at each point, by their intersections, a set of triply orthogonal directions, such that, when the axes of reference are

* No consistent name for the k_{rs} appears in the literature. The above nomenclature is offered as a suggestion.

taken along these triply orthogonal directions at any point, the components of Hooke's tensor are independent of position. The material is then said to possess *curvilinear anisotropy*.

For such material infinitesimal parallelepipeds bounded by three pairs of the triply-orthogonal surfaces will be identical in respect of their elastic properties. On the other hand identical and identically oriented parallelepipeds will have different properties at different points of the material.

The concepts of linear and curvilinear anisotropy can be generalized to include non-homogeneous materials in which the Hooke's tensor is a function of position.

A simple example of approximately curvilinear anisotropy is that of a block of wood cut in any shape from a tree trunk, the triply orthogonal surfaces being planes perpendicular to the axis of the trunk, axial planes and cylinders coaxial with the axis of the trunk.

1.82. The strain-energy function

From 1.62 (8) we have Castigliano's first theorem

$$\mathsf{S} = \frac{\partial U}{\partial \mathsf{D}}, \tag{1}$$

where U is the strain-energy function. Combining this with Hooke's law we have

$$\frac{\partial U}{\partial \mathsf{D}} = \mathsf{H}_{(4)} \cdot\cdot \mathsf{D} \tag{2}$$

and therefore, save for an arbitrary constant (which we take to be zero)

$$U = \frac{1}{2}(\mathsf{D} \cdot\cdot \mathsf{H}_{(4)} \cdot\cdot \mathsf{D}) \tag{3}$$

$$= \frac{1}{2}(\mathsf{D} \cdot\cdot \mathsf{S}) \qquad \text{from Hooke's law}. \tag{4}$$

$$= \frac{1}{2}(\mathsf{S} \cdot\cdot \mathsf{K}_{(4)} \cdot\cdot \mathsf{S}) \qquad \text{from 1.7 (5)}. \tag{5}$$

Thus the strain-energy function U can be expressed in terms of deformation or stress or in a mixed form.

When the tensors are represented in a coordinate system, U is a homogeneous quadratic of stress components or deformation components or is bilinear in the two sets.

It follows from (5) that

$$\frac{\partial U}{\partial \mathsf{S}} = \mathsf{K}_{(4)} \cdot\cdot \mathsf{S} = \mathsf{D} \tag{6}$$

which is Castigliano's second theorem.

If we take the representations

$$\mathsf{D} = e_{\alpha\beta}(\boldsymbol{i}_\alpha;\boldsymbol{i}_\beta), \qquad \mathsf{H}_{(4)} = H_{pq\,rs}(\boldsymbol{i}_p;\boldsymbol{i}_q;\boldsymbol{i}_r;\boldsymbol{i}_s), \tag{7}$$

we have

$$U = \frac{1}{2} H_{pq\,rs} e_{pq} e_{rs}. \tag{8}$$

while

$$\frac{\partial U}{\partial \mathsf{D}} = \frac{\partial U}{\partial e_{pq}} (\boldsymbol{i}_p;\boldsymbol{i}_q) = \widehat{x_p x_q} (\boldsymbol{i}_p;\boldsymbol{i}_q) \tag{9}$$

from (1). Therefore using (8)

$$\widehat{x_p x_q} = \frac{\partial U}{\partial e_{pq}} = \frac{1}{2} H_{pq\,rs}(e_{rs} + e_{sr}) = H_{pq\,rs} e_{rs}. \tag{10}$$

It is important to note that in performing the differentiation indicated in (9) we must regard e_{pq} and e_{qp} as formally distinct. From (8) we get

$$\begin{aligned} U = \frac{1}{2} H_{11\,11} e_{11}^2 &+ H_{11\,22}\, e_{11}\, e_{22} + H_{11\,33}\, e_{11}\, e_{33} \\ &+ 2H_{11\,23}\, e_{11}\, e_{23} + 2H_{11\,31}\, e_{11}\, e_{31} + 2H_{11\,12}\, e_{11}\, e_{12} \\ + \frac{1}{2} H_{22\,22} e_{22}^2 &+ H_{22\,33}\, e_{22}\, e_{33} \\ &+ 2H_{22\,23}\, e_{22}\, e_{23} + 2H_{22\,31}\, e_{22}\, e_{31} + 2H_{22\,12}\, e_{22}\, e_{12} \\ + \frac{1}{2} H_{33\,33} e_{33}^2 & \\ &+ 2H_{33\,23}\, e_{33}\, e_{23} + 2H_{33\,31}\, e_{33}\, e_{31} + 2H_{33\,12}\, e_{33}\, e_{12} \\ + 2H_{23\,23} e_{23}^2 &+ 4H_{23\,31}\, e_{23}\, e_{31} + 4H_{23\,12}\, e_{23}\, e_{12} \\ &+ 2H_{31\,31} e_{31}^2 + 4H_{31\,12}\, e_{31}\, e_{12} + 2H_{12\,12} e_{12}^2. \end{aligned}$$

Using 1.72 (5) we have the strain-energy function expressed in terms of deformation by

$$\begin{aligned} U = \frac{1}{2} h_{11} e_{11}^2 + h_{12} e_{11} e_{22} + h_{13} e_{11} e_{33} + 2h_{14} e_{11} e_{23} + 2h_{15} e_{11} e_{31} + 2h_{16} e_{11} e_{12} \\ + \frac{1}{2} h_{22} e_{22}^2 + h_{23} e_{22} e_{33} + 2h_{24} e_{22} e_{23} + 2h_{25} e_{22} e_{31} + 2h_{26} e_{22} e_{12} \\ + \frac{1}{2} h_{33} e_{33}^2 + 2h_{34} e_{33} e_{23} + 2h_{35} e_{33} e_{31} + 2h_{36} e_{33} e_{12} \\ + 2h_{44} e_{23}^2 + 4h_{45} e_{23} e_{31} + 4h_{46} e_{23} e_{12} \\ + 2h_{55} e_{31}^2 + 4h_{56} e_{31} e_{12} \\ + 2h_{66} e_{12}^2. \end{aligned} \tag{11}$$

To express U in terms of stress we introduce the temporary notation $t_{rs} = \widehat{x_r x_s}$. We then have

$$\begin{aligned} U = \frac{1}{2} k_{11} t_{11}^2 &+ k_{12} t_{11} t_{22} + k_{13} t_{11} t_{33} + k_{14} t_{11} t_{23} + k_{15} t_{11} t_{31} + k_{16} t_{11} t_{12} \\ &+ \frac{1}{2} k_{22} t_{22}^2 + k_{23} t_{22} t_{33} + k_{24} t_{22} t_{23} + k_{25} t_{22} t_{31} + k_{26} t_{22} t_{12} \\ &+ \frac{1}{2} k_{33} t_{33}^2 + k_{34} t_{33} t_{23} + k_{35} t_{33} t_{31} + k_{36} t_{33} t_{12} \\ &+ \frac{1}{2} k_{44} t_{23}^2 + k_{45} t_{23} t_{31} + k_{46} t_{23} t_{12} \\ &+ \frac{1}{2} k_{55} t_{31}^2 + k_{56} t_{31} t_{12} \\ &+ \frac{1}{2} k_{66} t_{12}^2 \quad (12) \end{aligned}$$

The mixed form can be developed from (4).

In certain cases of elastic symmetry, which will be discussed below, it will be found that by proper choice of axes of reference certain terms will drop out of the expression (11) so that instead of 21 independent elastic moduli a smaller number k will appear. The transformation 1.7 (14) now shows that if we change to other axes of reference although the number of terms in the expression may be increased, the number of *independent* moduli will still be k.

A similar remark applies to the inverse moduli of expression (12).

1.9. Elastic symmetry

If the internal structure of a material possesses symmetry of any kind, its elastic properties are observed to be symmetrical in a sense which we propose to make more precise.

Elastic symmetry will be shown to manifest itself more particularly in the possibility of simplifying the strain-energy function by a proper choice of axes of reference.

We consider the case of linear anisotropy for which (1.8) Hooke's tensor is the same at every point. In this case all identical and identically directed elements in the form of a rectangular parallelepiped are identical in respect of their elastic properties. This means that if P and Q are any two points of the material, referred to rectangular axes at P and to parallel translated rectangular axes at Q, Hooke's moduli of elasticity will be the same.

This leads us to consider sets of three non coplanar lines issuing from a point. Such a set will be called a *triad*. There is no loss of generality in considering only *orthogonal triads*.

Two such triads will be called *homothetic* when the corresponding lines in one are *parallel in the same sense* to corresponding lines in the other.

When issuing from the same point homothetic triads are coincident. Fig. 1.9 shows triads at P and Q which are homethetic. The triads at P and R are not homothetic, for the lines 1 and 3 are parallel and in the same sense but the lines 2 and 2′, while parallel, are in opposite senses.

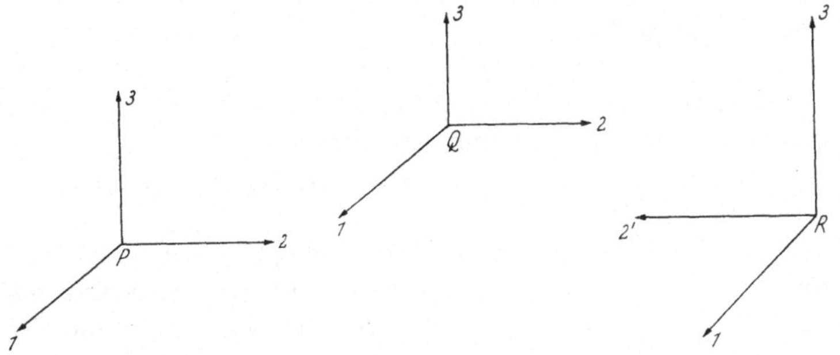

Fig. 1.0

Definition: *Equivalent triads* are triads referred to which, Hooke's moduli of elasticity are the same. *Equivalent directions* are the directions of corresponding lines in equivalent triads.

It follows from 1.82 that the form of the expression for the strain-energy is the same when referred to equivalent triads.

We see that in the case of linear anisotropy all homothetic triads are equivalent.

Definition: Material having linear anisotropy is said to possess *elastic symmetry* when there exist at the same point triads which are equivalent but not homothetic.

Thus in material having linear anisotropy which possesses elastic symmetry, at every point there are equivalent directions or rays. A figure made up of a set of equivalent directions is a geometrical figure exhibiting some kind of geometrical symmetry.

Definition: A geometrical figure which can be brought into coincidence with itself by an operation which changes the position of any of its points is said to possess *symmetry*.

Such operations are known as *covering operations* and a figure which can be brought into coincidence with itself by means of the operation is said to *allow* the operation.

The possible covering operations include

(1) *Rotation about an axis through a definite angle,* for example a prism whose cross-section is a regular hexagon allows rotation through the angle $2\pi/6$ about its longitudinal axis.

(2) *Rotation about an axis through an angle of arbitrary magnitude,* for example a circular cylinder allows rotation through any angle about its longitudinal axis.

A figure which allows rotation about an axis is said to have this axis as an *axis of symmetry*.

(3) *Reflexion in a plane*. For example an ellipsoid with unequal principal axes allows reflexion in a principal plane.

A figure which allows reflexion in a plane is said to have this plane as a *plane of symmetry*.

It can be shown that every covering operation which is neither a rotation nor a reflexion is equivalent to a combination of such operations.

Consider an ellipsoid of unequal semi-axes a, b, c. This ellipsoid allows a rotation of angle π about each principal axis, and a reflexion in each principal plane.

Suppose the ellipsoid to be cut in half along the principal plane which contains a, b; one half to be rotated relatively to the other through the angle $\pi/2$ about the axis c; and the two parts to be firmly united in this configuration.

The resulting figure allows rotation through the angle $\pi/2$ about the axis c followed by reflexion in the principal plane perpendicular to c, or vice versa. It does not allow the rotation alone or the reflexion alone.

A figure which allows the operation of rotation about an axis combined with reflexion in a plane perpendicular to the axis is said to have that axis as an *axis of alternating symmetry* as contrasted with the axis of *direct symmetry* defined by (1) and (2) above.

Covering operations form a group, for it is evident that two such operations performed in succession have the same effect as a single covering operation, or else the first and last positions of every point are identical. This *identical operation*, which alters the position of no point, is the *unit element* of the group.

In the analytical description of a covering operation two elements are involved, a triad of reference and the figure. Clearly the operation can be described either by keeping the triad fixed and altering the position of points of the figure, or by keeping the figure fixed and altering the triad. If we adopt the latter point of view, a covering operation can be described in terms of a linear transformation of coordinates.

1.92. Plane of elastic symmetry

Suppose the plane $x_3 = 0$ to be one of elastic symmetry. We pass from the triad whose unit vectors are $(\boldsymbol{i}_1, \boldsymbol{i}_2, \boldsymbol{i}_3)$ to the equivalent triad $(\boldsymbol{i}_1, \boldsymbol{i}_2, -\boldsymbol{i}_3)$ and therefore the position vector (x_1, x_2, x_3) becomes (x_1', x_2', x_3') where

$$x_1' = x_1, \quad x_2' = x_2, \quad x_3' = -x_3$$

while the displacement (u_1, u_2, u_3) becomes (u_1', u_2', u_3') where

$$u_1' = u_1, \quad u_2' = u_2, \quad u_3' = -u_3$$

and therefore from the definitions

$$
\begin{aligned}
&e'_{11} = e_{11}, \quad e'_{22} = e_{22}, \quad e'_{33} = e_{33}. \\
&e'_{23} = -e_{23}, \quad e'_{31} = -e_{31}, \quad e'_{12} = e_{12}.
\end{aligned} \tag{1}
$$

If the form of the strain-energy function, 1.82 (11), is to remain unaltered, the terms in its expression which contain e_{23} or e_{31} (but not their squares or products) must be absent, for such terms change sign in passing from the first triad to the second.

Therefore the moduli on and above the leading diagonal of the matrix of 1.72 (1) are

$$
\begin{matrix}
h_{11} & h_{12} & h_{13} & 0 & 0 & h_{16} \\
 & h_{22} & h_{23} & 0 & 0 & h_{26} \\
 & & h_{33} & 0 & 0 & h_{36} \\
 & & & h_{44} & h_{45} & 0 \\
 & & & & h_{55} & 0 \\
 & & & & & h_{66}
\end{matrix} \tag{2}
$$

Such a material has 13 moduli of elasticity.

If, in addition, $x_2 = 0$ is also a plane of elastic symmetry, we find by a similar argument that additionally

$$h_{16} = h_{26} = h_{36} = h_{45} = 0$$

and therefore the scheme of moduli becomes

$$
\begin{matrix}
h_{11} & h_{12} & h_{13} & 0 & 0 & 0 \\
 & h_{22} & h_{23} & 0 & 0 & 0 \\
 & & h_{33} & 0 & 0 & 0 \\
 & & & h_{44} & 0 & 0 \\
 & & & & h_{55} & 0 \\
 & & & & & h_{66}
\end{matrix} \tag{3}
$$

But this is also the condition that $x_1 = 0$ should be a plane of elastic symmetry. Thus if a homogeneous material has two perpendicular planes of elastic symmetry, it has also three such planes.

Material with three mutually perpendicular planes of elastic symmetry is called *orthotropic.*

Such material has 9 moduli of elasticity. A thin wooden board whose plane contains the axis of the tree trunk from which it is cut affords an example of approximate orthotropy.

1.94. Axis of elastic symmetry

Let the x_3-axis be one of elastic symmetry, and let

$$F = e_{11} + e_{22}, \; G = e_{22} - e_{11} + 2ie_{12}, \; H = e_{31} - ie_{23}, \; K = e_{33} \tag{1}$$

the notation for F and G being that of [MILNE-THOMSON (6)]. If we rotate about the axis x_3 through the angle θ, the above quantities transform to F', G', H', K' where

$$F' = F\,,\; G' = G e^{2i\theta}\,,\; H' = H e^{i\theta}\,,\; K' = K\,. \tag{2}$$

The proofs of these follow on the lines of those for the fundamental stress combinations Θ and Φ, [MILNE-THOMSON (6)] or from the transformation $x' = x\cos\theta + y\sin\theta$, $y' = -x\sin\theta + y\cos\theta$, $z' = z$.

Since the strain-energy function U is a real homogeneous quadratic function of the e_{rs} it is likewise a homogeneous quadratic function of F, G, H, K and their complex* conjugates F, $\bar{G}$, $\bar{H}$, K the first and last F and K being real numbers. Therefore we can write

$$\begin{aligned} U = {} & aF^2 + bK^2 + cFK + dG\bar{G} + eH\bar{H} \\ & + fG^2 + \bar{f}\bar{G}^2 + gGH + \bar{g}\bar{G}\bar{H} \\ & + hFG + kGK + lH^2 + mFH + nG\bar{H} + pHK \\ & + \bar{h}F\bar{G} + \bar{k}\bar{G}K + \bar{l}\bar{H}^2 + \bar{m}F\bar{H} + \bar{n}\bar{G}H + \bar{p}\bar{H}K\,, \end{aligned} \tag{3}$$

where $a, b, \ldots, p$ are certain constants, of which a, b, c, d, e must be real since they multiply real quantities.

Now rotate about the x_3-axis through the angle θ. The new form for U is obtained from (3) by writing F', G', H', K' for F, G, H, K, which gives, on use of (2),

$$\begin{aligned} U = {} & aF^2 + bK^2 + cFK + dG\bar{G} + eH\bar{H} \\ & + fG^2 e^{4i\theta} + \bar{f}\bar{G}^2 e^{-4i\theta} + gGH e^{3i\theta} + \bar{g}\bar{G}\bar{H} e^{-3i\theta} \\ & + (hFG + kGK + lH^2)\, e^{2i\theta} + (\bar{h}F\bar{G} + \bar{k}\bar{G}K + \bar{l}\bar{H}^2)\, e^{-2i\theta} \\ & + (mFH + nG\bar{H} + pHK)\, e^{i\theta} + (\bar{m}F\bar{H} + \bar{n}\bar{G}H + \bar{p}\bar{H}K)\, e^{-i\theta}. \end{aligned} \tag{4}$$

Subtracting (4) from (3) we have the identity

$$\begin{aligned} & fG^2(1 - e^{4i\theta}) + \bar{f}\bar{G}^2(1 - e^{-4i\theta}) + gGH(1 - e^{3i\theta}) + \bar{g}\bar{G}\bar{H}(1 - e^{-3i\theta}) \\ & \quad + L(1 - e^{2i\theta}) + \bar{L}(1 - e^{-2i\theta}) + M(1 - e^{i\theta}) + \bar{M}(1 - e^{-i\theta}) \equiv 0 \end{aligned} \tag{5}$$

where

$$L = hFG + kGK + lH^2\,,\; M = mFH + nG\bar{H} + pHK\,. \tag{6}$$

If (5) is to hold for every θ, we must have

$$f = 0,\quad g = 0,\quad L = 0,\quad M = 0\,. \tag{7}$$

Since the components e_{rs} are arbitrary we must have

$$f = 0, g = 0, h = 0, k = 0, l = 0, m = 0, n = 0, p = 0 \tag{8}$$

* We denote the complex conjugate of a complex number by a bar. Thus $z = x + iy$ and $\bar{z} = x - iy$ are complex conjugates.

and these are the conditions that the x_3-axis shall be an axis of symmetry for every value of the rotation angle θ. When the conditions (8) are satisfied we have from (1) and (3)

$$\begin{aligned} U = a(e_{11} + e_{22})^2 + b e_{33}^2 + c e_{33}(e_{11} + e_{22}) \\ + d[(e_{22} - e_{11})^2 + 4e_{12}^2] + e(e_{23}^2 + e_{31}^2)\,. \end{aligned} \tag{9}$$

Comparing this with 1.82 (11) we see that the following 12 moduli are zero

$$h_{14},\ h_{15},\ h_{16},\ h_{24},\ h_{25},\ h_{26},\ h_{34},\ h_{35},\ h_{36},\ h_{45},\ h_{46},\ h_{56} \tag{10}$$

while

$$\begin{aligned} \frac{1}{2} h_{11} = a + d, \quad h_{12} = 2(a - d), \quad \frac{1}{2} h_{22} = a + d\,, \\ h_{13} = c,\ h_{23} = c,\ \frac{1}{2} h_{33} = b,\ 2h_{44} = e,\ 2h_{55} = e,\ 2h_{66} = 4d \end{aligned} \tag{11}$$

whence we get

$$h_{11} = h_{22},\quad h_{13} = h_{23},\quad h_{44} = h_{55},\quad 2h_{66} = h_{11} - h_{12} \tag{12}$$

and the matrix of 1.72 (1) yields the scheme

$$\begin{array}{cccccc} h_{11} & h_{12} & h_{13} & 0 & 0 & 0 \\ & h_{11} & h_{13} & 0 & 0 & 0 \\ & & h_{33} & 0 & 0 & 0 \\ & & & h_{44} & 0 & 0 \\ & & & & h_{44} & 0 \\ & & & & & \frac{1}{2}(h_{11} - h_{12}) \end{array} \tag{13}$$

with 5 independent moduli.

The symmetry indicated by (13) is called *monotropy* or *transverse isotropy*, and the axis of symmetry (here the x_3-axis) is the *axis of monotropy*.

All orthogonal triads for which one arm is parallel to the axis of monotropy are equivalent and therefore all directions in any plane perpendicular to the axis of monotropy are equivalent. Thus the material is *isotropic* in respect of directions in planes perpendicular to the axis of monotropy.

Note that monotropy is a particular case of orthotropy, cf. 1.92 (3).

We observe that (5) holds without restriction when $\theta = 2\pi$ and this corresponds with the identical transformation.

We also note that two terms on the left hand side of (5) disappear when $\theta = \pi/2$, when $\theta = 2\pi/3$, and when $\theta = \pi$, and the conclusions (7) must then be modified.

If $\theta = \pi/2$ so that the body allows only rotation through a right angle, (7) is replaced by the less restrictive conditions $g = 0$, $L = 0$, $M = 0$ leading to

$$g = 0, h = 0, k = 0, l = 0, m = 0, n = 0, p = 0 \tag{14}$$

so that

$$U = a(e_{11} + e_{22})^2 + b e_{33}^2 + c e_{33}(e_{11} + e_{22}) + d[(e_{22} - e_{11})^2 + 4e_{12}^2] + e(e_{23}^2 + e_{31}^2) + f(e_{22} - e_{11} + 2ie_{12})^2 + \bar{f}(e_{22} - e_{11} - 2ie_{12})^2 .$$

Comparing this with 1.82 (11) we find that the following 10 moduli are zero

$$h_{14}, h_{15}, h_{24}, h_{25}, h_{34}, h_{35}, h_{36}, h_{45}, h_{46}, h_{56} \tag{15}$$

while

$$h_{11} = h_{22}, h_{13} = h_{23}, h_{44} = h_{55}, h_{26} = -h_{16} . \tag{16}$$

Thus when the material allows only the rotation $\pi/2$, we have the scheme

$$\begin{matrix} h_{11} & h_{12} & h_{13} & 0 & 0 & h_{16} \\ & h_{11} & h_{13} & 0 & 0 & -h_{16} \\ & & h_{33} & 0 & 0 & 0 \\ & & & h_{44} & 0 & 0 \\ & & & & h_{44} & 0 \\ & & & & & h_{66} \end{matrix} \tag{17}$$

with 7 independent moduli.

When $\theta = \pi$ so that the body allows only rotation through two right angles about the x_3-axis (7) is modified to $g = 0$, $M = 0$, leading by similar steps to the vanishing of 8 moduli, to no relations between the remaining moduli and the scheme

$$\begin{matrix} h_{11} & h_{12} & h_{13} & 0 & 0 & h_{16} \\ & h_{22} & h_{23} & 0 & 0 & h_{26} \\ & & h_{33} & 0 & 0 & h_{36} \\ & & & h_{44} & h_{45} & 0 \\ & & & & h_{55} & 0 \\ & & & & & h_{66} \end{matrix} \tag{18}$$

with 13 independent moduli, and a plane of elastic symmetry.

Similarly when the body allows only the rotation $2\pi/3$ about the x_3-axis, (5) reduces to $f = 0$, $L = 0$, $M = 0$. This implies the vanishing of the 6 moduli

$$h_{16}, h_{26}, h_{34}, h_{35}, h_{36}, h_{45} \tag{19}$$

and the relations

$$h_{11} = h_{22}, h_{14} = -h_{24} = h_{56}, h_{13} = h_{23}, h_{15} = -h_{25} = -h_{46}$$
$$h_{44} = h_{55}, h_{66} = \frac{1}{2}(h_{11} - h_{12}) . \tag{20}$$

Thus when $\theta = 2\pi/3$ we have the scheme

$$\begin{matrix} h_{11} & h_{12} & h_{13} & h_{14} & h_{15} & 0 \\ & h_{11} & h_{13} & -h_{14} & -h_{15} & 0 \\ & & h_{33} & 0 & 0 & 0 \\ & & & h_{44} & 0 & -h_{15} \\ & & & & h_{44} & h_{14} \\ & & & & & \frac{1}{2}(h_{11} - h_{12}) \end{matrix} \tag{21}$$

with 7 independent moduli.

Thus it appears that the only axes of rotational symmetry which anisotropic material allows are such that either *every angle* is allowable, or the allowable angles are one of

$$2\pi/n, n = 1, 2, 3, 4 \tag{22}$$

wherein $n = 1$ corresponds to the identical covering and is allowed by all materials.

For monotropic material we have $n = \infty$, which corresponds to the case in which every angle of rotation is allowable on the principle that any angle may be attained by sufficiently many small rotations. Nothing in the above argument asserts that material can have only one axis of rotational symmetry, indeed many crystals have several.

1.95. Axis of alternating symmetry

Let us perform the following operations

(1) Rotation about the x_3-axis through the angle θ, so that our coordinates change from (x_1, x_2, x_3) to (x_1', x_2', x_3'), $x_3' = x_3$.

(2) Reflexion in the plane $x_3 = 0$, so that our coordinates change from (x_1', x_2', x_3') to (x_1'', x_2'', x_3''), $x_1'' = x_1'$, $x_2'' = x_2'$, $x_3'' = -x_3'$.

If we denote by primes and double primes the corresponding changed values of the basic combinations 1.94 (1) we have, using 1.92 (1) and 1.94 (2)

$$F'' = F' = F, G'' = G' = Ge^{2i\theta},$$
$$H'' = -H' = -He^{i\theta}, K'' = K' = K . \tag{1}$$

The final transformation could therefore be obtained from 1.94 (2) by replacing therein the prime by a double prime and θ by $\theta + \pi$. Making the replacement of θ by $\theta + \pi$ in 1.94 (5) therefore leads to the condition that the x_3-axis may be one of alternating symmetry for the rotation θ about it combined with reflexion in the plane $x_3 = 0$ namely

$$\begin{aligned} & fG^2(1 - e^{4i\theta}) + \bar{f}\bar{G}^2(1 - e^{-4i\theta}) + gGH(1 + e^{3i\theta}) + \bar{g}\bar{G}\bar{H}(1 + e^{-3i\theta}) \\ & \quad + L(1 - e^{2i\theta}) + \bar{L}(1 - e^{-2i\theta}) + M(1 + e^{i\theta}) + \bar{M}(1 + e^{-i\theta}) \equiv 0. \end{aligned} \tag{2}$$

Therefore when the material allows rotation through every angle θ combined with reflexion the conditions are the same as 1.94 (8) and we have for the moduli the same scheme as for direct symmetry namely 1.94 (13).

$$\begin{matrix} h_{11} & h_{12} & h_{13} & 0 & 0 & 0 \\ & h_{11} & h_{13} & 0 & 0 & 0 \\ & & h_{33} & 0 & 0 & 0 \\ & & & h_{44} & 0 & 0 \\ & & & & h_{44} & 0 \\ & & & & & \frac{1}{2}(h_{11} - h_{12}) \end{matrix} \tag{3}$$

The rotations which cause two terms of (2) to disappear are now $\pi/2$, $\pi/3$, π.

If the material allows only the rotation $\pi/2$ combined with reflexion, (2) gives

$$g = 0, L = 0, M = 0$$

which is exactly the same as the condition for the rotation $\pi/2$ in direct symmetry and therefore we have the scheme 1.94 (17)

$$\begin{matrix} h_{11} & h_{12} & h_{13} & 0 & 0 & h_{16} \\ & h_{11} & h_{13} & 0 & 0 & -h_{16} \\ & & h_{33} & 0 & 0 & 0 \\ & & & h_{44} & 0 & 0 \\ & & & & h_{44} & 0 \\ & & & & & h_{66} \end{matrix} \tag{4}$$

When the material allows the rotation $\pi/3$ combined with reflexion the conditions given by (2) are

$$f = 0, L = 0, M = 0$$

which are the same as for the case of direct symmetry with rotation $2\pi/3$ and the scheme of moduli is that of 1.94 (21) namely

$$\begin{matrix} h_{11} & h_{12} & h_{13} & h_{14} & h_{15} & 0 \\ & h_{11} & h_{13} & -h_{14} & -h_{15} & 0 \\ & & h_{33} & 0 & 0 & 0 \\ & & & h_{44} & 0 & -h_{15} \\ & & & & h_{44} & h_{14} \\ & & & & & \frac{1}{2}(h_{11} - h_{12}) \end{matrix} \tag{5}$$

If the material allows only the rotation π followed by reflexion, (2) vanishes identically and therefore no conditions are imposed on the moduli. This operation, called *central perversion* [LOVE], corresponds to the transformation which replaces (x_1, x_2, x_3) by $(-x_1, -x_2, -x_3)$. The origin is then a *centre of symmetry*. Thus the existence of a centre of symmetry does not affect the elastic properties. It appears that for an axis of alternating symmetry the allowable single rotations are 2π, π, $\pi/2$, $\pi/3$ or $2\pi/n$, $n = 1, 2, 4, 6$. Combining this result with 1.94 (22) we see that allowable single rotations for an axis of symmetry are

$$2\pi/n, \, n = 1, 2, 3, 4, 6$$

the value 6 being peculiar to alternating symmetry, and the value 3 to direct symmetry.

The geometrical symmetry of crystals may be described by reference to the group of covering operations. Thus classified, crystals fall into 32 classes.

One class allows of the identical operation alone; the corresponding figure has no symmetry.

A second class allows besides the identical operation, the operation of central perversion only.

A third class besides the identical operation allows the operation of reflexion in a plane only.

There are a further 24 classes for which a *principal axis* exists, that is to say every axis of symmetry other than the principal axis is at right angles to the principal axis, and every plane of symmetry either passes through the principal axis or is at right angles to that axis.

The 5 remaining classes are characterized by the presence of 4 axes of symmetry for which the rotation is $2\pi/3$ and which are equally inclined to one another like the diagonals of a cube.

Not all these classes of geometrical symmetry correspond with distinct classes of elastic symmetry, indeed from the point of view of elastic symmetry a division into 9 classes suffices*.

1.98. Isotropic material

Material is said to be *elastically isotropic* when all orthogonal triads whether homothetic or not are equivalent.

For such material every plane is a plane of symmetry and every axis is an axis of symmetry for which the rotation may be of any amount. Now 1.92 (3) exhibits the scheme of moduli when there are three perpendicular planes of symmetry so that this scheme must apply to isotropic material. But the scheme of 1.94 (13) must also apply and therefore for

* For details of the elastic symmetry of crystals see [LOVE].

complete symmetry the scheme for isotropy must be

$$\begin{matrix} h_{11} & h_{12} & h_{12} & 0 & 0 & 0 \\ & h_{11} & h_{12} & 0 & 0 & 0 \\ & & h_{11} & 0 & 0 & 0 \\ & & & \frac{1}{2}(h_{11}-h_{12}) & 0 & 0 \\ & & & & \frac{1}{2}(h_{11}-h_{12}) & 0 \\ & & & & & \frac{1}{2}(h_{11}-h_{12}) \end{matrix} \qquad (1)$$

with two independent moduli.

The strain-energy-function is therefore

$$U = \frac{1}{2} h_{11}(e_{11}^2 + e_{22}^2 + e_{33}^2) + h_{12}(e_{22}e_{33} + e_{33}e_{11} + e_{11}e_{22}) + (h_{11} - h_{12})(e_{23}^2 + e_{31}^2 + e_{12}^2)\,.$$

If we write

$$h_{11} = \lambda + 2\mu,\ h_{12} = \lambda\,,$$

the quantities λ, μ, each having the dimensions of a stress, are known as Lamé's constants for the isotropic material and the strain-energy function is

$$U = \frac{1}{2}\lambda(e_{11} + e_{22} + e_{33})^2 + \mu\{e_{11}^2 + e_{22}^2 + e_{33}^2 + 2e_{23}^2 + 2e_{31}^2 + 2e_{12}^2\}\,. \qquad (3)$$

From 1.72 the h-matrix is

$$\begin{pmatrix} \lambda+2\mu & \lambda & \lambda & 0 & 0 & 0 \\ \lambda & \lambda+2\mu & \lambda & 0 & 0 & 0 \\ \lambda & \lambda & \lambda+2\mu & 0 & 0 & 0 \\ 0 & 0 & 0 & \mu & 0 & 0 \\ 0 & 0 & 0 & 0 & \mu & 0 \\ 0 & 0 & 0 & 0 & 0 & \mu \end{pmatrix} \qquad (4)$$

and therefore its inverse, the k-matrix, is

$$\begin{pmatrix} \frac{1}{E} & \frac{-\eta}{E} & \frac{-\eta}{E} & 0 & 0 & 0 \\ \frac{-\eta}{E} & \frac{1}{E} & \frac{-\eta}{E} & 0 & 0 & 0 \\ \frac{-\eta}{E} & \frac{-\eta}{E} & \frac{1}{E} & 0 & 0 & 0 \\ 0 & 0 & 0 & \frac{1}{\mu} & 0 & 0 \\ 0 & 0 & 0 & 0 & \frac{1}{\mu} & 0 \\ 0 & 0 & 0 & 0 & 0 & \frac{1}{\mu} \end{pmatrix} \qquad (5)$$

Here $\eta = \lambda/[2(\lambda+\mu)]$ is Poisson's ratio and $E = 2\mu(1+\eta)$ is Young's modulus.

EXAMPLES I

1. If A is a 2-tensor, show that in a cartesian representation we can write

$$\begin{aligned}\mathsf{A} = a_{11}(\boldsymbol{i}_1;\boldsymbol{i}_1) + a_{12}(\boldsymbol{i}_1;\boldsymbol{i}_2) + a_{13}(\boldsymbol{i}_1;\boldsymbol{i}_3)\\ + a_{21}(\boldsymbol{i}_2;\boldsymbol{i}_1) + a_{22}(\boldsymbol{i}_2;\boldsymbol{i}_2) + a_{23}(\boldsymbol{i}_2;\boldsymbol{i}_3)\\ + a_{31}(\boldsymbol{i}_3;\boldsymbol{i}_1) + a_{32}(\boldsymbol{i}_3;\boldsymbol{i}_2) + a_{33}(\boldsymbol{i}_3;\boldsymbol{i}_3) = a_{rs}(\boldsymbol{i}_r;\boldsymbol{i}_s)\end{aligned}$$

and that the conjugate tensor is

$$\mathsf{A}_c = a_{rs}(\boldsymbol{i}_s;\boldsymbol{i}_r) = a_{sr}(\boldsymbol{i}_r;\boldsymbol{i}_s)\ .$$

2. In the notation of Ex. 1 verify that the tensor $\mathsf{A} - \mathsf{A}_c$ is skew symmetric.

3. If A is a 2-tensor and $\boldsymbol{r}$ a position vector, show that the equation

$$\mathsf{A}\cdot\boldsymbol{r} = \lambda\boldsymbol{r}$$

is, in general, satisfied by exactly three eigen-values λ_1, λ_2, λ_3 of the scalar λ.

4. In Ex. 3 define the first, second, and third scalar invariants of A by

$$\mathsf{A}_I = \lambda_1 + \lambda_2 + \lambda_3,\ \mathsf{A}_{II} = \lambda_2\lambda_3 + \lambda_3\lambda_1 + \lambda_1\lambda_2,\ \mathsf{A}_{III} = \lambda_1\lambda_2\lambda_3$$

and hence prove the Cayley-Hamilton identity

$$\mathsf{A}_{III} - \mathsf{A}_{II}\lambda + \mathsf{A}_I\lambda^2 - \lambda^3 = (\mathsf{A} - \lambda\mathsf{I})_{III},$$

where I is the idemfactor.

5. Use Ex. 4 and the notation of Ex. 1 to show that the right hand side of the Cayley-Hamilton identity can be written

$$\begin{vmatrix} a_{11}-\lambda & a_{12} & a_{13}\\ a_{21} & a_{22}-\lambda & a_{23}\\ a_{31} & a_{32} & a_{33}-\lambda\end{vmatrix}$$

and that

$$\mathsf{A}_{III} = \begin{vmatrix} a_{11} & a_{12} & a_{13}\\ a_{21} & a_{22} & a_{23}\\ a_{31} & a_{32} & a_{33}\end{vmatrix}$$

6. In the first state, three infinitesimal vectors $d\boldsymbol{P}_0$, $d\boldsymbol{P}_0'$, $d\boldsymbol{P}_0''$ issue from $\boldsymbol{P}_0$ and in the second state, they are deformed into three infinitesimal vectors $d\boldsymbol{P}$, $d\boldsymbol{P}'$, $d\boldsymbol{P}''$ issuing from $\boldsymbol{P}$. Show that these vector triplets define parallelepipeds of volumes dV_0, dV respectively, where

$$dV_0 = [d\boldsymbol{P}_0, d\boldsymbol{P}_0', d\boldsymbol{P}_0''],\quad dV = [d\boldsymbol{P}, d\boldsymbol{P}', d\boldsymbol{P}''],$$

and the square brackets denote triple scalar products.

7. In Ex. 6 write $d\boldsymbol{P}_0 = d\boldsymbol{P}\cdot\mathsf{E}$, $\mathsf{E} = \dfrac{\partial;\boldsymbol{P}_0}{\partial\boldsymbol{P}} = \mathsf{I} - \dfrac{\partial;\boldsymbol{q}}{\partial\boldsymbol{P}}$.

Prove that

$$\frac{dV_0}{dV} = \mathsf{E}_{III} = \left(\mathsf{I} - \frac{\partial;\boldsymbol{q}}{\partial \boldsymbol{P}}\right)_{III}.$$

8. If in Ex. 7, $dV_0/dV = 1 - \varepsilon$, use the Cayley-Hamilton identity (Ex. 4) to prove that

$$(1-\varepsilon)^2 = 1 - 2\mathsf{D}_I^E + 4\mathsf{D}_{II}^E - 8\mathsf{D}_{III}^E.$$

9. If A, B, C are any three dyads, prove that

$$(\mathsf{A}\cdot\mathsf{B})\cdot\cdot\,\mathsf{C} = (\mathsf{B}\cdot\mathsf{C})\cdot\cdot\,\mathsf{A} = (\mathsf{C}\cdot\mathsf{A})\cdot\cdot\,\mathsf{B}$$
$$= \mathsf{A}\cdot\cdot\,(\mathsf{B}\cdot\mathsf{C}) = \mathsf{B}\cdot\cdot\,(\mathsf{C}\cdot\mathsf{A}) = \mathsf{C}\cdot\cdot\,(\mathsf{A}\cdot\mathsf{B}),$$

in other words the product is unchanged by cyclic interchange of the letters and any distribution of the signs (·) and (· ·).

Prove that the same results hold when A, B, C are 2-tensors.

10. Find the norm of the 2-tensor of Ex. 1 in terms of the a_{rs} and show that if the a_{rs} are real, so is the norm as defined in the footnote of 1.32.

11. Use 1.32 (7) to obtain the compatibility equations for the deformation tensor $\mathsf{D} = e_{rs}(\boldsymbol{i}_r;\boldsymbol{i}_s)$ expressed in cartesian coordinates, proving that only six distinct equations are obtained.

12. Show that for hydrostatic pressure p the stress tensor is $-p\mathsf{I}$, where I is the idemfactor.

13. If S is the stress tensor at a fixed point P and if $\boldsymbol{r}$ is the position vector of Q referred to P, show that

$$\boldsymbol{r}\cdot\mathsf{S}\cdot\boldsymbol{r} = c$$

for different values of the constant c is a family of homothetic quadric surfaces (called *stress quadrics*).

14. Show that the normal stress across any plane through the centre of the stress quadric (Ex. 13) is inversely proportional to the square of that radius vector of the quadric which is normal to the plane.

15. Show that when a stress quadric (Ex. 13) is referred to its principal axes, the tangential tractions across the principal planes of the quadric are zero.

Use this property to show that at any point of the material there exist three orthogonal planes across each of which the traction is purely normal. These normal tractions are called *principal stresses*.

16. Show that if the state of stress is purely normal across every plane at a given point of the material, the stress quadric is a sphere.

17. The state of stress at a point is one of simple tension, that is to say, the traction across one plane through the point is normal to the plane and the traction across any perpendicular plane vanishes. Prove that the stress quadric reduces to two parallel planes.

Show that of the three principal stresses only one is different from zero.

18. If the body-vector is zero, show that the equations of equilibrium 1.4 (7) are identically satisfied by

$$\widehat{xx} = \frac{\partial^2 \chi_3}{\partial y^2} + \frac{\partial^2 \chi_2}{\partial R^2}, \; \widehat{yy} = \frac{\partial^2 \chi_1}{\partial R^2} + \frac{\partial^2 \chi_3}{\partial x^2}, \; \widehat{zz} = \frac{\partial^2 \chi_2}{\partial x^2} + \frac{\partial^2 \chi_1}{\partial y^2},$$

$$\widehat{yz} = -\frac{\partial^2 \chi_1}{\partial y \partial R}, \; \widehat{zx} = -\frac{\partial^2 \chi_2}{\partial R \partial x}, \; \widehat{xy} = -\frac{\partial^2 \chi_3}{\partial x \partial y}$$

where χ_1, χ_2, χ_3 are arbitrary (differentiable) functions (Maxwell's stress functions).

19. Show that, if the body-vector is zero, the equations of equilibrium 1.4 (7) are identically satisfied by

$$\widehat{xx} = \frac{\partial^2 \psi_1}{\partial y \partial R}, \; \widehat{yy} = \frac{\partial^2 \psi_2}{\partial R \partial x}, \; \widehat{zz} = \frac{\partial^2 \psi_3}{\partial x \partial y},$$

$$\widehat{yz} = -\frac{1}{2}\frac{\partial}{\partial x}\left(-\frac{\partial \psi_1}{\partial x} + \frac{\partial \psi_2}{\partial y} + \frac{\partial \psi_3}{\partial R}\right),$$

$$\widehat{zx} = -\frac{1}{2}\frac{\partial}{\partial y}\left(\frac{\partial \psi_1}{\partial x} - \frac{\partial \psi_2}{\partial y} + \frac{\partial \psi_3}{\partial R}\right),$$

$$\widehat{xy} = -\frac{1}{2}\frac{\partial}{\partial R}\left(\frac{\partial \psi_1}{\partial x} + \frac{\partial \psi_2}{\partial y} - \frac{\partial \psi_3}{\partial R}\right),$$

where ψ_1, ψ_2, ψ_3 are arbitrary (differentiable) functions (Morera's stress functions).

20. Prove 1.7 (8) for a general tensor and in particular verify that

$$\mathsf{J}_{(4)} \cdot\cdot\, \mathsf{J}_{(4)} = \mathsf{J}_{(4)}\,.$$

21. Use 1.7 (14) to obtain the scheme of elastic moduli when $x_3 = 0$ is plane of elastic symmetry.

22. Use 1.82 (4) to develop the strain-energy function in the mixed form $U = \frac{1}{2}\delta_{\alpha s}\delta_{r\beta}\widehat{x_r x_s} e_{\alpha\beta}$.

23. Obtain the transformation (1.94)

$$F' = F, \; G' = G e^{2i\theta}, \; H' = H e^{i\theta}, \; K' = K$$

for rotation through the angle θ about the x_3-axis.

Chapter II

Stress functions and complex stresses

2.0. Introductory notions

Consider a cylindrical or prismatic rod of any form of cross-section bounded by plane ends perpendicular to the generators.

Suppose the rod to be in equilibrium under the action of external forces applied to the end sections and suppose further that body force is absent.

When the external force is parallel to the generators we have the problem of *extension* (or compression), and when the external force is perpendicular to the generators we have the problem of *flexure*. If the external forces reduce to twisting couples about a longitudinal axis we have the *torsion* problem, while the problem of *bending by couples* arises when the external forces reduce to couples about an axis perpendicular to the generators.

Naturally all the above sets of forces may be applied simultaneously to give a mixed problem of extension, flexure, torsion and bending. Within the framework of the linearized theory of elasticity the stresses which arise from the simultaneous application of these sets may be arrived at by adding the results obtained when each set is conceived to act in the absence of all the others. As we shall be concerned only with the linearized theory it is often convenient to consider each of the above problems in isolation.

When the rod is sufficiently long the principle of de St. Venant asserts that the state of stress in the middle parts of the rod is independent of the precise manner in which the end loads are applied. Suppose then that we know the state of stress in the middle parts and let S_1, S_2 be two cross-sections of the rod situated in these middle parts. Clearly if we imagine the rod to be cut across at S_1 and S_2, the part between S_1 and S_2 would still be in equilibrium under the action across these faces of a stress distribution which exactly equilibrates the stress system existing over S_1 and S_2 before the rod was cut. By the rules of statics at chosen base points O_1 in S_1 and O_2 in S_2 the equilibrating distributions would reduce to a force $\boldsymbol{F}_1$ and a couple $\boldsymbol{M}_1$ applied to S_1 at O_1 and to a force $\boldsymbol{F}_2$ and a couple $\boldsymbol{M}_2$ applied to S_2 at O_2.

We should therefore have the exact solution for the stress distribution in a rod bounded by S_1 and S_2 in equilibrium when acted upon by the system $(\boldsymbol{F}_1, \boldsymbol{M}_1)$ at O_1 and the system $(\boldsymbol{F}_2, \boldsymbol{M}_2)$ at O_2, provided that we suppose these systems to be applied not as concentrated forces or couples, but as the stress distributions from which they were deduced.

These considerations are the basis of de St. Venant's *semi-inverse method*. Suppose, for example, the problem is to find the stress distribution in a rod in equilibrium under a prescribed total load $(\boldsymbol{F}_1, \boldsymbol{M}_1)$ at the end S_1 applied at the point O_1, the other end S_2 of the rod being fixed, and the lateral faces being free of applied force. We seek for a stress distribution which will satisfy the condition of zero stress on the lateral faces and which will give the required *resultant* $(\boldsymbol{F}_1, \boldsymbol{M}_1)$ over S_1. We can then assert that this is the exact solution of the particular problem for which the load on S_1 is applied not by concentrated forces or couples but by the distribution over S_1 which has just been found.

Take rectangular cartesian axes x, y, R of which the R-axis is parallel to the generators (fig. 2.0 (i)). Consider the matrix of the stress components partitioned as shown in (1)

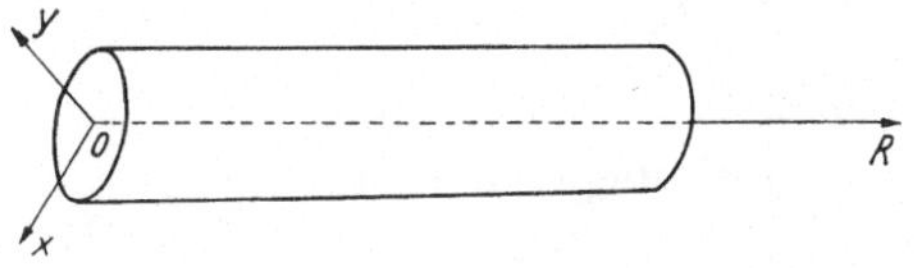

Fig. 2.0 (i)

$$\left[\begin{array}{cc|c} \widehat{xx} & \widehat{xy} & \widehat{xz} \\ \widehat{yx} & \widehat{yy} & \widehat{yz} \\ \hline \widehat{zx} & \widehat{zy} & \widehat{zz} \end{array}\right] \qquad (1)$$

where the symmetry relations

$$\widehat{yz} = \widehat{zy}, \ \widehat{zx} = \widehat{xz}, \ \widehat{xy} = \widehat{yx} \qquad (2)$$

are satisfied.

When the matrix

$$\begin{bmatrix} \widehat{zx} & \widehat{zy} \end{bmatrix} = 0 \qquad (3)$$

and the matrix

$$\begin{bmatrix} \widehat{xx} & \widehat{xy} \\ \widehat{yx} & \widehat{yy} \end{bmatrix} \text{ is independent of } R \qquad (4)$$

we have *plane* problems [MILNE-THOMSON (6)].

Problems for *isotropic* rods in which the matrix

$$\begin{bmatrix} \widehat{zx} & \widehat{zy} \end{bmatrix} \text{ is independent of } R \qquad (5)$$

and for which the matrix

$$\begin{bmatrix} \widehat{xx} & \widehat{xy} \\ \widehat{yx} & \widehat{yy} \end{bmatrix} = 0 \qquad (6)$$

have been called by Filon *antiplane* problems (see fig. 2.0 (ii)). The solution of such problems by the semi-inverse method is based on the assumptions (5) and (6). It is also found to be characteristic of such problems that isotropic material which occupies planes perpendicular to the generators before deformation in general becomes warped after

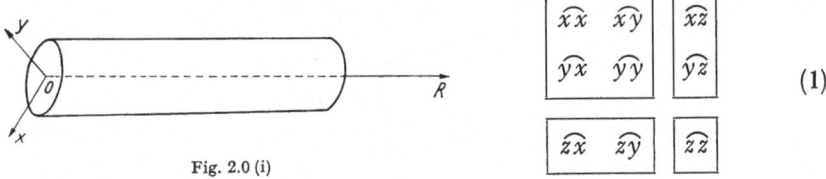

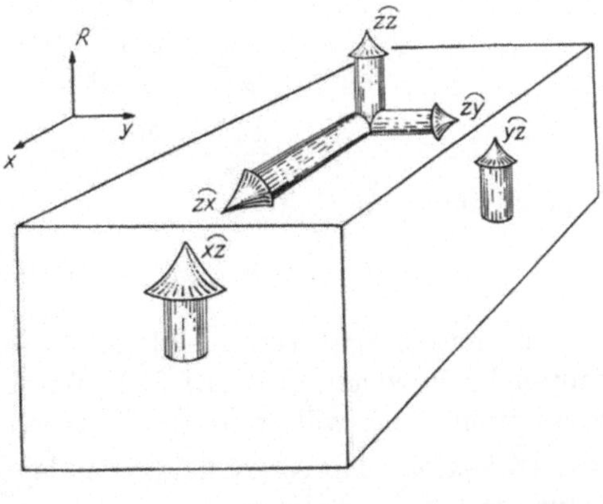

Fig. 2.0 (ii)

deformation, so as no longer to occupy planes perpendicular to the generators, or even to occupy planes at all.

When the material is no longer isotropic it is found that the assumption (6) is too restrictive but can be replaced by

$$\boxed{\begin{matrix} \widehat{xx} & \widehat{xy} \\ \widehat{yx} & \widehat{yy} \end{matrix}} \text{ is independent of } R\,. \tag{7}$$

We shall therefore generalize FILON's definition of antiplane systems in the manner described in the following section.

2.01. Definition of an antiplane system

We shall call a system *antiplane* when there exists a fixed plane Π such that the following conditions are satisfied.

(i) The only stress component which can depend upon distance from Π is the stress component normal to Π.

(ii) Particles, which before deformation lie in planes parallel to Π, cease to do so after deformation.

We shall call Π the *antiplane* of the system.

The antiplane Π can be replaced by any parallel plane Π', for the distances of a point from Π and Π' differ by a constant, so that a function of distance from one is equally a function of distance from the other.

Let us take orthogonal cartesian axes x, y, R of which the R-axis is perpendicular to the antiplane.

The equations of motion 1.4 (7) then reduce to

$$\begin{aligned} \frac{\partial \widehat{xx}}{\partial x} + \frac{\partial \widehat{yx}}{\partial y} &= b_1 \\ \frac{\partial \widehat{xy}}{\partial x} + \frac{\partial \widehat{yy}}{\partial y} &= b_2 \\ \frac{\partial \widehat{xz}}{\partial x} + \frac{\partial \widehat{yz}}{\partial y} &= b_3 - \frac{\partial \widehat{zz}}{\partial R}\,. \end{aligned} \tag{3}$$

It now follows that

$$b_1,\ b_2,\ b_3 - \partial \widehat{zz}/\partial R \text{ are all independent of } R\,. \tag{4}$$

This means that the antiplane stress situation can exist only in body-fields in which (4) is satisfied. We shall complete this result by an assumption which will apply to all our subsequent work.

Assumption. The body-field is independent of distance from the antiplane.

It will then follow from (4) that $\partial\widehat{zz}/\partial R$ is independent of R and therefore that

$$\widehat{zz} = Rf(x, y) + g(x, y) \tag{5}$$

where the functions f and g are independent of R.

All the foregoing applies independently of elastic response. In the case of infinitesimal elastic deformation we have from 1.7 (5)

$$\mathsf{D} = \mathsf{K}_{(4)} \cdot\cdot \mathsf{S}.$$

Therefore the deformation tensor D is at most a linear function of R, and if S is independent of R, so is D.

It is clear from the definition of antiplane stress that plane elastic systems as defined in Milne-Thomson (6) are limiting cases of antiplane systems for which $\widehat{zz}$ is independent of distance from Π and particles in planes parallel to Π undergo zero displacement in the direction normal to Π.

Other examples of limiting cases are the extension of an isotropic cylinder by normal stress uniformly distributed over the end section; and an isotropic cylinder whose cross-section is a circle twisted by a shear distribution proportional to the distance r from the centre of an end and tangential to the circumference of the circle of radius r about the centre of the end.

The purpose of condition (ii) of the definition of antiplane stress is to insist that, in general, there must be some type of warping after deformation of planes parallel to the antiplane, and it is just this condition which becomes nullified in the limiting cases mentioned above.

2.1. Stress functions and fundamental stress combinations

From 2.01 we have the equations of motion

$$\frac{\partial\widehat{xx}}{\partial x} + \frac{\partial\widehat{xy}}{\partial y} = b_1 \tag{1}$$

$$\frac{\partial\widehat{xy}}{\partial x} + \frac{\partial\widehat{yy}}{\partial y} = b_2 \tag{2}$$

$$\frac{\partial\widehat{xz}}{\partial x} + \frac{\partial\widehat{yz}}{\partial y} = b_3 - \frac{\partial\widehat{zz}}{\partial R}. \tag{3}$$

By the assumption introduced in 2.01, b_1, b_2, b_3 are all independent of R.

Now introduce the complex variables

$$z = x + iy, \quad \bar{z} = x - iy$$

so that

$$\frac{\partial}{\partial x} = \frac{\partial}{\partial z} + \frac{\partial}{\partial \bar{z}}, \quad \frac{\partial}{\partial y} = i\frac{\partial}{\partial z} - i\frac{\partial}{\partial \bar{z}}. \tag{4}$$

Substitute in (1) and (2) and form the combination $b_1 - i b_2$. Then we find that

$$\frac{\partial}{\partial z}(\widehat{xx} + \widehat{yy}) - \frac{\partial}{\partial \bar{z}}(\widehat{yy} - \widehat{xx} + 2i\,\widehat{xy}) = b_1 - i b_2\,, \tag{5}$$

while substitution in (3) gives

$$\frac{\partial}{\partial z}(\widehat{\bar{x}\bar{z}} + i\widehat{\bar{y}\bar{z}}) + \frac{\partial}{\partial \bar{z}}(\widehat{xz} - i\widehat{yz}) = B_3\,, \tag{6}$$

where

$$B_3 = b_3 - \frac{\partial \widehat{zz}}{\partial R} = B_3(x, y)\,. \tag{7}$$

From 2.01, B_3 is independent of R.

We call

$$\Theta = \widehat{xx} + \widehat{yy} \tag{8}$$

$$\Phi = \widehat{yy} - \widehat{xx} + 2i\,\widehat{xy} \tag{9}$$

$$\Psi = \widehat{xz} - i\,\widehat{yz} \tag{10}$$

the *fundamental stress combinations*. All the stress components can be expressed by means of Θ, Φ, Ψ and their complex conjugates. Equations (5) and (6) now become

$$\frac{\partial \Theta}{\partial z} - \frac{\partial \Phi}{\partial \bar{z}} = b_1 - i b_2 \tag{11}$$

$$\frac{\partial \overline{\Psi}}{\partial z} + \frac{\partial \Psi}{\partial \bar{z}} = B_3\,. \tag{12}$$

Let us denote by Θ_0, Φ_0, Ψ_0 particular integrals of these equations. Then

$$\frac{\partial \Theta_0}{\partial z} - \frac{\partial \Phi_0}{\partial \bar{z}} = b_1 - i b_2 \tag{13}$$

$$\frac{\partial \overline{\Psi}_0}{\partial z} + \frac{\partial \Psi_0}{\partial \bar{z}} = B_3 \tag{14}$$

and therefore

$$\frac{\partial}{\partial z}(\Theta - \Theta_0) - \frac{\partial}{\partial \bar{z}}(\Phi - \Phi_0) = 0\,,$$

$$\frac{\partial}{\partial z}(\overline{\Psi} - \overline{\Psi}_0) + \frac{\partial}{\partial \bar{z}}(\Psi - \Psi_0) = 0\,.$$

If we introduce real-valued functions χ and ψ, called *stress functions*, the above equations have the solutions

$$\Theta = \Theta_0 + 4\frac{\partial^2 \chi}{\partial z \partial \bar{z}}\,, \quad \Phi = \Phi_0 + 4\frac{\partial^2 \chi}{\partial z^2}\,, \quad \Psi = \Psi_0 + 2i\frac{\partial \psi}{\partial z}\,. \tag{15}$$

The determination of the stress components is thus reduced to finding suitable stress functions. If we write

$$\Theta_0 = \widehat{xx}_0 + \widehat{yy}_0\,, \Phi_0 = \widehat{yy}_0 - \widehat{xx}_0 + 2i\,\widehat{xy}_0\,, \Psi_0 = \widehat{xz}_0 - i\,\widehat{yz}_0\,, \tag{16}$$

we have from (15)

$$\widehat{xx} = \frac{\partial^2 \chi}{\partial y^2} + \widehat{xx}_0, \ \widehat{yy} = \frac{\partial^2 \chi}{\partial x^2} + \widehat{yy}_0, \ \widehat{xy} = -\frac{\partial^2 \chi}{\partial x \partial y} + \widehat{xy}_0$$
$$\widehat{xz} = \frac{\partial \psi}{\partial y} + \widehat{xz}_0, \ \widehat{yz} = -\frac{\partial \psi}{\partial x} + \widehat{yz}_0. \tag{17}$$

2.12. Case where the body-vector derives from a potential

When the body-vector is derivable from a potential V_0

$$b_1 = \frac{\partial V_0}{\partial x}, \quad b_2 = \frac{\partial V_0}{\partial y}, \quad b_3 = \frac{\partial V_0}{\partial R} \tag{1}$$

we note that V_0 can only depend linearly on R, otherwise the right-hand members of 2.1 (11), (12) would depend on R which is impossible since Θ, Φ, Ψ are independent of R.

Thus we must have

$$V_0 = V(x, y) + R V_1(x, y) = V + R V_1$$

so that

$$b_1 - i b_2 = \frac{\partial V}{\partial x} - i \frac{\partial V}{\partial y} + R \left(\frac{\partial V_1}{\partial x} - i \frac{\partial V_1}{\partial y} \right).$$

Since $b_1 - i b_2$ is independent of R we must have

$$\frac{\partial V_1}{\partial x} = 0, \qquad \frac{\partial V_1}{\partial y} = 0$$

and therefore V_1 must be a constant, say c. Therefore

$$V_0 = V(x, y) + Rc \tag{2}$$

and therefore

$$B_3 = c - \frac{\partial \widehat{zz}}{\partial R}. \tag{3}$$

Since $b_1 - i b_2 = \frac{\partial V}{\partial x} - i \frac{\partial V}{\partial y} = 2 \frac{\partial V}{\partial z}$ (4)

we can satisfy 2.1 (13) by taking

$$\Theta_0 = 2V, \ \Phi_0 = 0, \tag{5}$$

so that in this case

$$\Theta = 4 \frac{\partial^2 \chi}{\partial z \partial \bar{z}} + 2V, \qquad \Phi = 4 \frac{\partial^2 \chi}{\partial z^2}. \tag{6}$$

The foregoing results do not depend on the assumption of elastic response of the material. They hold also for plastic and other types of response.

For Ψ_0 see 2.24.

2.2. The strain coefficients for antiplane stress

If we introduce cartesian coordinates x_1, x_2, x_3 with unit vectors $\boldsymbol{i}_1$, $\boldsymbol{i}_2$, $\boldsymbol{i}_3$, the deformation tensor is

$$\mathsf{D} = e_{pq}(\boldsymbol{i}_p; \boldsymbol{i}_q)$$

and the compatibility equation 1.32 (7) can be written

$$\boldsymbol{i}_\alpha \frac{\partial}{\partial x_\alpha} \wedge e_{pq}(\boldsymbol{i}_p;\boldsymbol{i}_q) \wedge \boldsymbol{i}_\beta \frac{\partial}{\partial x_\beta} = 0$$

or

$$(\boldsymbol{i}_\alpha \wedge \boldsymbol{i}_p);\,(\boldsymbol{i}_q \wedge \boldsymbol{i}_\beta)\,\frac{\partial^2 e_{pq}}{\partial x_\alpha \partial x_\beta} = 0\,.$$

Given α, β, p, q we get four combinations

$$(\boldsymbol{i}_\alpha \wedge \boldsymbol{i}_p);\,(\boldsymbol{i}_q \wedge \boldsymbol{i}_\beta),\ (\boldsymbol{i}_p \wedge \boldsymbol{i}_\alpha);\,(\boldsymbol{i}_q \wedge \boldsymbol{i}_\beta),\ (\boldsymbol{i}_p \wedge \boldsymbol{i}_\alpha);\,(\boldsymbol{i}_\beta \wedge \boldsymbol{i}_q),\ (\boldsymbol{i}_\alpha \wedge \boldsymbol{i}_p);\,(\boldsymbol{i}_\beta \wedge \boldsymbol{i}_q)$$

which carry alternating signs so that the compatibility equations are

$$\frac{\partial^2 e_{pq}}{\partial x_\alpha \partial x_\beta} - \frac{\partial^2 e_{\alpha q}}{\partial x_\beta \partial x_p} + \frac{\partial^2 e_{\alpha\beta}}{\partial x_p \partial x_q} - \frac{\partial^2 e_{\beta p}}{\partial x_\alpha \partial x_q} = 0\,.$$

Of these 81 equations only 6 are not identities. Reverting to the coordinates (x, y, R), unit vectors $\boldsymbol{i}, \boldsymbol{j}, \boldsymbol{k}$, and the strain coefficients

$$e_{xx},\ e_{yy},\ e_{zz},\ e_{yz},\ e_{zx},\ e_{xy}$$

we have the following two groups of compatibility equations

$$\left.\begin{aligned}
&\frac{\partial^2 e_{yy}}{\partial R^2} - 2\frac{\partial^2 e_{yz}}{\partial y \partial R} + \frac{\partial^2 e_{zz}}{\partial y^2} = 0\\
&\frac{\partial^2 e_{zz}}{\partial x^2} - 2\frac{\partial^2 e_{zx}}{\partial x \partial R} + \frac{\partial^2 e_{xx}}{\partial R^2} = 0\\
&\frac{\partial^2 e_{zx}}{\partial y \partial R} - \frac{\partial^2 e_{xy}}{\partial R^2} + \frac{\partial^2 e_{yz}}{\partial x \partial R} - \frac{\partial^2 e_{zz}}{\partial x \partial y} = 0
\end{aligned}\right\} \tag{1}$$

$$\frac{\partial^2 e_{xx}}{\partial y^2} - 2\frac{\partial^2 e_{xy}}{\partial x \partial y} + \frac{\partial^2 e_{yy}}{\partial x^2} = 0 \tag{2}$$

$$\frac{\partial^2 e_{xy}}{\partial x \partial R} - \frac{\partial^2 e_{xx}}{\partial y \partial R} + \frac{\partial^2 e_{zx}}{\partial x \partial y} - \frac{\partial^2 e_{yz}}{\partial x^2} = 0 \tag{3}$$

$$\frac{\partial^2 e_{yy}}{\partial x \partial R} - \frac{\partial^2 e_{xy}}{\partial y \partial R} + \frac{\partial^2 e_{zx}}{\partial y^2} - \frac{\partial^2 e_{yz}}{\partial x \partial y} = 0\,. \tag{4}$$

Take $R = 0$ as the antiplane of the system.

Then of the stress components $\widehat{zz}$ alone depends on R and then linearly so that we can write the stress tensor in the form

$$\mathsf{S} = \mathsf{S}_0(x, y) + R f(x, y)\,(\boldsymbol{k};\boldsymbol{k}), \tag{5}$$

where $\mathsf{S}_0(x, y)$ and $f(x, y)$ are independent of R.

Since $\mathsf{D} = \mathsf{K}_{(4)} \cdot\cdot\, \mathsf{S} = \mathsf{K}_{(4)} \cdot\cdot\, \mathsf{S}_0 + R f(x, y)\,(\mathsf{K}_{(4)} \cdot\cdot\, \boldsymbol{k};\boldsymbol{k})$ we see that D is linear in R so that

$$e_{\alpha\beta} = e^0_{\alpha\beta}(x, y) + R f(x, y)\, K_{\alpha\beta 33} \tag{6}$$

where $e^0_{\alpha\beta}(x, y)$ is independent of R.

It now follows from (1), by differentiation with respect to R that

$$\frac{\partial^2}{\partial x^2}\left(\frac{\partial e_{zz}}{\partial R}\right) = \frac{\partial^2}{\partial y^2}\left(\frac{\partial e_{zz}}{\partial R}\right) = \frac{\partial^2}{\partial x \partial y}\left(\frac{\partial e_{zz}}{\partial R}\right) = 0\,.$$

Therefore $\partial e_{zz}/\partial R$ is a linear function of x, y only so that we can write

$$\frac{\partial e_{zz}}{\partial R} = A_1 x + B_1 y + C_1 m \tag{7}$$

where A_1, B_1, C_1, m are constants. If we suppose m to have the dimensions of a length, A_1, B_1, C_1 will have the dimensions of (length)$^{-2}$. Integrating (7) we get

$$e_{zz} = (A_1 x + B_1 y + C_1 m)\, R + e_{zz}^0 , \tag{8}$$

where e_{zz}^0 is independent of R.

Comparing this with (6) we see that

$$K_{3333} f(x, y) = k_{33} f(x, y) = (A_1 x + B_1 y + C_1 m)$$

which shows that the stress tensor is

$$\mathsf{S} = \mathsf{S}_0(x, y) + \frac{1}{k_{33}} (A_1 x + B_1 y + C_1 m)\, R\, (\boldsymbol{k}; \boldsymbol{k}) , \tag{9}$$

while from (6)

$$e_{\alpha\beta} = e_{\alpha\beta}^0 + \frac{K_{\alpha\beta 33}}{k_{33}} (A_1 x + B_1 y + C_1 m)\, R . \tag{10}$$

Formula (10) shows that, in general, all the strain coefficients depend linearly on R, whereas of the stress components $\widehat{zz}$ alone depends on R. From (9) we have

$$\widehat{zz} = \widehat{zz}_0 + \frac{1}{k_{33}} (A_1 x + B_1 y + C_1 m)\, R , \tag{11}$$

where $\widehat{zz}_0$ is independent of R.

2.24. The stress combination Ψ

From 2.1 (3) and 2.2 (11) we have

$$\frac{\partial \widehat{zx}}{\partial x} + \frac{\partial \widehat{yz}}{\partial y} + \frac{1}{k_{33}} (A_1 x + B_1 y + C_1 m) = b_3 , \tag{1}$$

where if the body-vector is derived from the potential V_0, cf. 2.12 (2), which we shall write in the form

$$V_0 = V(x, y) + \frac{\varepsilon m R}{k_{33}} \tag{2}$$

we have $b_3 = \varepsilon m / k_{33}$. Therefore (1) can be written

$$\frac{\partial}{\partial x} \left[\widehat{zx} + \frac{1}{2 k_{33}} \{A_1 x^2 + (C_1 - \varepsilon)\, m x\} \right] + \frac{\partial}{\partial y} \left[\widehat{yz} + \frac{1}{2 k_{33}} \{B_1 y^2 + (C_1 - \varepsilon)\, m y\} \right] = 0 ,$$

and so, introducing the stress function $\psi(x, y) = \psi$, we have

$$\widehat{zx} = \frac{\partial \psi}{\partial y} - \frac{1}{2k_{33}} \{A_1 x^2 + (C_1 - \varepsilon) m x\}$$

$$\widehat{yz} = -\frac{\partial \psi}{\partial x} - \frac{1}{2k_{33}} \{B_1 y^2 + (C_1 - \varepsilon) m y\} \tag{3}$$

whence

$$\Psi = \widehat{xz} - i\widehat{yz} = 2i\frac{\partial \psi}{\partial z} - \frac{1}{8k_{33}} \{A_1(z+\bar{z})^2 + iB_1(z-\bar{z})^2 + 4(C_1-\varepsilon) m\bar{z}\}.$$

Thus if we write

$$A_1 + iB_1 = \beta, \tag{4}$$

we have

$$\Psi = \widehat{xz} - i\widehat{yz} = 2i\frac{\partial \psi}{\partial z} - \frac{1}{8k_{33}} \{\beta(z^2+\bar{z}^2) + 2\bar{\beta}z\bar{z} + 4(C_1-\varepsilon) m\bar{z}\}. \tag{5}$$

Thus in 2.1 (15) we can take

$$\Psi_0 = -\frac{1}{8k_{33}} \{\beta(z^2+\bar{z}^2) + 2\bar{\beta}z\bar{z} + 4(C_1-\varepsilon) m\bar{z}\}. \tag{6}$$

2.3. The displacement

We continue to take $R = 0$ as the antiplane. Then from 2.2 (10)

$$e_{\alpha\beta} = e^0_{\alpha\beta} + \frac{K_{\alpha\beta 33}}{k_{33}} (A_1 x + B_1 y + C_1 m) R, \tag{1}$$

so that for example

$$e_{xx} = e^0_{xx} + \frac{k_{13}}{k_{33}} (A_1 x + B_1 y + C_1 m) R$$

$$2e_{yz} = 2e^0_{yz} + \frac{k_{43}}{k_{33}} (A_1 x + B_1 y + C_1 m) R. \tag{2}$$

For brevity write

$$\nu_s = \frac{k_{s3}}{k_{33}}, \quad \nu_3 = 1. \tag{3}$$

Then from the definition of the deformation tensor in terms of displacement (u, v, w) we have

$$\left.\begin{aligned} &\frac{\partial u}{\partial x} - \nu_1(A_1 x + B_1 y + C_1 m) R = e^0_{xx} \\ &\frac{\partial v}{\partial y} - \nu_2(A_1 x + B_1 y + C_1 m) R = e^0_{yy} \\ &\frac{\partial u}{\partial y} + \frac{\partial v}{\partial x} - \nu_6(A_1 x + B_1 y + C_1 m) R = 2e^0_{xy} \end{aligned}\right\} \tag{4}$$

$$\left.\begin{aligned} &\frac{\partial w}{\partial R} = (A_1 x + B_1 y + C_1 m) R + e^0_{zz} \\ &\frac{\partial u}{\partial R} = \nu_5(A_1 x + B_1 y + C_1 m) R + 2e^0_{zx} - \frac{\partial w}{\partial x} \\ &\frac{\partial v}{\partial R} = \nu_4(A_1 x + B_1 y + C_1 m) R + 2e^0_{yz} - \frac{\partial w}{\partial y} \end{aligned}\right\} \tag{5}$$

Integrating $(5)_1$ with respect to R we get

$$w = \frac{1}{2}(A_1 x + B_1 y + C_1 m) R^2 + e^0_{zz} R + w_0(x, y), \tag{6}$$

here $w_0 = w_0(x, y)$ is an arbitrary function independent of R.

We then have from $(5)_2$ and $(5)_3$

$$\frac{\partial u}{\partial R} = -\frac{1}{2} A_1 R^2 + \left[\nu_5(A_1 x + B_1 y + C_1 m) - \frac{\partial e^0_{zz}}{\partial x}\right] R + 2 e^0_{zx} - \frac{\partial w_0}{\partial x}$$

$$\frac{\partial v}{\partial R} = -\frac{1}{2} B_1 R^2 + \left[\nu_4(A_1 x + B_1 y + C_1 m) - \frac{\partial e^0_{zz}}{\partial y}\right] R + 2 e^0_{yz} - \frac{\partial w_0}{\partial y}.$$

Integrating these equations with respect to R we get

$$u = -\frac{1}{6} A_1 R^3 + \frac{1}{2} P R^2 + Q R + u_0(x, y), \tag{7}$$

$$v = -\frac{1}{6} B_1 R^3 + \frac{1}{2} P' R^2 + Q' R + v_0(x, y), \tag{8}$$

where $u_0 = u_0(x, y)$, $v_0 = v_0(x, y)$ are arbitrary functions independent of R, and where for brevity we have written

$$P = \nu_5(A_1 x + B_1 y + C_1 m) - \frac{\partial e^0_{zz}}{\partial x}, \quad Q = 2 e^0_{zx} - \frac{\partial w_0}{\partial x} \tag{9}$$

$$P' = \nu_4(A_1 x + B_1 y + C_1 m) - \frac{\partial e^0_{zz}}{\partial y}, \quad Q' = 2 e^0_{yz} - \frac{\partial w_0}{\partial y}. \tag{10}$$

The values of u, v given by (7) and (8) must satisfy (4) identically for all values of R.

Substituting we get

$$R^2\left(\frac{1}{2}\frac{\partial P}{\partial x}\right) + \left[\frac{\partial Q}{\partial x} - \nu_1(A_1 x + B_1 y + C_1 m)\right] R + \frac{\partial u_0}{\partial x} - e^0_{xx} \equiv 0$$

$$R\left(\frac{1}{2}\frac{\partial P'}{\partial y}\right) + \left[\frac{\partial Q'}{\partial y} - \nu_2(A_1 x + B_1 y + C_1 m)\right] R + \frac{\partial v_0}{\partial y} - e^0_{yy} \equiv 0$$

$$R^2\left(\frac{1}{2}\frac{\partial P}{\partial y} + \frac{1}{2}\frac{\partial P'}{\partial x}\right) + \left[\frac{\partial Q}{\partial y} + \frac{\partial Q'}{\partial x} - \nu_6(A_1 x + B_1 y + C_1 m)\right] R$$
$$+ \frac{\partial v_0}{\partial x} + \frac{\partial u_0}{\partial y} - 2 e^0_{xy} \equiv 0.$$

In these identities we equate to zero the coefficients of R^2, R, and the term independent of R. Then

$$\frac{\partial^2 e^0_{zz}}{\partial x^2} = \nu_5 A_1, \quad \frac{\partial^2 e^0_{zz}}{\partial y^2} = \nu_4 B_1, \quad 2\,\frac{\partial^2 e^0_{zz}}{\partial x \partial y} = \nu_4 A_1 + \nu_5 B_1 \tag{11}$$

$$\frac{\partial Q}{\partial x} - \nu_1(A_1 x + B_1 y + C_1 m) = 0, \quad \frac{\partial Q'}{\partial y} - \nu_2(A_1 x + B_1 y + C_1 m) = 0$$

$$\frac{\partial Q}{\partial y} + \frac{\partial Q'}{\partial x} - \nu_6(A_1 x + B_1 y + C_1 m) = 0 \tag{12}$$

$$\frac{\partial u_0}{\partial x} = e^0_{xx}, \quad \frac{\partial v_0}{\partial y} = e^0_{yy}, \quad \frac{\partial u_0}{\partial y} + \frac{\partial v_0}{\partial x} = 2 e^0_{xy}. \tag{13}$$

From (11) it appears that e^0_{zz} is a quadratic function of x and y so that, as is easily verified by differentiation,

$$e^0_{zz} = \frac{1}{2}[\nu_5 A_1 x^2 + (\nu_4 A_1 + \nu_5 B_1) xy + \nu_4 B_1 y^2] + A_2 x + B_2 y + C_2 m, \tag{14}$$

where A_2, B_2, C_2 are constants each of the dimensions (length)$^{-1}$.

From (12) we see that Q and Q' are quadratic functions of x, y which are easily proved or verified to be of the form

$$\begin{aligned} Q = \frac{1}{2}[\nu_1 A_1 x^2 + 2\nu_1 B_1 xy + (\nu_6 B_1 - \nu_2 A_1) y^2] \\ + \nu_1 C_1 m x + \frac{1}{2}\nu_6 C_1 m y - \tau y + \omega_2, \end{aligned} \tag{15}$$

$$\begin{aligned} Q' = \frac{1}{2}[(\nu_6 A_1 - \nu_1 B_1) x^2 + 2\nu_2 A_1 xy + \nu_2 B_1 y^2] \\ + \frac{1}{2}\nu_6 C_1 m x + \nu_2 C_1 m y + \tau x - \omega_1, \end{aligned} \tag{16}$$

where ω_1, ω_2, τ are real constants and from (9) and (10) we now get

$$P = \frac{1}{2}(\nu_5 B_1 - \nu_4 A_1) y + \nu_5 C_1 m - A_2 \tag{17}$$

$$P' = -\frac{1}{2}(\nu_5 B_1 - \nu_4 A_1) x + \nu_4 C_1 m - B_2. \tag{18}$$

Finally let us introduce an arbitrary rigid body motion by writing

$$\begin{aligned} u_0 &= u_1(x, y) - \omega_3 y + \alpha \\ v_0 &= v_1(x, y) + \omega_3 x + \beta \\ w_0 &= w_1(x, y) + \omega_1 y - \omega_2 x + \gamma. \end{aligned}$$

We then find by substitution of these values into (6), (7), (8) together with the values of P, Q, P', Q',

$$\begin{aligned} u = u_1 - \frac{1}{6}A_1 R^3 + \frac{1}{4}[(\nu_5 B_1 - \nu_4 A_1) y + 2(\nu_5 C_1 m - A_2)] R^2 \\ + \left[\frac{1}{2}\nu_1 A_1 x^2 + \nu_1 B_1 xy + \frac{1}{2}(\nu_6 B_1 - \nu_2 A_1) y^2\right. \\ \left. + \nu_1 C_1 m x + \frac{1}{2}\nu_6 C_1 m y - \tau y\right] R + \alpha + \omega_2 R - \omega_3 y \end{aligned} \tag{19}$$

$$\begin{aligned} v = v_1 - \frac{1}{6}B_1 R^3 + \frac{1}{4}[-(\nu_5 B_1 - \nu_4 A_1) x + 2(\nu_4 C_1 m - B_2)] R^2 \\ + \left[\frac{1}{2}(\nu_6 A_1 - \nu_1 B_1) x^2 + \nu_2 A_1 xy + \frac{1}{2}\nu_2 B_1 y^2\right. \\ \left. + \frac{1}{2}\nu_6 C_1 m x + \nu_2 C_1 m y + \tau x\right] R + \beta + \omega_3 x - \omega_1 R. \end{aligned} \tag{20}$$

$$\begin{aligned} w = w_1 + \frac{1}{2}(A_1 x + B_1 y + C_1 m) R^2 \\ + \left[\frac{1}{2}\nu_5 A_1 x^2 + \frac{1}{2}(\nu_4 A_1 + \nu_5 B_1) xy + \frac{1}{2}B_1 \nu_4 y^2 + A_2 x + B_2 y + C_2 m\right] R \\ + \gamma - \omega_2 x + \omega_1 y. \end{aligned} \tag{21}$$

In the right-hand sides of (19) and (20) put all quantities equal to zero except $\tau y R$, $\tau x R$. Then

$$u + iv = -\tau y R + i\tau x R = iz\tau R .$$

Therefore τR is an angle of rotation and thus τ is to be interpreted as an angle of twist per unit length measured along a line perpendicular to the antiplane.

Here (α, β, γ) is an arbitrary translation and $(\omega_1, \omega_2, \omega_3)$ an arbitrary small rotation so that the last three terms on the right-hand side of the expressions for u, v, w are the components of an arbitrary rigid body movement.

We note that, since $e_{zz} = \partial w/\partial R$,

$$\begin{aligned} e_{zz} = {} & (A_1 x + B_1 y + C_1 m) R \\ & + \frac{1}{2}\nu_5 A_1 x^2 + \frac{1}{2}(\nu_4 A_1 + \nu_5 B_1) xy + \frac{1}{2} B_1 \nu_4 y^2 \\ & + A_2 x + B_2 y + C_2 m . \end{aligned} \tag{22}$$

We can sum up these results as follows

$$\left.\begin{aligned} u &= u_1 - \frac{1}{6} A_1 R^3 + \frac{1}{2} P R^2 + Q_1 R + \alpha + \omega_2 R - \omega_3 y , \\ v &= v_1 - \frac{1}{6} B_1 R^3 + \frac{1}{2} P' R^2 + Q_1' R + \beta + \omega_3 x - \omega_1 R , \\ w &= w_1 + \frac{1}{2} L R^2 + T R + \gamma + \omega_1 y - \omega_2 x , \end{aligned}\right\} \tag{23}$$

where

$$\widehat{zz} = \widehat{zz}_0 + \frac{1}{k_{33}} (A_1 x + B_1 y + C_1 m) R = \widehat{zz}_0 + \frac{1}{k_{33}} L R \tag{24}$$

$$L = A_1 x + B_1 y + C_1 m \tag{25}$$

$$\begin{aligned} T = e^0_{zz} = {} & \frac{1}{2}\{\nu_5 A_1 x^2 + (\nu_4 A_1 + \nu_5 B_1) xy + \nu_4 B_1 y^2\} \\ & + A_2 x + B_2 y + C_2 m \end{aligned} \tag{26}$$

$$P = \frac{1}{2}(\nu_5 B_1 - \nu_4 A_1) y + \nu_5 C_1 m - A_2 \tag{27}$$

$$P' = -\frac{1}{2}(\nu_5 B_1 - \nu_4 A_1) x + \nu_4 C_1 m - B_2 \tag{28}$$

$$\begin{aligned} Q_1 = {} & \frac{1}{2}[\nu_1 A_1 x^2 + 2\nu_1 B_1 xy + (\nu_6 B_1 - \nu_2 A_1) y^2] \\ & + \nu_1 C_1 m x + \frac{1}{2}\nu_6 C_1 m y - \tau y \end{aligned} \tag{29}$$

$$\begin{aligned} Q_1' = {} & \frac{1}{2}[(\nu_6 A_1 - \nu_1 B_1) x^2 + 2\nu_2 A_1 xy + \nu_2 B_1 y^2] \\ & + \frac{1}{2}\nu_6 C_1 mx + \nu_2 C_1 m y + \tau x , \end{aligned} \tag{30}$$

m is a given length,

$$\nu_s = k_{3s}/k_{33}\,. \tag{31}$$

If the stresses are independent of R, $A_1 = B_1 = C_1 = 0$ and

$$\left.\begin{aligned} u &= u_1 - \frac{1}{2} A_2 R^2 - \tau y R + \alpha + \omega_2 R - \omega_3 y \\ v &= v_1 - \frac{1}{2} B_2 R^2 + \tau x R + \beta + \omega_3 x - \omega_1 R \\ w &= w_1 + (A_2 x + B_2 y + C_2 m) R + \gamma - \omega_2 x + \omega_1 y\,. \end{aligned}\right\} \tag{32}$$

2.32. The elimination of $\widehat{zz}$

Since

$$e_{zz} = k_{31}\widehat{xx} + k_{32}\widehat{yy} + k_{33}\widehat{zz} + k_{34}\widehat{yz} + k_{35}\widehat{zx} + k_{36}\widehat{xy}, \tag{1}$$

we have

$$\widehat{zz} = \frac{e_{zz}}{k_{33}} - \nu_1\widehat{xx} - \nu_2\widehat{yy} - \nu_4\widehat{yz} - \nu_5\widehat{zx} - \nu_6\widehat{xy}\,. \tag{2}$$

Now from Hooke's law

$$e_{xx} = k_{11}\widehat{xx} + k_{12}\widehat{yy} + k_{13}\widehat{zz} + k_{14}\widehat{yz} + k_{15}\widehat{zx} + k_{16}\widehat{xy}\,. \tag{3}$$

Substitute for $\widehat{zz}$ from (2). Then

$$\begin{aligned} e_{xx} = {} & (k_{11} - k_{13}\nu_1)\widehat{xx} + (k_{12} - k_{13}\nu_2)\widehat{yy} + (k_{14} - k_{13}\nu_4)\widehat{yz} \\ & + (k_{15} - k_{13}\nu_5)\widehat{zx} + (k_{16} - k_{13}\nu_6)\widehat{xy} + \frac{k_{13}}{k_{33}} e_{zz}\,. \end{aligned} \tag{4}$$

Write

$$\left.\begin{aligned} l_{11} &= k_{11} - k_{13}\nu_1 = k_{11} - \frac{k_{13}}{k_{33}} k_{13} \\ l_{12} &= k_{12} - k_{13}\nu_2 = k_{12} - \frac{k_{23}}{k_{33}} k_{13} \\ l_{14} &= k_{14} - k_{13}\nu_4 = k_{14} - \frac{k_{43}}{k_{33}} k_{13} \\ l_{15} &= k_{15} - k_{13}\nu_5 = k_{15} - \frac{k_{53}}{k_{33}} k_{13} \\ l_{16} &= k_{16} - k_{13}\nu_6 = k_{16} - \frac{k_{63}}{k_{33}} k_{13} \end{aligned}\right\} \tag{5}$$

and generally

$$l_{rs} = k_{rs} - k_{r3}\frac{k_{3s}}{k_{33}}, \quad l_{r3} = l_{3r} = 0\,.$$

We shall call the l_{rs} the *modified inverse moduli.*

With this notation we have, on the lines of (4),

$$\left.\begin{aligned} \frac{\partial u}{\partial x} &= e_{xx} = l_{11}\widehat{xx} + l_{16}\widehat{xy} + l_{12}\widehat{yy} + l_{14}\widehat{yz} + l_{15}\widehat{zx} + \nu_1 e_{zz} \\ \frac{\partial v}{\partial y} &= e_{yy} = l_{21}\widehat{xx} + l_{26}\widehat{xy} + l_{22}\widehat{yy} + l_{24}\widehat{yz} + l_{25}\widehat{zx} + \nu_2 e_{zz} \\ \frac{\partial v}{\partial x} + \frac{\partial u}{\partial y} &= 2e_{xy} = l_{61}\widehat{xx} + l_{66}\widehat{xy} + l_{62}\widehat{yy} + l_{64}\widehat{yz} + l_{65}\widehat{zx} + \nu_6 e_{zz} \end{aligned}\right\} \tag{6}$$

$$\left.\begin{aligned} \frac{\partial v}{\partial R} + \frac{\partial w}{\partial y} &= 2e_{yz} = l_{41}\widehat{xx} + l_{46}\widehat{xy} + l_{42}\widehat{yy} + l_{44}\widehat{yz} + l_{45}\widehat{zx} + \nu_4 e_{zz} \\ \frac{\partial u}{\partial R} + \frac{\partial w}{\partial x} &= 2e_{zx} = l_{51}\widehat{xx} + l_{56}\widehat{xy} + l_{52}\widehat{yy} + l_{54}\widehat{yz} + l_{55}\widehat{zx} + \nu_5 e_{zz} \end{aligned}\right\} \tag{7}$$

2.4. The strain-energy function

Take the mixed form for the strain-energy U where 1.82 (4)

$$2U = e_{xx}\widehat{xx} + e_{yy}\widehat{yy} + e_{zz}\widehat{zz} + 2e_{yz}\widehat{yz} + 2e_{zx}\widehat{zx} + 2e_{xy}\widehat{xy}\,. \tag{1}$$

Substituting in this the expressions of 2.32 (6), (7) and using 2.32 (1) we find that

$$2U = 2U_1 + \frac{(e_{zz})^2}{k_{33}} \tag{2}$$

where

$$\begin{aligned} 2U_1 = l_{11}(\widehat{xx})^2 &+ 2l_{12}\widehat{xx}\,\widehat{yy} + 2l_{14}\widehat{xx}\,\widehat{yz} + 2l_{15}\widehat{xx}\,\widehat{zx} + 2l_{16}\widehat{xx}\,\widehat{xy} \\ &+ l_{22}(\widehat{yy})^2 + 2l_{24}\widehat{yy}\,\widehat{yz} + 2l_{25}\widehat{yy}\,\widehat{zx} + 2l_{26}\widehat{yy}\,\widehat{xy} \\ &+ l_{44}(\widehat{yz})^2 + 2l_{45}\widehat{yz}\,\widehat{zx} + 2l_{46}\widehat{yz}\,\widehat{xy} \\ &+ l_{55}(zx)^2 + 2l_{56}\widehat{zx}\,\widehat{xy} \\ &+ l_{66}(\widehat{xy})^2 \end{aligned} \tag{3}$$

which is precisely what would arise from 1.82 (12) by writing therein $k_{3s} = 0$ and then putting l_{rs} for k_{rs}. We also note that

$$e_{zz} = k_{31}\widehat{xx} + k_{32}\widehat{yy} + k_{33}\widehat{zz} + k_{34}\widehat{yz} + k_{35}\widehat{zx} + k_{36}\widehat{xy}\,. \tag{4}$$

2.5. The elimination of the displacements

From 2.3 writing

$$L = A_1 x + B_1 y + C_1 m \tag{1}$$

we have

$$\left.\begin{aligned}\frac{\partial u_1}{\partial x} &= e_{xx} - \nu_1 LR\\ \frac{\partial v_1}{\partial y} &= e_{yy} - \nu_2 LR\\ \frac{\partial v_1}{\partial x} + \frac{\partial u_1}{\partial y} &= 2e_{xy} - \nu_6 LR\end{aligned}\right\}\tag{2}$$

$$\left.\begin{aligned}\frac{\partial w_1}{\partial y} &= 2e_{yz} - \nu_4 LR - Q_1'\\ \frac{\partial w_1}{\partial x} &= 2e_{zx} - \nu_5 LR - Q_1\end{aligned}\right\}\tag{3}$$

where Q_1 and Q_1' are given by 2.3 (29) (30).

Also from 2.3 (1), (14)

$$e_{zz} = e_{zz}^0 + LR\,,\tag{4}$$

$$e_{zz}^0 = \frac{1}{2}\left[\nu_5 A_1 x^2 + (\nu_4 A_1 + \nu_5 B_1)\,xy + \nu_4 B_1 y^2\right] + A_2 x + B_2 y + C_2 m\,.\tag{5}$$

Therefore from 2.32 (6), (7)

$$\left.\begin{aligned}\frac{\partial u_1}{\partial x} &= l_{11}\widehat{xx} + l_{16}\widehat{xy} + l_{12}\widehat{yy} + l_{14}\widehat{yz} + l_{15}\widehat{zx} + \nu_1 e_{zz}^0\\ \frac{\partial v_1}{\partial y} &= l_{21}\widehat{xx} + l_{26}\widehat{xy} + l_{22}\widehat{yy} + l_{24}\widehat{yz} + l_{25}\widehat{zx} + \nu_2 e_{zz}^0\\ \frac{\partial v_1}{\partial x} + \frac{\partial u_1}{\partial y} &= l_{61}\widehat{xx} + l_{66}\widehat{xy} + l_{62}\widehat{yy} + l_{64}\widehat{yz} + l_{65}\widehat{zx} + \nu_6 e_{zz}^0\end{aligned}\right\}\tag{6}$$

$$\left.\begin{aligned}\frac{\partial w_1}{\partial y} &= l_{41}\widehat{xx} + l_{46}\widehat{xy} + l_{42}\widehat{yy} + l_{44}\widehat{yz} + l_{45}\widehat{zx} + \nu_4 e_{zz}^0 - Q_1'\\ \frac{\partial w_1}{\partial x} &= l_{51}\widehat{xx} + l_{56}\widehat{xy} + l_{52}\widehat{yy} + l_{54}\widehat{yz} + l_{55}\widehat{zx} + \nu_5 e_{zz}^0 - Q_1\end{aligned}\right\}\tag{7}$$

We now eliminate the displacements by the identities

$$\frac{\partial^2}{\partial x^2}\left(\frac{\partial v_1}{\partial y}\right) + \frac{\partial^2}{\partial y^2}\left(\frac{\partial u_1}{\partial x}\right) - \frac{\partial^2}{\partial x \partial y}\left(\frac{\partial v_1}{\partial x} + \frac{\partial u_1}{\partial y}\right) = 0$$

$$\frac{\partial}{\partial x}\left(\frac{\partial w_1}{\partial y}\right) - \frac{\partial}{\partial y}\left(\frac{\partial w_1}{\partial x}\right) = 0$$

which give

$$\left.\begin{aligned}&\left(l_{21}\frac{\partial^2}{\partial x^2} - l_{61}\frac{\partial^2}{\partial x \partial y} + l_{11}\frac{\partial^2}{\partial y^2}\right)\widehat{xx}\\ &+\left(l_{22}\frac{\partial^2}{\partial x^2} - l_{62}\frac{\partial^2}{\partial x \partial y} + l_{12}\frac{\partial^2}{\partial y^2}\right)\widehat{yy}\\ &+\left(l_{26}\frac{\partial^2}{\partial x^2} - l_{66}\frac{\partial^2}{\partial x \partial y} + l_{16}\frac{\partial^2}{\partial y^2}\right)\widehat{xy}\\ &+\left(l_{24}\frac{\partial^2}{\partial x^2} - l_{64}\frac{\partial^2}{\partial x \partial y} + l_{14}\frac{\partial^2}{\partial y^2}\right)\widehat{yz}\\ &+\left(l_{25}\frac{\partial^2}{\partial x^2} - l_{65}\frac{\partial^2}{\partial x \partial y} + l_{15}\frac{\partial^2}{\partial y^2}\right)\widehat{zx}\\ &+\left(\nu_2\frac{\partial^2}{\partial x^2} - \nu_6\frac{\partial^2}{\partial x \partial y} + \nu_1\frac{\partial^2}{\partial y^2}\right)e_{zz}^0 = 0\end{aligned}\right\}\tag{8}$$

and

$$\left.\begin{aligned}&\left(l_{41}\frac{\partial}{\partial x}-l_{51}\frac{\partial}{\partial y}\right)\widehat{xx}+\left(l_{46}\frac{\partial}{\partial x}-l_{56}\frac{\partial}{\partial y}\right)\widehat{xy}\\&+\left(l_{42}\frac{\partial}{\partial x}-l_{52}\frac{\partial}{\partial y}\right)\widehat{yy}+\left(l_{44}\frac{\partial}{\partial x}-l_{54}\frac{\partial}{\partial y}\right)\widehat{yz}+\left(l_{45}\frac{\partial}{\partial x}-l_{55}\frac{\partial}{\partial y}\right)\widehat{zx}\\&+\left(\nu_4\frac{\partial}{\partial x}-\nu_5\frac{\partial}{\partial y}\right)e_{zz}^0-\frac{\partial Q_1'}{\partial x}+\frac{\partial Q_1}{\partial y}=0\,.\end{aligned}\right\}\quad(9)$$

Also we have

$$\begin{aligned}&\left(\nu_2\frac{\partial^2}{\partial x^2}-\nu_6\frac{\partial^2}{\partial x\partial y}+\nu_1\frac{\partial^2}{\partial y^2}\right)e_{zz}^0\\&=\nu_2[\nu_5A_1]-\frac{1}{2}\nu_6[\nu_4A_1+\nu_5B_1]+\nu_1[\nu_4B_1]\\&=A_1\left(\nu_2\nu_5-\frac{1}{2}\nu_4\nu_6\right)+B_1\left(\nu_1\nu_4-\frac{1}{2}\nu_5\nu_6\right)=E_1\,,\end{aligned}\qquad(10)$$

say, and

$$\begin{aligned}&\left(\nu_4\frac{\partial}{\partial x}-\nu_5\frac{\partial}{\partial y}\right)e_{zz}^0-\frac{\partial Q_1'}{\partial x}+\frac{\partial Q_1}{\partial y}\\&=\nu_4\left[\nu_5A_1x+\frac{1}{2}(\nu_4A_1+\nu_5B_1)y+A_2\right]\\&-\nu_5\left[\frac{1}{2}(\nu_4A_1+\nu_5B_1)x+\nu_4B_1y+B_2\right]\\&-[(\nu_6A_1-\nu_1B_1)x+\nu_2A_1y+\tau]\\&+[\nu_1B_1x+(\nu_6B_1-\nu_2A_1)y-\tau]\\&=\left\{\frac{1}{2}\nu_4\nu_5A_1-\frac{1}{2}\nu_5^2B_1-\nu_6A_1+2\nu_1B_1\right\}x\\&+\left\{\frac{1}{2}\nu_4^2A_1-\frac{1}{2}\nu_4\nu_5B_1+\nu_6B_1-2\nu_2A_1\right\}y\\&+(\nu_4A_2-\nu_5B_2-2\tau)\\&=\alpha_1x+\beta_1y+\gamma_1\end{aligned}\qquad(11)$$

say.

2.52. The equations satisfied by the stress functions

In terms of stress functions, 2.12, 2.24, we have

$$\widehat{xx}=\frac{\partial^2\chi}{\partial y^2}+V,\quad\widehat{xy}=-\frac{\partial^2\chi}{\partial x\partial y},\quad\widehat{yy}=\frac{\partial^2\chi}{\partial x^2}+V$$

$$\widehat{zx}=\frac{\partial\psi}{\partial y}-\frac{1}{2k_{33}}[A_1x^2+(C_1-\varepsilon)mx]$$

$$\widehat{yz}=-\frac{\partial\psi}{\partial x}-\frac{1}{2k_{33}}[B_1y^2+(C_1-\varepsilon)my]\,.$$

Substitute in 2.5 (8), (9). Then from 2.5 (8)

$$
\begin{aligned}
&\left(l_{21}\frac{\partial^2}{\partial x^2}-l_{61}\frac{\partial^2}{\partial x\partial y}+l_{11}\frac{\partial^2}{\partial y^2}\right)\left(\frac{\partial^2\chi}{\partial y^2}+V\right)\\
&+\left(l_{22}\frac{\partial^2}{\partial x^2}-l_{62}\frac{\partial^2}{\partial x\partial y}+l_{12}\frac{\partial^2}{\partial y^2}\right)\left(\frac{\partial^2\chi}{\partial x^2}+V\right)\\
&-\left(l_{26}\frac{\partial^2}{\partial x^2}-l_{66}\frac{\partial^2}{\partial x\partial y}+l_{16}\frac{\partial^2}{\partial y^2}\right)\left(\frac{\partial^2\chi}{\partial x\partial y}\right)\\
&-\left(l_{24}\frac{\partial^2}{\partial x^2}-l_{64}\frac{\partial^2}{\partial x\partial y}+l_{14}\frac{\partial^2}{\partial y^2}\right)\left[\frac{\partial\psi}{\partial x}+\frac{1}{2k_{33}}\{B_1y^2+(C_1-\varepsilon)my\}\right]\\
&+\left(l_{25}\frac{\partial^2}{\partial x^2}-l_{65}\frac{\partial^2}{\partial x\partial y}+l_{15}\frac{\partial^2}{\partial y^2}\right)\left[\frac{\partial\psi}{\partial y}-\frac{1}{2k_{33}}\{A_1x^2+(C_1-\varepsilon)mx\}\right]\\
&+E_1=0\,,
\end{aligned}
$$

or

$$
\begin{aligned}
&l_{22}\frac{\partial^4\chi}{\partial x^4}-2l_{26}\frac{\partial^4\chi}{\partial x^3\partial y}+(2l_{12}+l_{66})\frac{\partial^4\chi}{\partial x^2\partial y^2}-2l_{16}\frac{\partial^4\chi}{\partial x\partial y^3}+l_{11}\frac{\partial^4\chi}{\partial y^4}\\
&-l_{24}\frac{\partial^3\psi}{\partial x^3}+(l_{64}+l_{25})\frac{\partial^3\psi}{\partial x^2\partial y}-(l_{65}+l_{14})\frac{\partial^3\psi}{\partial x\partial y^2}+l_{15}\frac{\partial^3\psi}{\partial y^3}\\
&+(l_{21}+l_{22})\frac{\partial^2V}{\partial x^2}-(l_{61}+l_{62})\frac{\partial^2V}{\partial x\partial y}+(l_{11}+l_{12})\frac{\partial^2V}{\partial y^2}\\
&-\frac{1}{k_{33}}(l_{25}A_1+l_{14}B_1)+E_1=0
\end{aligned}
\qquad (1)
$$

where E_1 is given by 2.5 (10).

Again 2.5 (9) gives

$$
\begin{aligned}
&\left(l_{41}\frac{\partial}{\partial x}-l_{51}\frac{\partial}{\partial y}\right)\left(\frac{\partial^2\chi}{\partial y^2}+V\right)\\
&-\left(l_{46}\frac{\partial}{\partial x}-l_{56}\frac{\partial}{\partial y}\right)\frac{\partial^2\chi}{\partial x\partial y}\\
&+\left(l_{42}\frac{\partial}{\partial x}-l_{52}\frac{\partial}{\partial y}\right)\left(\frac{\partial^2\chi}{\partial x^2}+V\right)\\
&-\left(l_{44}\frac{\partial}{\partial x}-l_{54}\frac{\partial}{\partial y}\right)\left(\frac{\partial\psi}{\partial x}+\frac{1}{2k_{33}}[B_1y^2+(C_1-\varepsilon)my]\right)\\
&+\left(l_{45}\frac{\partial}{\partial x}-l_{55}\frac{\partial}{\partial y}\right)\left(\frac{\partial\psi}{\partial y}-\frac{1}{2k_{33}}[A_1x^2+(C_1-\varepsilon)mx]\right)\\
&+\alpha_1x+\beta_1y+\gamma_1=0
\end{aligned}
$$

or

$$
\begin{aligned}
&l_{42}\frac{\partial^3\chi}{\partial x^3}-(l_{46}+l_{52})\frac{\partial^3\chi}{\partial x^2\partial y}+(l_{41}+l_{56})\frac{\partial^3\chi}{\partial x\partial y^2}-l_{51}\frac{\partial^3\chi}{\partial y^3}\\
&-l_{44}\frac{\partial^2\psi}{\partial x^2}+2l_{45}\frac{\partial^2\psi}{\partial x\partial y}-l_{55}\frac{\partial^2\psi}{\partial y^2}\\
&+(l_{41}+l_{42})\frac{\partial V}{\partial x}-(l_{51}+l_{52})\frac{\partial V}{\partial y}\\
&+\frac{l_{45}}{k_{33}}(-A_1x+B_1y)+\alpha_1x+\beta_1y+\gamma_1=0\,,
\end{aligned}
\qquad (2)
$$

where $\alpha_1x+\beta_1y+\gamma_1$ is given by 2.5 (11).

Introduce the operators

$$D^{(4)} = l_{22}\frac{\partial^4}{\partial x^4} - 2l_{26}\frac{\partial^4}{\partial x^3 \partial y} + (2l_{12} + l_{66})\frac{\partial^4}{\partial x^2 \partial y^2} - 2l_{16}\frac{\partial^4}{\partial x \partial y^3} + l_{11}\frac{\partial^4}{\partial y^4} \quad (3)$$

$$D^{(3)} = l_{24}\frac{\partial^3}{\partial x^3} - (l_{64} + l_{25})\frac{\partial^3}{\partial x^2 \partial y} + (l_{65} + l_{14})\frac{\partial^3}{\partial x \partial y^2} - l_{15}\frac{\partial^3}{\partial y^3} \quad (4)$$

$$D^{(2)} = l_{44}\frac{\partial^2}{\partial x^2} - 2l_{45}\frac{\partial^2}{\partial x \partial y} + l_{55}\frac{\partial^2}{\partial y^2} \quad (5)$$

$$H^{(2)} = (l_{21} + l_{22})\frac{\partial^2}{\partial x^2} - (l_{61} + l_{62})\frac{\partial^2}{\partial x \partial y} + (l_{11} + l_{12})\frac{\partial^2}{\partial y^2} \quad (6)$$

$$H^{(1)} = (l_{41} + l_{42})\frac{\partial}{\partial x} - (l_{51} + l_{52})\frac{\partial}{\partial y}\,. \quad (7)$$

Then

$$D^{(4)}\chi - D^{(3)}\psi + H^{(2)}V + F = 0\,, \quad (8)$$

where

$$F = E_1 - \frac{1}{k_{33}}(l_{25}A_1 + l_{14}B_1) \quad (9)$$

$$D^{(3)}\chi - D^{(2)}\psi + H^{(1)}V + \alpha_2 x + \beta_2 y + \gamma_2 = 0\,, \quad (10)$$

where

$$\alpha_2 x + \beta_2 y + \gamma_2 = \alpha_1 x + \beta_1 y + \gamma_1 + \frac{l_{45}}{k_{33}}(-A_1 x + B_1 y)\,. \quad (11)$$

Thus (8) and (10) are the equations satisfied by the stress functions.

2.56. Notation for indefinite integration

We now introduce a convenient notation. Just as we often write $dW(z)/dz = W'(z)$, thereby expressing differentiation with respect to the argument (in this case z) by a prime to the right of W, so we shall express indefinite integration with respect to the argument by a prime to the left of W. Thus

$$'W(z) = \int W(z)\,dz \quad (1)$$

The operator can be repeated

$$''W(z) = \int 'W(z)\,dz \quad (2)$$

and we note that, for example,

$$('W(z))' = W(z)\,,\ (''W(z))' = 'W(z) \quad (3)$$

2.6. The complex stresses

Consider equations 2.52 (8), (10).

Let χ_0, ψ_0 be particular solutions so that

$$D^{(4)}\chi_0 - D^{(3)}\psi_0 + H^{(2)}V + F = 0 \quad (1)$$

$$D^{(3)}\chi_0 - D^{(2)}\psi_0 + H^{(1)}V + \alpha_2 x + \beta_2 y + \gamma_2 = 0\,. \quad (2)$$

Then we have

$$D^{(4)}(\chi - \chi_0) - D^{(3)}(\psi - \psi_0) = 0 \tag{3}$$

$$D^{(3)}(\chi - \chi_0) - D^{(2)}(\psi - \psi_0) = 0\,. \tag{4}$$

$$\text{Put } \chi - \chi_0 = D^{(2)}\Omega,\ \psi - \psi_0 = D^{(3)}\Omega\,. \tag{5}$$

Then (4) is satisfied identically and (3) gives

$$[D^{(4)}D^{(2)} - D^{(3)}D^{(3)}]\,\Omega = 0\,. \tag{6}$$

This is a linear partial differential equation of the sixth order with constant coefficients.

In (6) write $\exp(x + \lambda y)$ for Ω. This will satisfy (6) if

$$f(\lambda) = d^{(4)}(\lambda)\,d^{(2)}(\lambda) - [d^{(3)}(\lambda)]^2 = 0\,, \tag{7}$$

where

$$d^{(4)}(\lambda) = l_{11}\lambda^4 - 2l_{16}\lambda^3 + (2l_{12} + l_{66})\lambda^2 - 2l_{26}\lambda + l_{22} \tag{8}$$

$$d^{(3)}(\lambda) = -l_{15}\lambda^3 + (l_{14} + l_{65})\lambda^2 - (l_{25} + l_{46})\lambda + l_{24} \tag{9}$$

$$d^{(2)}(\lambda) = l_{55}\lambda^2 - 2l_{45}\lambda + l_{44}\,. \tag{10}$$

We call $f(\lambda)$ the *characteristic function* and $f(\lambda) = 0$ the *characteristic equation*. Let λ_1, λ_2, λ_3, λ_4, λ_5, λ_6 be the six roots of the characteristic equation. In the general case, which we assume for the present,

$$l_{11}l_{55} - l_{15}^2 \neq 0 \tag{11}$$

so that we can write

$$f(\lambda) = [l_{11}l_{55} - l_{15}^2]\,(\lambda - \lambda_1)\,(\lambda - \lambda_2)\,(\lambda - \lambda_3)\,(\lambda - \lambda_4)\,(\lambda - \lambda_5)\,(\lambda - \lambda_6)\,. \tag{12}$$

Equation (6) can now be written in the form

$$D_1D_2D_3D_4D_5D_6 = 0\,,\ D_\nu = \frac{\partial}{\partial y} - \lambda_\nu\frac{\partial}{\partial x}\,, \tag{13}$$

which is solved by writing

$$D_6\Omega = \phi_5\,,\ D_5D_6\Omega = \phi_4\,,\ D_4D_5D_6\Omega = \phi_3\,,\ D_3D_4D_5D_6\Omega = \phi_2\,,$$

$$D_2D_3D_4D_5D_6\Omega = \phi_1\,,$$

so that

$$D_1\phi_1 = 0\,,\ D_2\phi_2 = \phi_1\,,\ D_3\phi_3 = \phi_2\,,\ D_4\phi_4 = \phi_3\,,\ D_5\phi_5 = \phi_4\,,\ D_6\Omega = \phi_5\,.$$

The solution of $D_1\phi_1 = 0$ or $\left(\frac{\partial}{\partial y} - \lambda_1\frac{\partial}{\partial x}\right)\phi_1 = 0$ is

$$\phi_1 = f_1(x + \lambda_1 y)$$

as is readily verified by differentiation, f_1 being an arbitrary function. Therefore $D_2\phi_2 = f_1(x + \lambda_1 y)$ has the solution

$$\phi_2 = \frac{'f_1(x + \lambda_1 y)}{\lambda_1 - \lambda_2} + f_2(x + \lambda_2 y) .$$

Here a prime to the left of a function indicates indefinite integration (2.56).

Proceeding in this way we find that

$$\Omega = \sum_{\nu=1}^{6} \Omega_\nu(z_\nu), \; z_\nu = x + \lambda_\nu y , \tag{14}$$

where the Ω_ν are arbitrary functions, and

$$z_\nu = \frac{1}{2} z(1 - i\lambda_\nu) + \frac{1}{2}\bar{z}(1 + i\lambda_\nu) . \tag{15}$$

Therefore from (5)

$$\chi - \chi_0 = \sum_{\nu=1}^{6} D^{(2)}\Omega_\nu(z_\nu), \; \psi - \psi_0 = \sum_{\nu=1}^{6} D^{(3)}\Omega_\nu(z_\nu)$$

or using (9) and (10)

$$\chi - \chi_0 = \sum_{\nu=1}^{6} d^{(2)}(\lambda_\nu)\,\Omega''(z_\nu), \; \psi - \psi_0 = \sum_{\nu=1}^{6} d^{(3)}(\lambda_\nu)\,\Omega_\nu'''(z_\nu) .$$

Now write

$$\Omega_\nu(z_\nu) = \frac{1}{4}\,\frac{''''W_\nu(z_\nu)}{d^{(2)}(\lambda_\nu)}, \quad \mu_\nu = \frac{d^{(3)}(\lambda_\nu)}{d^{(2)}(\lambda_\nu)} . \tag{16}$$

Then

$$\chi - \chi_0 = \sum_{\nu=1}^{6} \frac{1}{4}\,''W_\nu(z_\nu), \; \psi - \psi_0 = \sum_{\nu=1}^{6} \frac{1}{4}\,\mu_\nu\,'W_\nu(z_\nu) . \tag{17}$$

Since χ and ψ are real valued, so is Ω. Therefore the roots of the characteristic equation are either real or conjugate complex. We shall show below, 2.62, that none are real. Therefore we can write them in conjugate pairs

$$\lambda_\nu = \alpha_\nu + i\beta_\nu, \quad \bar{\lambda}_\nu = \alpha_\nu - i\beta_\nu, \quad \beta_\nu > 0, \; \nu = 1,2,3, \tag{18}$$

where α_ν and β_ν are real and where, without loss of generality, we can so number them that $\beta_\nu > 0$.

Thus the pairs of roots are

$$\lambda_1, \bar{\lambda}_1; \; \lambda_2, \bar{\lambda}_2; \; \lambda_3, \bar{\lambda}_3$$

and the corresponding variables z_ν occur in the conjugate pairs

$$z_1, \bar{z}_1; \; z_2, \bar{z}_2; \; z_3, \bar{z}_3$$

and therefore since $\chi - \chi_0$ is real-valued the functions $W_\nu(z_\nu)$ occur in the conjugate pairs*

$$W_1(z_1),\ \overline{W}_1(\bar{z}_1);\ W_2(z_2),\ \overline{W}_2(\bar{z}_2);\ W_3(z_3),\ \overline{W}_3(\bar{z}_3)\,.$$

It therefore follows from (17) that

$$\chi - \chi_0 = \frac{1}{4}\sum_{\nu=1}^{3}\left[{}^{\prime\prime}W_\nu(z_\nu) + {}^{\prime\prime}\overline{W}_\nu(\bar{z}_\nu)\right], \tag{19}$$

$$\psi - \psi_0 = \frac{1}{4}\sum_{\nu=1}^{3}\left[\mu_\nu{}^{\prime}W_\nu(z_\nu) + \bar{\mu}_\nu{}^{\prime}\overline{W}_\nu(\bar{z}_\nu)\right]. \tag{20}$$

Conformably with the notation of [MILNE-THOMSON (6)] we shall call $W_1(z_1)$, $W_2(z_2)$, $W_3(z_3)$ the *complex stresses*. Thus in the general case of antiplane elastic stress we have three complex stresses in contrast with the two required for the case of plane problems.

2.62. The characteristic equation has no real roots

Inasmuch as the strain-energy function represents stored and recoverable energy, when the system is in stable equilibrium the strain-energy function must be positive semi-definite, that is to say greater than zero for any system of values assigned to the stresses which are not simultaneously zero.

We shall assume that a system undergoing elastic deformation is in stable equilibrium provided that none of the stress components exceeds some positive value ε.

Proof. To prove that $f(\lambda) = 0$ has no real roots where $f(\lambda)$ is defined by 2.6 (7), write in the strain energy function 2.4 (2)

$$\widehat{xx} = \frac{c^2}{N},\ \widehat{yy} = \frac{1}{N},$$

$$\widehat{zz} = -\frac{1}{k_{33}N}\left(k_{13}c^2 - k_{36}c + k_{23} + k_{35}\mu c - k_{34}\mu\right)$$

$$\widehat{yz} = -\frac{\mu}{N},\ \widehat{zx} = \frac{c\mu}{N},\ \widehat{xy} = -\frac{c}{N},$$

where $\mu = -d^{(3)}(c)/d^{(2)}(c)$. We can always choose N large enough to make all stress components less than ε.

Then from 2.4 (4)

$$N e_{zz} = k_{31}c^2 + k_{32} - \left(k_{13}c^2 - k_{36}c + k_{23} + k_{35}\mu c - k_{34}\mu\right)$$
$$- k_{34}\mu + k_{35}c\mu - k_{36}c = 0$$

and therefore from 2.4 (2)

$$U = U_1$$

* From $f(z)$ we can form the complex conjugate $\overline{f(z)} = \bar{f}(\bar{z})$, and from these the associated functions; $f(\bar{z})$ by writing $\bar{z}$ for z in $f(z)$, and $\bar{f}(z)$ by writing z for $\bar{z}$ in $\bar{f}(\bar{z})$.

so that

$$\begin{aligned}2N^2U_1 = {} & l_{11}c^4 + 2l_{12}c^2 - 2l_{14}\mu c^2 + 2l_{15}c^3\mu - 2l_{16}c^3\\ & + l_{22} - 2l_{24}\mu + 2l_{25}c\mu - 2l_{26}c\\ & + l_{44}\mu^2 - 2l_{45}c\mu^2 + 2l_{46}c\mu\\ & + l_{55}c^2\mu^2 - 2l_{56}c^2\mu + l_{66}c^2\\ = {} & l_{11}c^4 - 2l_{16}c^3 + (2l_{12} + l_{66})c^2 - 2l_{26}c + l_{22}\\ & + 2\mu\{l_{15}c^3 - (l_{14} + l_{56})c^2 + (l_{25} + l_{46})c - l_{24}\}\\ & + \mu^2\{l_{44} - 2l_{45}c + l_{55}c^2\}\\ = {} & d^{(4)}(c) - 2\mu d^{(3)}(c) + \mu^2 d^{(2)}(c)\\ = {} & d^{(4)}(c) - \frac{[d^{(3)}(c)]^2}{d^{(2)}(c)} = \frac{f(c)}{d^{(2)}(c)}\,.\end{aligned}$$

Thus

$$U_1 = \frac{1}{2N^2}\frac{f(c)}{d^{(2)}(c)}\,.$$

Therefore

$$\frac{f(c)}{d^{(2)}(c)} > 0$$

and so $f(c)$ and $d^{(2)}(c)$ have the same sign.

We show that $d^{(2)}(c) > 0$ by writing in 2.4 (2)

$$\widehat{xx} = \widehat{yy} = \widehat{xy} = 0\,,$$

$$\widehat{yz} = -\frac{1}{N}\,,\ \widehat{xz} = \frac{c}{N}\,,\ \widehat{zz} = \frac{1}{k_{33}N}(k_{34} - k_{35}c)$$

which give $e_{zz} = 0$ and

$$2U = 2U_1 = \frac{l_{44}}{N^2} - 2l_{45}\frac{c}{N^2} + l_{55}\frac{c^2}{N^2} = \frac{d^{(2)}(c)}{N^2}$$

so that $d^{(2)}(c) > 0$.

Therefore finally $f(c) > 0$ for all values of c, so that the characteristic equation $f(c) = 0$ has no real roots. Q. E. D.

2.7. Expression of the fundamental stress combinations in terms of the complex stresses

Writing

$$\gamma_\nu = \frac{1}{2}(1 - i\lambda_\nu), \qquad \delta_\nu = \frac{1}{2}(1 + i\lambda_\nu) \tag{1}$$

we have from 2.6 (15)

$$z_\nu = \gamma_\nu z + \delta_\nu \bar{z}, \qquad \bar{z}_\nu = \bar{\delta}_\nu z + \bar{\gamma}_\nu \bar{z} \tag{2}$$

whence we find

$$\frac{\partial z_\nu}{\partial z} = \gamma_\nu, \qquad \frac{\partial z_\nu}{\partial \bar{z}} = \delta_\nu, \qquad \frac{\partial \bar{z}_\nu}{\partial z} = \bar{\delta}_\nu, \qquad \frac{\partial \bar{z}_\nu}{\partial \bar{z}} = \bar{\gamma}_\nu\,. \tag{3}$$

Therefore from 2.1 (15) and 2.6 (19), (20)

$$\widehat{xx} + \widehat{yy} = \Theta = \sum_{\nu=1}^{3} [\gamma_\nu \delta_\nu W_\nu(z_\nu) + \bar{\gamma}_\nu \bar{\delta}_\nu \overline{W}_\nu(\bar{z}_\nu)] + \Theta_0 + 4\frac{\partial^2 \chi_0}{\partial z \partial \bar{z}} \quad (4)$$

$$\widehat{yy} - \widehat{xx} + 2i\,\widehat{xy} = \Phi = \sum_{\nu=1}^{3} [\gamma_\nu^2 W_\nu(z) + \bar{\delta}_\nu^2 \overline{W}_\nu(\bar{z}_\nu)] + \Phi_0 + 4\frac{\partial^2 \chi_0}{\partial z^2}, \quad (5)$$

$$\widehat{xz} - i\,\widehat{yz} = \Psi = \sum_{\nu=1}^{3} \frac{1}{2} i\,[\mu_\nu \gamma_\nu W_\nu(z) + \bar{\mu}_\nu \bar{\delta}_\nu \overline{W}_\nu(\bar{z}_\nu)] + \Psi_0 + 2i\frac{\partial \psi_0}{\partial z}. \quad (6)$$

When the body vector derives from a potential, 2.12 (1),

$$V_0 = V + \varepsilon m R/k_{33} \quad (7)$$

we can take

$$\Theta_0 = 2V, \quad \Phi_0 = 0. \quad (8)$$

Also from 2.24 (6)

$$\Psi_0 = -\frac{1}{8k_{33}}\{(A_1 + iB_1)(z^2 + \bar{z}^2) + 2(A_1 - iB_1)z\bar{z} + 4(C_1 - \varepsilon)m\bar{z}\} \quad (9)$$

where A_1, B_1, C_1 are defined by 2.2 (7).

The individual stress components can be expressed by

$$\begin{aligned} \widehat{xx} &= \frac{1}{2}\Theta - \frac{1}{4}(\Phi + \bar{\Phi}), \quad \widehat{yy} = \frac{1}{2}\Theta + \frac{1}{4}(\Phi + \bar{\Phi}) \\ \widehat{xy} &= -\frac{1}{4}i(\Phi - \bar{\Phi}), \quad \widehat{zx} = \frac{1}{2}(\Psi + \bar{\Psi}), \quad \widehat{yz} = \frac{1}{2}i(\Psi - \bar{\Psi}). \end{aligned} \quad (10)$$

Again from 2.1 (17), when the body-vector derives from a potential we have

$$\left.\begin{aligned} \widehat{xx} &= \sum_{\nu=1}^{3}\left[\frac{1}{4}\lambda_\nu^2 W_\nu(z_\nu) + \frac{1}{4}\bar{\lambda}_\nu^2 \overline{W}_\nu(\bar{z}_\nu)\right] + V + \frac{\partial^2 \chi_0}{\partial y^2}, \\ \widehat{yy} &= \sum_{\nu=1}^{3}\left[\frac{1}{4} W_\nu(z_\nu) + \frac{1}{4}\overline{W}_\nu(\bar{z}_\nu)\right] + V + \frac{\partial^2 \chi_0}{\partial x^2}, \\ \widehat{xy} &= -\sum_{\nu=1}^{3}\left[\frac{1}{4}\lambda_\nu W_\nu(z_\nu) + \frac{1}{4}\bar{\lambda}_\nu \overline{W}_\nu(\bar{z}_\nu)\right] - \frac{\partial^2 \chi_0}{\partial x \partial y}, \end{aligned}\right\} \quad (11)$$

$$\left.\begin{aligned} \widehat{zx} &= \sum_{\nu=1}^{3}\left[\frac{1}{4}\lambda_\nu \mu_\nu W_\nu(z_\nu) + \frac{1}{4}\bar{\lambda}_\nu \bar{\mu}_\nu \overline{W}_\nu(\bar{z}_\nu)\right] \\ &\quad - \frac{1}{2k_{33}}[A_1 x^2 + (C_1 - \varepsilon)m x] + \frac{\partial \psi_0}{\partial y}, \\ \widehat{yz} &= -\sum_{\nu=1}^{3}\left[\frac{1}{4}\mu_\nu W_\nu(z_\nu) + \frac{1}{4}\bar{\mu}_\nu \overline{W}_\nu(\bar{z}_\nu)\right] \\ &\quad - \frac{1}{2k_{33}}[B_1 y^2 + (C_1 - \varepsilon)m y] - \frac{\partial \psi_0}{\partial x}. \end{aligned}\right\} \quad (12)$$

2.72. The displacement in terms of the complex stresses

The displacement components (u, v, w) given by 2.3 (19)—(21) involve terms dependent on the applicate R, a rigid body displacement and the components (u_1, v_1, w_1) which are independent of R. These latter can be expressed in terms of the complex stresses by the following procedure.

Substitute in (6) and (7) of 2.5 the stress components given by 2.7 (11), (12). We then get

$$\frac{\partial u_1}{\partial x} = \sum_{\nu=1}^{3} [L_{1\nu} W_\nu(z_\nu) + \bar{L}_{1\nu} \overline{W}_\nu(\bar{z}_\nu)] + U_1 \tag{1}$$

$$\frac{\partial v_1}{\partial y} = \sum_{\nu=1}^{3} [L_{2\nu} W_\nu(z_\nu) + \bar{L}_{2\nu} \overline{W}_\nu(\bar{z}_\nu)] + U_2 \tag{2}$$

$$\frac{\partial v_1}{\partial x} + \frac{\partial u_1}{\partial y} = \sum_{\nu=1}^{3} [L_{6\nu} W_\nu(z_\nu) + \bar{L}_{6\nu} \overline{W}_\nu(\bar{z}_\nu)] + U_6 \tag{3}$$

$$\frac{\partial w_1}{\partial y} = \sum_{\nu=1}^{3} [L_{4\nu} W_\nu(z_\nu) + \bar{L}_{4\nu} \overline{W}_\nu(\bar{z}_\nu)] + U_4 - Q_1' \tag{4}$$

$$\frac{\partial w_1}{\partial x} = \sum_{\nu=1}^{3} [L_{5\nu} W_\nu(z_\nu) + \bar{L}_{5\nu} \overline{W}_\nu(\bar{z}_\nu)] + U_5 - Q_1 \tag{5}$$

where

$$4 L_{k\nu} = l_{k1} \lambda_\nu^2 - l_{k6} \lambda_\nu + l_{k2} - l_{k4} \mu_\nu + l_{k5} \lambda_\nu \mu_\nu \qquad (k = 1,2,4,5,6) \tag{6}$$

$$\begin{aligned} U_k = {} & (l_{k1} + l_{k2}) V + l_{k1} \frac{\partial^2 \chi_0}{\partial y^2} - l_{k6} \frac{\partial^2 \chi_0}{\partial x \partial y} + l_{k2} \frac{\partial^2 \chi_0}{\partial x^2} \\ & - l_{k4} \left[\frac{\partial \psi_0}{\partial x} + \frac{1}{2 k_{33}} \{B_1 y^2 + (C_1 - \varepsilon) m y\} \right] \\ & + l_{k5} \left[\frac{\partial \psi_0}{\partial y} - \frac{1}{2 k_{33}} \{A_1 x^2 + (C_1 - \varepsilon) m x\} \right] + \nu_k e_{zz}^0 \\ & (k = 1, 2, 4, 5, 6) \, . \end{aligned} \tag{7}$$

Here Q_1, Q_1', e_{zz}^0 are given by 2.3 (29), (30), (14), and

$$\nu_k = \frac{k_{3k}}{k_{33}}. \tag{8}$$

Let $u^{(0)}$, $v^{(0)}$, $w^{(0)}$ be particular integrals of the equations

$$\begin{aligned} & \frac{\partial u_1}{\partial x} = U_1, \; \frac{\partial v_1}{\partial y} = U_2, \; \frac{\partial v_1}{\partial x} + \frac{\partial u_1}{\partial y} = U_6 \\ & \frac{\partial w_1}{\partial y} = U_4 - Q_1', \; \frac{\partial w_1}{\partial x} = U_5 - Q_1 \, . \end{aligned} \tag{9}$$

Then the values of u_1, v_1, w_1 are given by

$$u_1 = \sum_{\nu=1}^{3} [L_{1\nu}' W_\nu(z_\nu) + \bar{L}_{1\nu}' \overline{W}_\nu(\bar{z}_\nu)] + u^{(0)} \tag{10}$$

$$v_1 = \sum_{\nu=1}^{3} \left[\frac{L_{2\nu}}{\lambda_\nu}' W_\nu(z_\nu) + \frac{\bar{L}_{2\nu}}{\bar{\lambda}_\nu}' \overline{W}_\nu(\bar{z}_\nu)\right] + v^{(0)} \tag{11}$$

$$w_1 = \sum_{\nu=1}^{3} [L_{5\nu}' W_\nu(z_\nu) + \bar{L}_{5\nu}' \overline{W}_\nu(\bar{z}_\nu)] + w^{(0)}\,. \tag{12}$$

To see this we note first that for (10) and (11) to satisfy (3) we must have

$$\frac{\partial u^{(0)}}{\partial y} + \frac{\partial v^{(0)}}{\partial x} = U_6$$

which is true in virtue of (9), and we must also have

$$\frac{L_{2\nu}}{\lambda_\nu} + \lambda_\nu L_{1\nu} = L_{6\nu}$$

which on being developed is identical with the characteristic equation $f(\lambda_\nu) = 0$ of 2.6 (12) of which λ_ν is a root.

For w_1 to satisfy (4) and (5) we must have

$$\frac{\partial w^{(0)}}{\partial y} = U_4 - Q_1, \quad \frac{\partial w^{(0)}}{\partial x} = U_5 - Q_1$$

which are true in virtue of (9), and we must also have

$$\frac{\partial}{\partial x}(L_{4\nu} W_\nu) = \frac{\partial}{\partial y}(L_{5\nu} W_\nu)$$

which implies that $\lambda_\nu L_{5\nu} = L_{4\nu}$, and this reduces to an identity if we take into account 2.3 (9), (10) and the definition of μ_ν in 2.3 (16). The complete displacement (u, v, w) is then obtained by substituting the above values in 2.3 (19), (20), (21).

2.73. Induced circuits

Let C be any circuit drawn in the material in any plane parallel to the antiplane

$$z = x + iy, \qquad R = 0\,. \tag{1}$$

Since

$$z_\nu = x + \lambda_\nu y = (x + \alpha_\nu y) + i\beta_\nu y = x_\nu + i y_\nu \tag{2}$$

it follows that in the z_ν-plane, $z_\nu = x_\nu + i y_\nu$, we have an affine transformation of the circuit C into a circuit C_ν effected by the relations

$$x_\nu = x + \alpha_\nu y, \qquad y_\nu = \beta_\nu y\,. \tag{3}$$

We call C_ν the *induced circuit* induced by the affine transformation (3) operating on the *inducing circuit* C.

Also the Jacobian $\frac{\partial(x_\nu, y_\nu)}{\partial(x, y)} = \beta_\nu$ and since by hypothesis, 2.6 (18), $\beta_\nu > 0$, the sense of description of C_ν is the same as that of C.

Thus induced circuits have the same sense of description as the circuit which induces them, fig. 2.73.

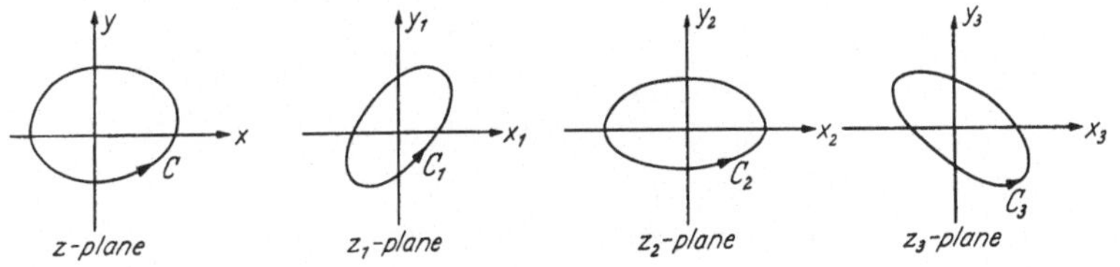

Fig. 2.73

2.74. The cyclic properties of the complex stresses

Let C be any circuit drawn in the material in a plane parallel to the antiplane and let $C_\nu (\nu = 1, 2, 3)$ be the circuits induced, 2.73, by C in the planes $z_\nu (\nu = 1, 2, 3)$.

Let the *cyclic function* [MILNE-THOMSON (6)] or change of the complex stress $W_\nu(z_\nu)$, when z is taken round C and therefore z_ν round C_ν, be denoted by

$$[W_\nu(z_\nu)]_{C_\nu} = k_\nu \,. \tag{1}$$

Since the stress components are necessarily one-valued their cyclic functions are zero with respect to the circuit C whatever the connectivity of the region occupied by the material.

We shall express this fact in terms of the expressions 2.7 (10), (11) with the hypothesis that V, and the derivatives of χ_0 and ψ_0 are one-valued. We then get, using (1), from the expressions for $\widehat{yy}$, $\widehat{xy}$, $\widehat{xx}$, $\widehat{yz}$, $\widehat{zx}$ respectively.

$$\left.\begin{aligned}
&k_1 + \bar{k}_1 + k_2 + \bar{k}_2 + k_3 + \bar{k}_3 &= 0\\
&\lambda_1 k_1 + \bar{\lambda}_1 \bar{k}_1 + \lambda_2 k_2 + \bar{\lambda}_2 \bar{k}_2 + \lambda_3 k_3 + \bar{\lambda}_3 \bar{k}_3 &= 0\\
&\lambda_1^2 k_1 + \bar{\lambda}_1^2 \bar{k}_1 + \lambda_2^2 k_2 + \bar{\lambda}_2^2 \bar{k}_2 + \lambda_3^2 k_3 + \bar{\lambda}_3^2 \bar{k}_3 &= 0\\
&\mu_1 k_1 + \bar{\mu}_1 \bar{k}_1 + \mu_2 k_2 + \bar{\mu}_2 \bar{k}_2 + \mu_3 k_3 + \bar{\mu}_3 \bar{k}_3 &= 0\\
&\lambda_1 \mu_1 k_1 + \bar{\lambda}_1 \bar{\mu}_1 \bar{k}_1 + \lambda_2 \mu_2 k_2 + \bar{\lambda}_2 \bar{\mu}_2 \bar{k}_2 + \lambda_3 \mu_3 k_3 + \bar{\lambda}_3 \bar{\mu}_3 \bar{k}_3 &= 0
\end{aligned}\right\} \tag{2}$$

Solving these linear equations we get

$$\frac{k_1}{\Delta_1} = \frac{-\bar{k}_1}{\Delta_1'} = \frac{k_2}{\Delta_2} = \frac{-\bar{k}_2}{\Delta_2'} = \frac{k_3}{\Delta_3} = \frac{-\bar{k}_3}{\Delta_3'} = K \,, \tag{3}$$

where for example

$$\Delta_1 = \begin{vmatrix} 1 & 1 & 1 & 1 & 1 \\ \bar{\lambda}_1 & \lambda_2 & \bar{\lambda}_2 & \lambda_3 & \bar{\lambda}_3 \\ \bar{\lambda}_1^2 & \lambda_2^2 & \bar{\lambda}_2^2 & \lambda_3^2 & \bar{\lambda}_3^2 \\ \bar{\mu}_1 & \mu_2 & \bar{\mu}_2 & \mu_3 & \bar{\mu}_3 \\ \bar{\lambda}_1\bar{\mu}_1 & \lambda_2\mu_2 & \bar{\lambda}_2\bar{\mu}_2 & \lambda_3\mu_3 & \bar{\lambda}_3\bar{\mu}_3 \end{vmatrix} \tag{4}$$

$$\Delta_1' = \begin{vmatrix} 1 & 1 & 1 & 1 & 1 \\ \lambda_1 & \lambda_2 & \bar{\lambda}_2 & \lambda_3 & \bar{\lambda}_3 \\ \lambda_1^2 & \lambda_2^2 & \bar{\lambda}_2^2 & \lambda_3^2 & \bar{\lambda}_3^2 \\ \mu_1 & \mu_2 & \bar{\mu}_2 & \mu_3 & \bar{\mu}_3 \\ \lambda_1\mu_1 & \lambda_2\mu_2 & \bar{\lambda}_2\bar{\mu}_2 & \lambda_3\mu_3 & \bar{\lambda}_3\bar{\mu}_3 \end{vmatrix} \tag{5}$$

Comparing (4) and (5) we see that $\Delta_1' = \bar{\Delta}_1$ and so from (3) $k_1/\Delta_1 = -\bar{k}_1/\bar{\Delta}_1$. This means that k_ν/Δ_ν is a pure imaginary ($\nu = 1, 2, 3$). Therefore we can write $K = 2\pi i \gamma$ where γ is real, so that

$$\frac{k_1}{\Delta_1} = \frac{k_2}{\Delta_2} = \frac{k_3}{\Delta_3} = 2\pi i \gamma \tag{6}$$

$$\Delta_2 = \begin{vmatrix} 1 & 1 & 1 & 1 & 1 \\ \lambda_1 & \bar{\lambda}_1 & \bar{\lambda}_2 & \lambda_3 & \bar{\lambda}_3 \\ \lambda_1^2 & \bar{\lambda}_1^2 & \bar{\lambda}_2^2 & \lambda_3^2 & \bar{\lambda}_3^2 \\ \mu_1 & \bar{\mu}_1 & \bar{\mu}_2 & \mu_3 & \bar{\mu}_3 \\ \lambda_1\mu_1 & \bar{\lambda}_1\bar{\mu}_1 & \bar{\lambda}_2\bar{\mu}_2 & \lambda_3\mu_3 & \bar{\lambda}_3\bar{\mu}_3 \end{vmatrix} \tag{7}$$

$$\Delta_3 = \begin{vmatrix} 1 & 1 & 1 & 1 & 1 \\ \lambda_1 & \bar{\lambda}_1 & \lambda_2 & \bar{\lambda}_2 & \bar{\lambda}_3 \\ \lambda_1^2 & \bar{\lambda}_1^2 & \lambda_2^2 & \bar{\lambda}_2^2 & \bar{\lambda}_3^2 \\ \mu_1 & \bar{\mu}_1 & \mu_2 & \bar{\mu}_2 & \bar{\mu}_3 \\ \lambda_1\mu_1 & \bar{\lambda}_1\bar{\mu}_1 & \lambda_2\mu_2 & \bar{\lambda}_2\bar{\mu}_2 & \bar{\lambda}_3\bar{\mu}_3 \end{vmatrix} \tag{8}$$

Therefore

$$[W_\nu(z_\nu)]_{C_\nu} = 2\pi i \gamma \Delta_\nu, \quad \nu = 1, 2, 3\,. \tag{9}$$

It follows that

$$[{}'W_\nu(z_\nu)]_{C_\nu} = 2\pi i (\gamma z_\nu \Delta_\nu + A_\nu) \quad (\nu = 1, 2, 3) \tag{10}$$

where γ is an arbitrary real constant and A_ν is an arbitrary (complex) constant.

2.75. Boundary conditions for the complex stresses

Consider a cylindrical or prismatic beam the contour of whose cross-section is the plane curve C.

Let the loading on the lateral surface of the beam be (X_n, Y_n, Z_n) and let $(l, m, 0)$ be the direction cosines of the outward normal $\boldsymbol{n}$ to the surface, so that

$$l = \frac{dy}{ds}, \quad m = -\frac{dx}{ds}$$

where ds is an element of arc of C.

Then the stress boundary condition is

$$\mathsf{S}\cdot\boldsymbol{n} = \boldsymbol{i}X_n + \boldsymbol{j}Y_n + \boldsymbol{k}Z_n$$

or

$$l\,\widehat{xx} + m\,\widehat{xy} = X_n \tag{1}$$

$$l\,\widehat{xy} + m\,\widehat{yy} = Y_n \tag{2}$$

$$l\,\widehat{xz} + m\,\widehat{yz} = Z_n\,. \tag{3}$$

Since

$$\frac{d}{ds}\,{}'W_\nu(z_\nu) = W_\nu(z_\nu)\,\frac{dz_\nu}{ds} = W_\nu(z_\nu)\,(-m + \lambda_\nu l)$$

it follows from (1), (2) and 2.7 (11) that

$$\frac{1}{4}\sum_{\nu=1}^{3}\frac{d}{ds}\left[\lambda_\nu\,{}'W_\nu(z_\nu) + \bar{\lambda}_\nu\,{}'\overline{W}(\bar{z}_\nu)\right] = X_n - V\frac{dy}{ds} - \frac{d}{ds}\left(\frac{\partial\chi_0}{\partial y}\right) = X_{ne}, \tag{4}$$

$$\frac{1}{4}\sum_{\nu=1}^{3}\frac{d}{ds}\left[{}'W_\nu(z_\nu) + {}'\overline{W}(\bar{z}_\nu)\right] = -Y_n - V\frac{dx}{ds} - \frac{d}{ds}\left(\frac{\partial\chi_0}{\partial x}\right) = -Y_{ne}, \tag{5}$$

where (X_{ne}, Y_{ne}, Z_{ne}) is the *effective lateral load* corresponding to (X_n, Y_n, Z_n), and Z_{ne} is defined in (10) below.

Multiply (4) by i and add to (5). Then, 2.7 (1),

$$\frac{1}{2}\frac{d}{ds}\sum_{\nu=1}^{3}\left[\delta_\nu\,{}'W_\nu(z_\nu) + \bar{\gamma}_\nu\,{}'\overline{W}(\bar{z}_\nu)\right] = i(X_{ne} + iY_{ne}), \tag{6}$$

Now let the effective load (X_{ne}, Y_{ne}) be equivalent to effective normal and tangential loads $(\widehat{nn}_e, \widehat{ns}_e)$, fig. 2.75. Then if α is the inclination of the normal to the x-axis

$$X_{ne} + iY_{ne} = (\widehat{nn}_e + i\,\widehat{ns}_e)\,e^{i\alpha}.$$

But

$$e^{i\alpha}ds = -i\,dz\,.$$

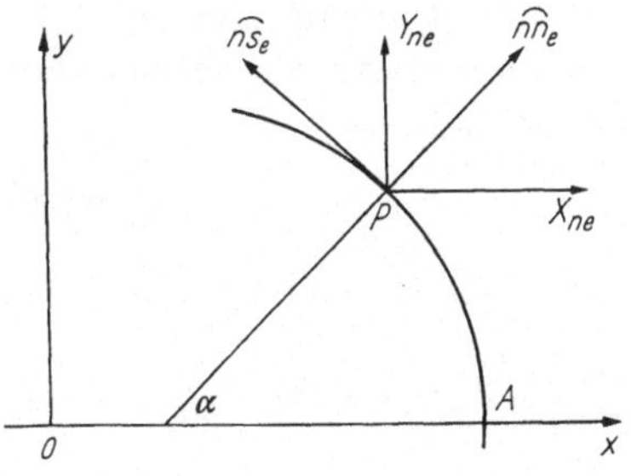

Fig. 2.75

Therefore

$$i(X_{ne} + iY_{ne})\,ds = (\widehat{nn}_e + i\,\widehat{ns}_e)\,dz\,. \tag{7}$$

Substitute in (6) and integrate along the boundary from a fixed point A to a variable point P. Then

$$\sum_{\nu=1}^{3}\left[\delta_\nu\,{}'W_\nu(z_\nu) + \bar{\gamma}_\nu\,{}'\overline{W}_\nu(\bar{z}_\nu)\right] = 2\int_{AP}(\widehat{nn}_e + i\,\widehat{ns}_e)\,dz + \text{constant}\,. \tag{8}$$

The conjugate complex is then

$$\sum_{\nu=1}^{3} [\gamma_\nu' W_\nu(z_\nu) + \delta_\nu' \overline{W}_\nu(\bar{z}_\nu)] = 2 \int_{AP} (\widehat{nn}_e - i\,\widehat{ns}_e)\, d\bar{z} + \text{constant} . \tag{9}$$

Combining (3) with 2.7 (12) we get

$$\frac{1}{4} \sum_{\nu=1}^{3} \frac{d}{ds} [\mu_\nu' W_\nu(z_\nu) + \bar{\mu}_\nu' \overline{W}_\nu(\bar{z}_\nu)]$$

$$= Z_n - \frac{d\psi_0}{ds} + \frac{1}{2k_{33}} [A_1 x^2 + (C_1 - \varepsilon) m x] \frac{dy}{ds} \tag{10}$$

$$- \frac{1}{2k_{33}} [B_1 y^2 + (C_1 - \varepsilon) m y] \frac{dx}{ds} = Z_{ne}$$

and therefore integrating along the boundary

$$\sum_{\nu=1}^{3} [\mu_\nu' W_\nu(z_\nu) + \bar{\mu}_\nu' \overline{W}_\nu(\bar{z}_\nu)] = 4 \int_{AP} Z_{ne}\, ds + \text{constant}. \tag{11}$$

Equations (8), (9), (11) constitute one form of the stress boundary conditions.

We can obtain a more convenient form, free from integrations if we suppose z, z_1, z_2, z_3 to be functions of the parameter $\sigma = e^{i\theta}$ so that

$$z = m(\sigma),\ z_1 = m_1(\sigma),\ z_2 = m_2(\sigma),\ z_3 = m_3(\sigma)$$

$$\sigma = e^{i\theta},\ \bar{\sigma} = \frac{1}{\sigma}. \tag{12}$$

This situation will always arise in mapping the boundary curve C on the unit circumference.

Differentiate the above boundary conditions with respect to σ, noting that, say,

$$\frac{d}{d\sigma} \overline{m}_1(\bar{\sigma}) = - \frac{1}{\sigma^2} \overline{m}_1' \left(\frac{1}{\sigma}\right)$$

and divide through by $d\sigma/ds$. We then get the boundary conditions

$$\sum_{\nu=1}^{3} \left[\delta_\nu W_\nu(z_\nu)\, m_\nu'(\sigma) - \frac{1}{\sigma^2} \bar{\gamma}_\nu \overline{W}_\nu(\bar{z}_\nu) \overline{m}_\nu' \left(\frac{1}{\sigma}\right)\right] = 2(\widehat{nn}_e + i\,\widehat{ns}_e)\, m'(\sigma) \tag{13}$$

$$\sum_{\nu=1}^{3} \left[\gamma_\nu W_\nu(z_\nu)\, m_\nu'(\sigma) - \frac{1}{\sigma^2} \delta_\nu \overline{W}_\nu(\bar{z}_\nu)\, \overline{m}_\nu' \left(\frac{1}{\sigma}\right)\right] = - \frac{2}{\sigma^2} (\widehat{nn}_e - i\,\widehat{ns}_e)\, \overline{m}' \left(\frac{1}{\sigma}\right) \tag{14}$$

$$\sum_{\nu=1}^{3} \left[\mu_\nu W_\nu(z_\nu)\, m_\nu'(\sigma) - \frac{1}{\sigma^2} \bar{\mu}_\nu \overline{W}_\nu(\bar{z}_\nu)\, \overline{m}_\nu' \left(\frac{1}{\sigma}\right)\right] = 4 Z_{ne} \frac{ds}{d\sigma}. \tag{15}$$

As to $ds/d\sigma$ note that

$$\frac{dz}{d\sigma} = m'(\sigma), \qquad \frac{d\bar{z}}{d\sigma} = - \frac{1}{\sigma^2} \overline{m}' \left(\frac{1}{\sigma}\right) \tag{16}$$

and therefore that

$$\left(\frac{ds}{d\sigma}\right)^2 = \frac{dz}{d\sigma}\cdot\frac{d\bar{z}}{d\sigma} = -\frac{1}{\sigma^2}\,m'(\sigma)\,\bar{m}'\left(\frac{1}{\sigma}\right). \tag{17}$$

When the displacement (u^*, v^*, w^*) is given on the boundary the displacement boundary condition is

$$(u, v, w) = (u^*, v^*, w^*) \text{ on the boundary}. \tag{18}$$

The part of the displacement which does not depend on the applicate R is given in terms of the complex stresses by **2.72** (10)—(12).

2.76. The form of the complex stresses at infinity

When the material extends to infinity in the antiplane, as for example in the case of elastic material which occupies all space except for a cylindrical hole, we make the hypothesis that

all stresses are bounded at infinity. (1)

It is easily proved that in the presence of a body-field such as that due to gravity or rotation this condition can not be fulfilled [MILNE-THOMSON (6)]. We therefore have as a consequence of (1) in such cases

there is no body-field. (2)

While in a material with holes, many-valued displacements, due to dislocations [MILNE-THOMSON (6)], are possible, for simplicity we shall restrict our attention to those cases where there are no dislocations, that is to say

the displacements are one-valued (3)

As a consequence of (1), for sufficiently large values of $|z_\nu|$ there must be an expansion of the form

$$W_\nu(z_\nu) = a_{\nu 0} + \frac{a_{\nu 1}}{z_\nu} + \frac{a_{\nu 2}}{z_\nu^2} + \frac{a_{\nu 3}}{z_\nu^3} + \cdots \tag{4}$$

and therefore the form of $`W_\nu(z_\nu)$ is

$$`W_\nu(z_\nu) = a_{\nu 0} z_\nu + a_{\nu 1}\ln z_\nu - \frac{a_{\nu 2}}{z_\nu} - \frac{a_{\nu 3}}{2 z_\nu^2} - \cdots \tag{5}$$

no arbitrary added constant being necessary.

Now let C be any simple closed circuit drawn wholly in the material. The circuit C in the z-plane induces (2.73) a circuit C_ν in the z_ν-plane, and from (5) the cyclic function of $`W_\nu(z_\nu)$ is

$$[`W_\nu(z_\nu)]_{C_\nu} = 2\pi i\, a_{\nu 1} \text{ or } 0 \tag{6}$$

according as the point $z_\nu = 0$ is inside or outside C_ν.

For the displacements u_1, v_1, w_1 to be one-valued we must have

$$[u_1]_C = 0, \; [v_1]_C = 0, \; [w_1]_C = 0$$

and therefore from (6) and 2.72 (10)—(12)

$$\left.\begin{aligned}
&\sum_{\nu=1}^{3} (L_{1\nu} a_{\nu 1} - \bar{L}_{1\nu} \bar{a}_{\nu 1}) = 0 \\
&\sum_{\nu=1}^{3} \left(\frac{L_{2\nu}}{\lambda_\nu} a_{\nu 1} - \frac{\bar{L}_{2\nu}}{\bar{\lambda}_\nu} \bar{a}_{\nu 1}\right) = 0 \\
&\sum_{\nu=1}^{3} (L_{5\nu} a_{\nu 1} - \bar{L}_{5\nu} \bar{a}_{\nu 1}) = 0
\end{aligned}\right\} \tag{7}$$

Let (X, Y, Z) be the resultant force due to the effective loading acting upon the circuit C. Then integration of 2.75 (6) gives

$$2i(X + iY) = \left[\sum_{\nu=1}^{3} (\delta_\nu` W_\nu(z_\nu) + \bar{\gamma}_\nu` \bar{W}_\nu(\bar{z}_\nu)\right]_C = 2\pi i \sum_{\nu=1}^{3} (\delta_\nu a_{\nu 1} - \bar{\gamma}_\nu \bar{a}_{\nu 1}) \tag{8}$$

on the assumption that $z = 0$ lies within C.

Similarly from 2.75 (11)

$$4Z = \left[\sum_{\nu=1}^{3} \mu_\nu` W_\nu(z_\nu) + \bar{\mu}_\nu` \bar{W}_\nu(\bar{z}_\nu)\right]_C = 2\pi i \sum_{\nu=1}^{3} (\mu_\nu a_{\nu 1} - \bar{\mu}_\nu \bar{a}_{\nu 1}) \tag{9}$$

The three equations (7), equation (8) and its complex conjugate, together with equation (9) serve to determine $a_{\nu 1}$, $\bar{a}_{\nu 1}$ for $\nu = 1, 2, 3$.

When the resultant force due to the effective lateral loading is zero the six equations for determining $a_{\nu 1}$, $\bar{a}_{\nu 1}$ will, in general, have only the trivial solution $\bar{a}_{\nu 1} = a_{\nu 1} = 0$, so that the form of $W_\nu(z_\nu)$ is

$$a_{\nu 0} + \frac{a_{\nu 2}}{z_\nu^2} + \frac{a_{\nu 3}}{z_\nu^3} + \cdots$$

The constants $a_{\nu 0}$ depend upon the stress at infinity, in fact from 2.7 (11), (12) we have

$$\left.\begin{aligned}
4\widehat{xx}_\infty &= \sum_{\nu=1}^{3} [\lambda_\nu^2 a_{\nu 0} + \bar{\lambda}_\nu^2 \bar{a}_{\nu 0}] \\
4\widehat{yy}_\infty &= \sum_{\nu=1}^{3} (a_{\nu 0} + \bar{a}_{\nu 0}) \\
4\widehat{xy}_\infty &= -\sum_{\nu=1}^{3} (\lambda_\nu a_{\nu 0} + \bar{\lambda}_\nu \bar{a}_{\nu 0}) \\
4\widehat{xz}_\infty &= \sum_{\nu=1}^{3} (\lambda_\nu \mu_\nu a_{\nu 0} + \bar{\lambda}_\nu \bar{\mu}_\nu \bar{a}_{\nu 0}) \\
4\widehat{yz}_\infty &= -\sum_{\nu=1}^{3} (\mu_\nu a_{\nu 0} + \bar{\mu}_\nu \bar{a}_{\nu 0})
\end{aligned}\right\} \tag{10}$$

These are 5 equations for 6 unknowns. One may therefore be arbitrarily assigned or we might take a sixth relation, say that a_{10} is real or

$$a_{10} = \bar{a}_{10} \,. \tag{11}$$

When the stress components at infinity are all zero, only the trivial solution

$$a_{\nu 0} = \bar{a}_{\nu 0} = 0\,, \; \nu = 1, 2, 3$$

is available.

Thus if the resultant force and the stress components at infinity all vanish, the form of $W_\nu(z_\nu)$ at infinity is

$$W^*_\nu(z_\nu) = \frac{a_{\nu 2}}{z_\nu^2} + \frac{a_{\nu 3}}{z_\nu^3} + \cdots \tag{12}$$

Put in another way; when the material extends to infinity the general form of the complex stress is

$$W_\nu(z_\nu) = a_{\nu 0} + \frac{a_{\nu 1}}{z_\nu} + W^*(z_\nu)\,, \tag{13}$$

where $W^*(z_\nu)$ is holomorphic in the neighbourhood of infinity, and for sufficiently large values of $|z_\nu|$ has an expansion of the form (12). The constants $a_{\nu 0}$ are determined by the stress at infinity and the constants $a_{\nu 1}$ by one-valuedness of the displacement and by the resultant loading.

2.77. Determinateness of the complex stresses

We seek the form of the complex stresses which give zero stress when the body-field is zero. Calling these complex stresses $W_1^{(0)}(z_1)$, $W_2^{(0)}(z_2)$, $W_3^{(0)}(z_3)$ we have from 2.7 (11), (12)

$$W_1^{(0)}(z_1) + \overline{W}_1^{(0)}(\bar{z}_1) + W_2^{(0)}(z_2) + \overline{W}_2^{(0)}(\bar{z}_2) + W_3^{(0)}(z_3) + \overline{W}_3^{(0)}(\bar{z}_3) = 0$$

$$\lambda_1 W_1^{(0)}(z_1) + \bar{\lambda}_1 \overline{W}_1^{(0)}(\bar{z}_1) + \lambda_2 W_2^{(0)}(z_2) + \bar{\lambda}_2 \overline{W}_2^{(0)}(\bar{z}_2) + \lambda_3 W_3^{(0)}(z_3) + \bar{\lambda}_3 \overline{W}_3^{(0)}(\bar{z}_3) = 0$$

$$\lambda_1^2 W_1^{(0)}(z_1) + \bar{\lambda}_1^2 \overline{W}_1^{(0)}(\bar{z}_1) + \lambda_2^2 W_2^{(0)}(z_2) + \bar{\lambda}_2^2 \overline{W}_2^{(0)}(\bar{z}_2) + \lambda_3^2 W_3^{(0)}(z_3) + \bar{\lambda}_3^2 \overline{W}_3^{(0)}(\bar{z}_3) = 0$$

$$\mu_1 W_1^{(0)}(z_1) + \bar{\mu}_1 \overline{W}_1^{(0)}(\bar{z}_1) + \mu_2 W_2^{(0)}(z_2) + \bar{\mu}_2 \overline{W}_2^{(0)}(\bar{z}_2) + \mu_3 W_3^{(0)}(z_3) + \bar{\mu}_3 \overline{W}_3^{(0)}(\bar{z}_3) = 0$$

$$\lambda_1\mu_1 W_1^{(0)}(z_1) + \bar{\lambda}_1\bar{\mu}_1 \overline{W}_1^{(0)}(\bar{z}_1) + \lambda_2\mu_2 W_2^{(0)}(z_2) + \bar{\lambda}_2\bar{\mu}_2 \overline{W}_2^{(0)}(\bar{z}_2) + \lambda_3\mu_3 W_3^{(0)}(z_3) + \bar{\lambda}_3\bar{\mu}_3 \overline{W}_3^{(0)}(\bar{z}_3) = 0\,.$$

These equations are of the same form as those in 2.74 (2) but with k_ν replaced by $W_\nu(z_\nu)$.

Therefore we can use exactly the same argument to show that

$$\frac{W_1^{(0)}(z_1)}{\Delta_1} = \frac{W_2^{(0)}(z_2)}{\Delta_2} = \frac{W_3^{(0)}(z_3)}{\Delta_3} = 2\pi i A\,, \tag{1}$$

where A is a real constant corresponding to γ in 2.74 and Δ_1, Δ_2, Δ_3 are given respectively by 2.74 (4), (7), (8).

Thus

$$W_1^{(0)}(z_1) = 2\pi i A\,\Delta_1,\; W_2^{(0)}(z_2) = 2\pi i A\,\Delta_2,\; W_3^{(0)}(z_3) = 2\pi i A\,\Delta_3. \tag{2}$$

If therefore $W_1(z_1)$, $W_2(z_2)$, $W_3(z_3)$ are complex stresses which solve a problem for given boundary stresses, the general solution will be

$$W_1(z_1) + 2\pi i A\,\Delta_1,\; W_2(z_2) + 2\pi i A\,\Delta_2,\; W_3(z_3) + 2\pi i A\,\Delta_3\,,$$

for the added terms give rise to no additional stress.

2.8. Effective stress functions

We shall call *effective stress functions* the functions χ_e, ψ_e defined by

$$\chi_e = \frac{1}{4}\sum_{\nu=1}^{3}\left[{}^{\backprime\backprime}W_\nu(z_\nu) + {}^{\backprime\backprime}\overline{W}_\nu(\bar{z}_\nu)\right], \tag{1}$$

$$\psi_e = \frac{1}{4}\sum_{\nu=1}^{3}\left[\mu_\nu\,{}^{\backprime}W_\nu(z_\nu) + \bar{\mu}_\nu\,{}^{\backprime}\overline{W}_\nu(\bar{z}_\nu)\right]. \tag{2}$$

Then from 2.7 (4), (5), (6) we have

$$\begin{aligned} \Theta &= 4\frac{\partial^2 \chi_e}{\partial z\,\partial \bar{z}} + \Theta_0 + 4\frac{\partial^2 \chi_0}{\partial z\,\partial \bar{z}}, \\ \Phi &= 4\frac{\partial^2 \chi_e}{\partial z^2} + \Phi_0 + 4\frac{\partial^2 \chi_0}{\partial z^2}, \\ \Psi &= 2i\frac{\partial \psi_e}{\partial z} + \Psi_0 + 2i\frac{\partial \psi_0}{\partial z}. \end{aligned} \tag{3}$$

We can write (3) in the form

$$\begin{aligned} \Theta &= \Theta_e + \Theta_0 + 4\frac{\partial^2 \chi_0}{\partial z\,\partial \bar{z}} \\ \Phi &= \Phi_e + \Phi_0 + 4\frac{\partial^2 \chi_0}{\partial z^2} \\ \Psi &= \Psi_e + \Psi_0 + 2i\frac{\partial \psi_0}{\partial z} \end{aligned} \tag{4}$$

in terms of effective fundamental stress combinations Θ_e, Φ_e, Ψ_e.

Since a prior condition for the solution of any problem is the determination of the particular integrals Θ_0, Φ_0, Ψ_0 of 2.1 (13), (14) and χ_0, ψ_0 of 2.6 (1), (2), we must regard these particular integrals as known. Thus we see from (4) that the actual fundamental stress combinations are determined when the effective ones are and vice versa.

Thus without loss of generality we can work in terms of effective stress functions with a consequent appreciable simplification of the algebra.

2.9. The shear function

The k-matrix in the isotropic case is given by 1.98 (5) and therefore the modified inverse moduli l_{rs} of 2.32 are the elements of the matrix

$$\begin{bmatrix} \frac{1-\eta^2}{E} & \frac{-\eta-\eta^2}{E} & 0 & 0 & 0 & 0 \\ \frac{-\eta-\eta^2}{E} & \frac{1-\eta^2}{E} & 0 & 0 & 0 & 0 \\ 0 & 0 & 0 & 0 & 0 & 0 \\ 0 & 0 & 0 & \frac{1}{\mu} & 0 & 0 \\ 0 & 0 & 0 & 0 & \frac{1}{\mu} & 0 \\ 0 & 0 & 0 & 0 & 0 & \frac{1}{\mu} \end{bmatrix}$$

where η is Poisson's ratio and E is Young's modulus, 1.98.

Therefore the operators of 2.52 (3)—(7) become, with

$$\nabla^2 = \frac{\partial^2}{\partial x^2} + \frac{\partial^2}{\partial y^2}, \tag{1}$$

$$D^{(4)} = \frac{1-\eta^2}{E} \nabla^4, \quad D^{(3)} \equiv 0, \quad D^{(2)} = \frac{1}{\mu} \nabla^2$$
$$H^{(2)} = \frac{(1+\eta)(1-2\eta)}{E} \nabla^2, \quad H^{(1)} \equiv 0 \tag{2}$$

and therefore 2.52 (8), (10) give

$$\nabla^4 \chi + \frac{1-2\eta}{1-\eta} \nabla^2 V = 0, \tag{3}$$

$$\frac{1}{\mu} \nabla^2 \psi = 2\eta (A_1 y - B_1 x) - 2\tau \tag{4}$$

where A_1, B_1 are given by 2.2 (7).

Thus in the isotropic case χ and ψ satisfy completely independent equations*. The solution of (3) is obtained in [MILNE-THOMSON (6) 1.74] in the form

$$\chi = \frac{1}{4} \{\bar{z}\,{}'W(z) + z\,{}'\overline{W}(\bar{z}) + {}''w(z) + {}''\overline{w}(\bar{z})\} - \nu Q \tag{5}$$

where

$$\nu = (1-2\eta)/(1-\eta) \tag{6}$$

and

$$4 \frac{\partial^2 Q}{\partial z \partial \bar{z}} = V. \tag{7}$$

* This is also true when $R = 0$ is a plane of elastic symmetry (see 6.5).

The fundamental stress combinations Θ and Φ are then expressed in terms of the complex stresses $W(z)$, $w(z)$ by

$$\widehat{xx} + \widehat{yy} = \Theta = W(z) + \overline{W}(\bar{z}) + (2 - \nu) V \tag{8}$$

$$\widehat{yy} - \widehat{xx} + 2i\widehat{xy} = \Phi = \bar{z} W'(z) + w(z) - 4\nu \frac{\partial^2 Q}{\partial z^2} . \tag{9}$$

As to (4) this can be written in the form

$$\frac{4i\partial^2 \psi}{\mu \partial z \partial \bar{z}} = \eta (\bar{\beta} z - \beta \bar{z}) - 2i\tau, \; \beta = A_1 + i B_1$$

and therefore integrating with respect to $\bar{z}$

$$\frac{2i}{\mu} \frac{\partial \psi}{\partial z} = \frac{1}{2} \eta \bar{\beta} z \bar{z} - \frac{1}{4} \eta \beta \bar{z}^2 - i\tau \bar{z} + f'(z) \tag{10}$$

where $f'(z)$ is an arbitrary function.

Integrating once more with respect to z

$$\frac{2i}{\mu} \psi = \frac{1}{4} \eta \bar{\beta} z^2 \bar{z} - \frac{1}{4} \eta \beta \bar{z}^2 z - i\tau z \bar{z} + f(z) - \bar{f}(\bar{z}) \tag{11}$$

the last function being chosen to make the right-hand side imaginary, as is the left-hand side.

Combining (10) with 2.1 (15) and 2.24 (6) we find

$$\begin{aligned} \frac{1}{\mu} (\widehat{xz} - i\widehat{yz}) = \Phi(z) - \{i\tau + (C_1 - \varepsilon) m (1 + \eta)\} \bar{z} - \frac{1 + 2\eta}{4} \beta \bar{z}^2 \\ - \frac{1}{2} \bar{\beta} z \bar{z} \end{aligned} \tag{12}$$

where we have written

$$\Phi(z) = f'(z) - \frac{1}{4} (1 + \eta) \bar{\beta} z^2 .$$

We shall call $\Phi(z)$ the *shear function*, for when it is known the shears $\widehat{xz}$, $\widehat{yz}$ are determined from (12).

EXAMPLES II

1. When the antiplane is horizontal show that the potential of the body field can be taken in the form $V_0 = gR + V(x, y)$.

2. Obtain in detail the compatibility equations of 2.2 and show that they are completely equivalent to

$$\frac{\partial^2 e_{pq}}{\partial x_a^2} - \frac{\partial^2 e_{aq}}{\partial x_a \partial x_p} + \frac{\partial^2 e_{aa}}{\partial x_p \partial x_q} - \frac{\partial^2 e_{ap}}{\partial x_a \partial x_q} = 0 ,$$

where now the left-hand side is subject to the summation convention.

3. Obtain the formula 2.3 (14) for e_{zz}^0 and the expressions 2.3 (15) (16) for Q and Q'.

4. Show that the equation 2.1 (14)

$$\frac{\partial \Psi_0}{\partial z} + \frac{\partial \Psi_0}{\partial \bar{z}} = B_3$$

has a solution

$$\Psi_0 = -\frac{1}{2k_{33}} [(A_1 - iB_1) z\bar{z} + (C_1 - \varepsilon) m\bar{z}] .$$

5. Show that it is possible to write $\widehat{zx}$, $\widehat{yz}$ in the forms

$$\widehat{zx} = \frac{\partial \psi}{\partial y} - Q, \quad \widehat{yz} = -\frac{\partial \psi}{\partial x} - Q ,$$

where

$$Q = \frac{1}{2k_{33}} [A_1 x^2 + B_1 y^2 + (C_1 - \varepsilon) m (x + y)] .$$

6. Verify that the displacement components u, v, w of 2.3 do in fact satisfy the equations from which they are derived.

7. Find the displacement components when the stress tensor is independent of R.

8. Obtain formulae for change of the modified inverse moduli l_{rs} with rotation of cartesian axes of reference.

9. Find the matrix for the modified inverse moduli in the case of (i) an orthotropic (ii) a monotropic material.

10. A cylindrical anisotropic elastic rod of length L bounded by plane cross-sections is fixed with the upper cross-section horizontal and the R-axis taken vertically downwards from an origin in the upper cross-section. To the lower end of the rod is applied a normal stress T uniformly applied over the lower end. Prove that if gravity is neglected the conditions of equilibrium are satisfied by

$$\widehat{zz} = T, \widehat{xx} = \widehat{xy} = \widehat{yy} = \widehat{yz} = \widehat{zx} = 0 .$$

11. In the preceding example show that the strain coefficients are

$$e_{xx} = Tk_{13}, \quad e_{yy} = Tk_{23}, \quad e_{zz} = Tk_{33}$$
$$2e_{yz} = Tk_{34}, 2e_{zx} = Tk_{35}, 2e_{xy} = Tk_{36}$$

12. In the preceding example show that the displacements apart from a rigid body displacement are given by

$$u = T(k_{13}x + k_{36}y + k_{35}R), v = T(k_{23}y + k_{34}R), w = Tk_{33}R.$$

Prove further that if the displacement is zero at the origin and if, also at the origin,

$$\frac{\partial u}{\partial R} = \frac{\partial v}{\partial R} = \frac{\partial v}{\partial x} - \frac{\partial u}{\partial y} = 0$$

the rigid body movement is determined and

$$u = T\left(k_{13}x + \frac{1}{2}k_{36}y\right), v = T\left(\frac{1}{2}k_{36}x + k_{23}y\right)$$
$$w = T(k_{35}x + k_{34}y + k_{33}R) .$$

13. In Ex. 10 if we take gravity into account, show that the conditions of equilibrium are satisfied by

$$b_1 = 0,\, b_2 = 0,\, b_3 = -g\varrho$$

$$\widehat{zz} = g\varrho(L - R),\, \widehat{xx} = \widehat{xy} = \widehat{yy} = \widehat{yz} = \widehat{zx} = 0$$

and that with the same conditions of fixing at the origin as in Ex. 12, the displacements are

$$u = g\varrho\left\{-\frac{1}{2}k_{35}R^2 + k_{13}x(L-R) + \frac{1}{2}k_{36}y(L-R)\right\}$$

$$v = g\varrho\left\{-\frac{1}{2}k_{34}R^2 + k_{23}y(L-R) + \frac{1}{2}k_{36}x(L-R)\right\}$$

$$w = g\varrho\left\{\frac{1}{2}k_{13}R^2 + \frac{1}{2}k_{23}y^2 + \frac{1}{2}k_{36}xy + (k_{34}y + k_{35}x)L + \frac{1}{2}k_{33}R(2L-R)\right\}.$$

14. In Ex. 13 find the equations after deformation of the line which occupied the position $x = 0$, $y = 0$ before deformation.

15. In the notation of 2.4 calculate the strain energy when

$$\widehat{xx} = \frac{k^2}{N},\, \widehat{yy} = \frac{1}{N},\, \widehat{zz} = -\frac{1}{N}(\nu_1 k^2 - \nu_6 k + \nu_2)$$

$$\widehat{xy} = -k/N, \quad \widehat{zx} = \widehat{yz} = 0$$

in the form $U = d^{(4)}(k)/2N^2$.

Deduce that the equation

$$d^{(4)}(\lambda) = 0$$

can have no real roots.

16. In the notation of 2.72 prove that

$$\frac{L_{2S}}{\lambda_S} + \lambda_S L_{1S} = L_{6S},$$

$$\lambda_S L_{5S} = L_{4S}.$$

17. In 2.72 verify that the six equations (9) are consistent with the existence of a set of three particular solutions $u^{(0)}$, $v^{(0)}$, $w^{(0)}$ by proving that

$$\frac{\partial^2 U_2}{\partial x^2} - \frac{\partial^2 U_6}{\partial x \partial y} + \frac{\partial^2 U_1}{\partial y^2} = 0$$

$$\frac{\partial}{\partial x}(U_4 - Q_1') = \frac{\partial}{\partial y}(U_5 - Q_1).$$

18. Find the curve C_1 induced from the ellipse C, $x^2/a^2 + y^2/b^2 = 1$, by the affine mapping

$$x_1 = x + \alpha y,\, y_1 = \beta y.$$

19. If, in the preceding example, the ellipse C is mapped on the circumference γ of the unit circle $|\sigma| = 1$ in the σ-plane by

$$z = \frac{1}{2}(a+b)\,\sigma + \frac{1}{2}(a-b)\frac{1}{\sigma},$$

find the function which maps the induced curve C_1 on γ.

20. Expand the determinants Δ_1, Δ_2, Δ_3 of section 2.74 with respect to the minors of the two bottom rows.

Chapter III

Isotropic beams

In this chapter we consider the general antiplane problem for isotropic (prismatic or cylindrical) beams and investigate in particular the cases of extension, bending by couples and push.

3.1. The boundary conditions for a prismatic beam

Consider a finite beam of length L bounded by two cross-sections, the *ends*. We suppose the beam to be fixed at one end, the *root*, and to be kept in equilibrium by systems of forces applied to the free end, forces applied to the lateral surface, and body-force.

Take right handed axes of reference Ox, Oy in the plane of the free end and OR directed parallel to the generators and into the material.

We shall suppose the stress-system to be antiplane with $R = 0$ as the antiplane.

Let the loading applied to the lateral surface be (X_n, Y_n, Z_n) at a point where $\boldsymbol{n}$ is the outwardly directed unit normal vector. As in 1.4 we denote the body-vector by (b_1, b_2, b_3), independent of R.

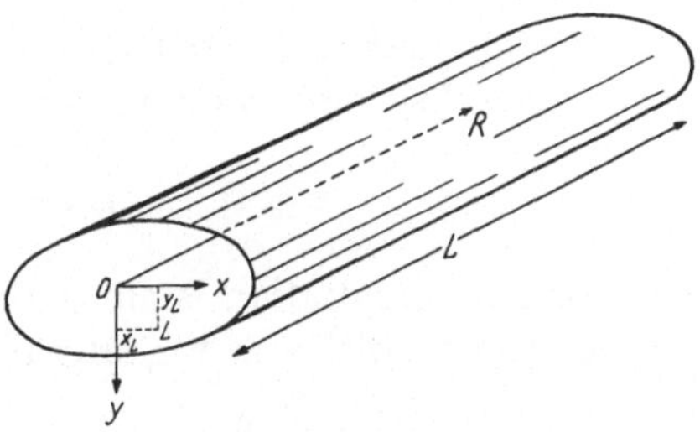

Fig. 3.1 (i)

Taking the origin as base-point, let the loading applied to the antiplane, $R = 0$, be statically equivalent to a force (P_1, P_2, P_3) and a couple (M_1, M_2, M_3), fig. 3.1 (ii). *This system of loading will be applied not as a concentrated system but as a distribution of tractions over the antiplane so arranged as to be consistent with antiplane stress.* The loading over the fixed end, $R = L$, of the beam will be such as to achieve equilibrium with the body field and the other applied loads and again to be consistent with antiplane stress.

The components of stress on the antiplane will be denoted by $\widehat{zx}$, $\widehat{zy}$, $\widehat{zz}_0$. Here $\widehat{zx}$, $\widehat{zy}$ are independent of R while $\widehat{zz}_0$ denotes the value of the stress $\widehat{zz}$ when $R = 0$. These components of traction are directed in the negative directions of the axes of reference, fig. 3.1 (ii).

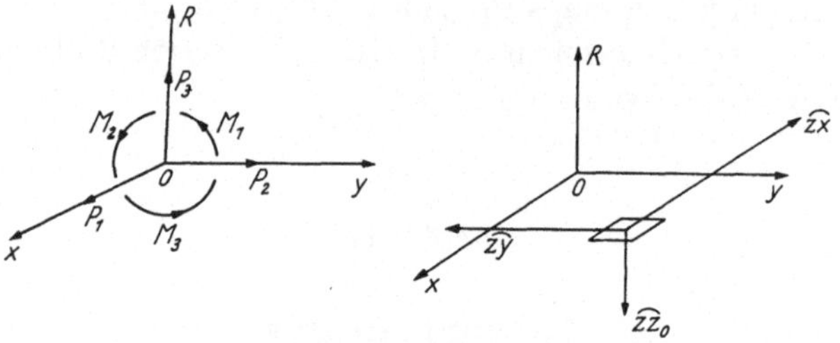

Fig. 3.1 (ii)

The statical equivalence of the load on the antiplane with the stress applied over it is expressed by resolution parallel to, and moments about, the axes. Thus

$$\int \widehat{zx}\, dS + P_1 = 0, \int \widehat{yz}\, dS + P_2 = 0 \tag{1}$$

$$\int (y \cdot \widehat{zx} - x \cdot \widehat{zy})\, dS = M_3 = -M \tag{2}$$

$$\int y \cdot \widehat{zz}_0\, dS = -M_1, \int x \cdot \widehat{zz}_0 = M_2 \tag{3}$$

$$\int \widehat{zz}_0 \cdot dS + P_3 = 0\,. \tag{4}$$

Here the integrals are taken over the whole of the cross-sectional area. At the lateral surface, the unit outward normal is

$$\boldsymbol{n} = l\boldsymbol{i} + m\boldsymbol{j}, \, l = \frac{\partial y}{\partial s}, \, m = -\frac{\partial x}{\partial s}, \tag{5}$$

where the differentiation is along the arc s of the periphery of a cross-section. The condition at the lateral surface is therefore

$$\mathsf{S} \cdot \boldsymbol{n} = \boldsymbol{i} X_n + \boldsymbol{j} Y_n + \boldsymbol{k} Z_n$$

or

$$l\widehat{xx} + m\,\widehat{xy} = X_n \tag{6}$$

$$l\widehat{xy} + m\,\widehat{yy} = Y_n \tag{7}$$

$$l\widehat{zx} + m\,\widehat{zy} = Z_n \tag{8}$$

Since $\widehat{xx}$, $\widehat{xy}$, $\widehat{yy}$, $\widehat{zx}$, $\widehat{zy}$ are by hypothesis independent of R, it follows that (X_n, Y_n, Z_n) must be independent of R and therefore *the same at every point of the same generator*. In general, however, (X_n, Y_n, Z_n) will change from one generator to another.

From this it appears that a necessary condition for antiplane stress in a beam, for which the antiplane is a given cross-section, is that the body-vector and the lateral surface loading shall be independent of distance from that cross-section.

When the resultant loading on one cross-section of the beam is given it will be found that sufficient arbitrariness will remain to allow us to determine that distribution of stress over this cross-section which will be statically equivalent to the given loading and which will lead to antiplane stress.

The requisite resultant loading on any other cross-section, consistent with antiplane stress, will then be determined by considerations of statical equilibrium.

In practice the end load will usually be applied in a manner different from that which will produce antiplane stress in the neighbourhood of the end. Nevertheless, if the beam is long enough, we can invoke the principle of de ST. VENANT to the effect that the state of stress sufficiently far from the place at which the load is applied will be practically the same for all statically equivalent loads.

3.2. The isotropic beam

From 2.9 it appears that in the case of isotropy the stress functions χ and ψ satisfy independent equations. It therefore appears from the boundary conditions 3.1 (6)—(8) that however the beam and its surface may be loaded consistently with antiplane stress, we have, 2.9 (3),

$$\nabla^4 \chi + \frac{1-2\eta}{1-\eta} \nabla^2 V = 0\,, \tag{1}$$

$$l\,\widehat{xx} + m\,\widehat{xy} = X_n,\; l\,\widehat{xy} + m\,\widehat{yy} = Y_n \text{ at the lateral surface} \tag{2}$$

where $(l, m, 0)$ are the direction cosines of the outward normal to the lateral surface, while

$$\frac{1}{\mu} \nabla^2 \psi = 2\eta (A_1 y - B_1 x) - 2\tau\,, \tag{3}$$

$$l\,\widehat{zx} + m\,\widehat{zy} = Z_n \text{ at the lateral surface}\,. \tag{4}$$

From 2.9 (12) we can replace (3) by its integrated form in terms of the shear function $\Phi(z)$ namely

$$\frac{\widehat{xz} - i\,\widehat{yz}}{\mu} = \Phi(z) - p\bar{z} - q\bar{z}^2 - r\,z\bar{z}\,, \tag{5}$$

where

$$p = i\tau + (1+\eta)(C_1 - \varepsilon)\,m\,,\; q = \frac{1+2\eta}{4}\beta\,,\; r = \frac{1}{2}\bar{\beta}\,,\; \beta = A_1 + iB_1\,, \tag{6}$$

together with the conditions 3.1 (1)—(4)

$$\int \widehat{zx}\, dS = -P_1, \int \widehat{yz}\, dS = -P_2, \int (y \cdot \widehat{zx} - x \cdot \widehat{zy})\, dS = M_3, \tag{7}$$

$$\int \widehat{zz}_0\, dS = -P_3, \int y \cdot \widehat{zz}_0\, dS = -M_1, \int x \cdot \widehat{zz}_0\, dS = M_2 \tag{8}$$

Thus the problem when the loading is given resolves itself into two quasi-independent problems as follows

(i) the problem arising from the boundary condition (2) namely the problem of finding $\widehat{xx}$, $\widehat{xy}$, $\widehat{yy}$ when the loading $(X_n, Y_n, 0)$ on the lateral surface and the part V of the body potential $V + \varepsilon Rm$ (see **2.24**) are given, and $\widehat{yz} = \widehat{zx} = 0$.

(ii) The problem arising when the force (P_1, P_2, P_3) and the moment (M_1, M_2, M_3) are applied to the free end, the loading on the lateral surface is $(0, 0, Z_n)$ and $\widehat{xx} = \widehat{xy} = \widehat{yy} = 0$. Into this problem body force enters only through ε.

We shall call problem (ii) the *pure antiplane problem* for isotropic material.

The two problems have been called quasi-independent for the following reason. Problem (i) is the plane elastic deformation isotropic case which has been fully treated in [MILNE-THOMSON (6)]. Assuming this plane deformation problem to have been solved to find $\widehat{xx}$, $\widehat{xy}$, $\widehat{yy}$ we have then a known longitudinal stress component $\widehat{zz}_p$ in the R-direction given by Hooke's law namely [MILNE-THOMSON (6), **1.7** (8)]

$$\widehat{zz}_p = \eta(\widehat{xx} + \widehat{yy}), \tag{9}$$

where suffix p refers to the solution arising from the plane deformation problem, which in this connection may be termed the *associated plane problem*.

For problem (ii) we proceed on the assumption that the conditions are as just stated. Assuming the problem to have been solved we find the longitudinal component of stress $\widehat{zz}_a$, and the shears $\widehat{zx}$, $\widehat{zy}$, where suffix a refers to the pure antiplane case $\widehat{xx} = \widehat{xy} = \widehat{yy} = 0$.

The complete solution will therefore consist of $\widehat{xx}$, $\widehat{xy}$, $\widehat{yy}$, $\widehat{zx}$, $\widehat{zy}$ as found together with

$$\widehat{zz} = \widehat{zz}_a + \widehat{zz}_p. \tag{10}$$

Thus the problems are independent except insofar as (10) intervenes. Further from **2.3** (22) we have for e_{zz} the form

$$e_{zz} = (A_1 x + B_1 y + C_1 m) R + A_2 x + B_2 y + C_2 m,$$

since in the isotropic case $\nu_4 = \nu_5 = 0$. Also by Hooke's law

$$\widehat{zz} = E e_{zz} + \eta(\widehat{xx} + \widehat{yy}) = E e_{zz} + \widehat{zz}_p$$

from (9). Therefore from (10)

$$\widehat{zz}_a = E e_{zz} = E(A_1 x + B_1 y + C_1 m) R + E(A_2 x + B_2 y + C_2 m) \tag{11}$$

and therefore

$$\widehat{zz} = E(A_1 x + B_1 y + C_1 m) R + E(A_2 x + B_2 y + C_2 m) + \widehat{zz}_p. \tag{12}$$

Remembering that $\widehat{zz}_p$ is independent of R we see that

$$\widehat{zz} = E(A_1 x + B_1 y + C_1 m) R + \widehat{zz}_0, \tag{13}$$

where

$$\widehat{zz}_0 = E(A_2 x + B_2 y + C_2 m) + \widehat{zz}_p \tag{14}$$

is the value of $\widehat{zz}$ on the antiplane.

Note also that for pure antiplane stress $\widehat{zz}_p = 0$ and therefore for pure antiplane stress $\widehat{zz}$ is a linear form in R whose coefficients are linear forms in x and y.

3.22. The moment equations

With the axes of reference and notation of 3.1 we suppose the free end of the beam to be acted upon by a wrench which, with the origin as base point, consists of a force (P_1, P_2, P_3) applied at the origin and a couple whose components are (M_1, M_2, M_3) along the axes. This system is statically equivalent to a force (P_1, P_2, P_3) acting at the *load point* $L(x_L, y_L, 0)$ together with a couple of components $(M_1 - y_L P_3, M_2 + x_L P_3, M_3 - x_L P_2 + y_L P_1)$ along the axes. If the force is zero, we have the couple alone, if the couple is zero, we have the force alone. Any intermediate conditions for which some components of the force or couple vanish are possible.

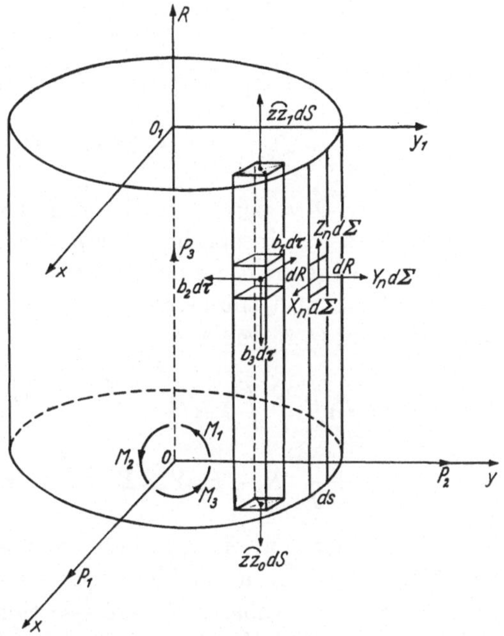

Fig. 3.22

We suppose a load (X_n, Y_n, Z_n) to act on the lateral surface of the beam and the body-vector to be (b_1, b_2, b_3). The *effective body-force* per unit volume is then $(-b_1, -b_2, -b_3)$.

Consider the portion of the beam between the free end $R = 0$ and an arbitrary cross-section $R = R_1$ across which the components of the stress vector will be denoted by $\widehat{zx}$, $\widehat{zy}$, $\widehat{zz}_1$, for $\widehat{zx}$, $\widehat{zy}$ are independent of R_1. From 3.2 (12) we have

$$\widehat{zz}_1 = L_1 R_1 + L_2 + \widehat{zz}_p\,, \tag{1}$$

where

$$L_1 = E(A_1 x + B_1 y + C_1 m)\,,\; L_2 = E(A_2 x + B_2 y + C_2 m)\,, \tag{2}$$

$$\widehat{zz}_p = \eta(\widehat{xx} + \widehat{yy})\,. \tag{3}$$

The term $\widehat{zz}_p$ arises from the associated plane problem (3.2) and is zero when such a problem is absent.

Thus we can write

$$\widehat{zz}_1 = L_1 R_1 + \widehat{zz}_0\,, \tag{4}$$

where

$$\widehat{zz}_0 = L_2 + \widehat{zz}_p \tag{5}$$

is independent of R.

From 3.2 (7) (8) we have

$$\int \widehat{zx}\, dS = -P_1,\; \int \widehat{yz}\, dS = -P_2,\; \int (y\cdot\widehat{xz} - x\cdot\widehat{yz})\, dS = M_3 \tag{6}$$

$$\int \widehat{zz}_0\, dS = -P_3,\; \int y\cdot\widehat{zz}_0\, dS = -M_1,\; \int x\cdot\widehat{zz}_0\, dS = M_2\,. \tag{7}$$

Draw O_1x_1, O_1y_1 parallel to the x- and y-axes through the point O_1 in which the R-axis meets the plane $R = R_1$.

The portion of the beam between $R = 0$ and $R = R_1$ is in equilibrium under the forces just enumerated and the stress across $R = R_1$. Taking moments about O_1x_1 we have

$$\begin{aligned} M_1 + R_1 P_2 - \int y b_3\, d\tau - \int (R_1 - R)\, b_2\, d\tau + \int y Z_n\, d\Sigma \\ + \int (R_1 - R)\, Y_n\, d\Sigma + \int y\cdot\widehat{zz}_1\, dS = 0\,. \end{aligned} \tag{8}$$

Here dS is an element of area of the cross-section , $d\Sigma$ is an element of area of the lateral surface and $d\tau$ is an element of volume. The integrals are taken over the whole of the cross-section, lateral surface or volume as the case may be. Write $d\tau = dR\, dS$, $d\Sigma = dR\, ds$, where ds is an element of arc of the contour of the cross-section and note that

$$\int_0^{R_1} (R_1 - R)\, dR = \frac{1}{2} R_1^2,$$

while b_1, b_2, b_3, Y_n, Z_n are independent of R. Then the above moment equation can be written

$$M_1 + R_1 P_2 - R_1 \int y b_3 \, dS - \frac{1}{2} R_1^2 \int b_2 \, dS + R_1 \oint y Z_n \, ds$$
$$+ \frac{1}{2} R_1^2 \oint Y_n ds + \int y (L_1 R_1 + \widehat{zz}_0) \, dS = 0 \, .$$

Equating to zero the coefficients of the several powers of R_1 we get the second of (7) together with

$$\begin{aligned} P_2 - \int y b_3 dS + \int y L_1 dS + \oint y Z_n ds &= 0 \, , \\ -\int b_2 dS + \oint Y_n ds &= 0; \end{aligned} \tag{9}$$

the last equation expresses that the body-force resolved parallel to the y-axis is in equilibrium with the force on the lateral surface resolved in the same direction.

Similarly taking moments about $O_1 y_1$ we get the third of (7) together with

$$\begin{aligned} P_1 - \int x b_3 \, dS + \int x L_1 \, dS + \oint x \cdot Z_n ds &= 0 \, , \\ -\int b_1 \, dS + \oint X_n ds &= 0 \, . \end{aligned} \tag{10}$$

It appears from the above equations that the components of the body-field parallel to the antiplane serves only to equilibrate the force parallel to the antiplane applied to the lateral surface.

When the body-field is due to gravity only and the antiplane is horizontal we see that

$$\oint X_n ds = 0 \, , \qquad \oint Y_n ds = 0 \tag{11}$$

in other words the forces, parallel to the antiplane, applied to the lateral surface between any pair of transverse planes are in equilibrium.

3.24. Determination of the longitudinal stress component $\widehat{zz}$

We have from 3.2 (12)

$$\widehat{zz} = E(A_1 x + B_1 y + C_1 m) R + E(A_2 x + B_2 y + C_2 m) + \widehat{zz}_p \, , \tag{1}$$

where $\widehat{zz}_p$ refers to the associated plane problem.

For the equilibrium of the portion of the beam between $R = 0$ and $R = R_1$, resolution parallel to OR gives

$$P_3 + \int \widehat{zz}_1 \, dS - \int b_3 \, d\tau + \int Z_n d\Sigma = 0 \tag{2}$$

in the notation of fig. 3.22 and from 3.22 (7), (5)

$$P_3 + \int \widehat{zz}_0 \, dS = 0 \text{ and } \widehat{zz}_1 = L_1 R + \widehat{zz}_0 .$$

Also we can write in the notation of 3.22

$$d\tau = dR\,dS, \; d\Sigma = dR\,ds$$

and therefore (2) gives, using 3.22 (4),

$$E \int (A_1 x + B_1 y + C_1 m)\, dS + \oint Z_n ds - \int b_3 dS = 0 . \qquad (3)$$

Also from 3.22 (9), (10)

$$\left.\begin{aligned} E \int y(A_1 x + B_1 y + C_1 m)\, dS + P_2 + \oint y Z_n ds - \int y b_3 dS = 0 \\ E \int x(A_1 x + B_1 y + C_1 m)\, dS + P_1 + \oint x Z_n ds - \int x b_3\, dS = 0 \end{aligned}\right\} \qquad (4)$$

Now let

$$A = \int y^2\, dS, \; B = \int x^2\, dS, \; H = \int xy\, dS \qquad (5)$$

be the second moments of area of the cross-section of the beam. Let S be the area and $(x_G, y_G, 0)$ be the coordinates of the centroid of the free end.

Then from (3) and (4)

$$\left.\begin{aligned} S A_1 x_G + S B_1 y_G + S C_1 m + k_1 &= 0 \\ A_1 H + B_1 A + S C_1 m y_G + k_2 &= 0 \\ A_1 B + B_1 H + S C_1 m x_G + k_3 &= 0 \end{aligned}\right\} \qquad (6)$$

where

$$\left.\begin{aligned} k_1 &= \frac{1}{E} \oint Z_n ds - \frac{1}{E} \int b_3\, dS \\ k_2 &= \frac{P_2}{E} + \frac{1}{E} \oint y Z_n\, ds - \frac{1}{E} \int y b_3\, dS \\ k_3 &= \frac{P_1}{E} + \frac{1}{E} \oint x Z_n\, ds - \frac{1}{E} \int x b_3\, dS \end{aligned}\right\} \qquad (7)$$

These are known quantities and so the linear equations (6) determine A_1, B_1, C_1. Into these equations the associated plane problem does not enter.

The quantities A, B, H are moments referred to the axes of reference. If we denote by A_G, B_G, H_G the corresponding moments referred to parallel axes through the centroid G we have

$$A = A_G + y_G^2 S, \; B = B_G + x_G^2 S, \; H = H_G + x_G y_G S . \qquad (8)$$

Eliminating $C_1 m$ we then get

$$A_1 H_G + B_1 A_G + k_2 - y_G k_1 = 0$$
$$A_1 B_G + B_1 H_G + k_3 - x_G k_1 = 0$$

and therefore

$$\left.\begin{aligned} A_1 &= \frac{k_1 (x_G A_G - y_G H_G) + k_2 H_G - k_3 A_G}{A_G B_G - H_G^2} \\ B_1 &= \frac{k_1 (y_G B_G - x_G H_G) - k_2 B_G + k_3 H_G}{A_G B_G - H_G^2} \end{aligned}\right\} \tag{9}$$

To determine A_2, B_2, C_2, we have similarly from 3.22 (2), (5), (7)

$$\left.\begin{aligned} S A_2 x_G + S B_2 y_G + S C_2 m + l_1 &= 0\,, \\ A_2 H + B_2 A + S C_2 m y_G + l_2 &= 0\,, \\ A_2 B + B_2 H + S C_2 m x_G + l_3 &= 0\,, \end{aligned}\right\} \tag{10}$$

where

$$\left.\begin{aligned} l_1 &= \frac{P_3}{E} + \frac{1}{E}\int \widehat{zz}_p \, dS \\ l_2 &= \frac{M_1}{E} + \frac{1}{E}\int y \cdot \widehat{zz}_p \, dS \\ l_3 &= -\frac{M_2}{E} + \frac{1}{E}\int x \cdot \widehat{zz}_p \, dS \end{aligned}\right\} \tag{11}$$

and thus A_2, B_2, C_2 are found from the linear equations (10) in terms of known quantities.

Note that the body-field does not enter (10), (11). Equations analogous to (9) may be written down by substituting l for k.

3.25. The area theorem

STOKES's theorem for a curve C closed by a diaphragm S can be enunciated in the form [MILNE-THOMSON (5)]

$$\int_C d\mathbf{C} \circ X = \int_S \left(d\mathbf{S}_\wedge \frac{\partial}{\partial \mathbf{P}} \right) \circ X\,, \tag{1}$$

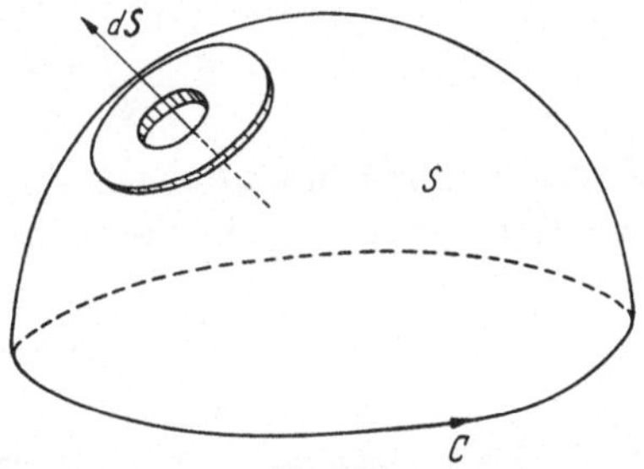

Fig. 3.25

where X is any continuous function of position, $d\mathbf{C}$, $d\mathbf{S}$ are properly directed elements of arc of C and surface of S and the small circle represents multiplication which may be scalar, vector or dyadic. The symbol $\wedge$ denotes vector multiplication. The situation is shown in fig. 3.25.

Now let C be a simple closed curve in the xy-plane, and let the diaphragm S be the portion of the xy-plane enclosed by C. Then if

$\boldsymbol{i}, \boldsymbol{j}, \boldsymbol{k}$ are orthogonal unit vectors along the axes and X is a scalar function $F(x, y)$, STOKES's theorem becomes

$$\oint (\boldsymbol{i}\, dx + \boldsymbol{j}\, dy)\, F(x, y) = \int_S \boldsymbol{k}_\wedge \left(\boldsymbol{i} \frac{\partial F}{\partial x} + \boldsymbol{j} \frac{\partial F}{\partial y}\right) dS. \tag{2}$$

Observing that $\boldsymbol{j} = \boldsymbol{k}_\wedge \boldsymbol{i}$ and that the operational equivalence

$$\boldsymbol{k}_\wedge = i \tag{3}$$

yields the complex variable, (2) becomes

$$\oint_C (dx + i\, dy)\, F(x, y) = i \int \left(\frac{\partial F}{\partial x} + i \frac{\partial F}{\partial y}\right) dS = 2i \int \frac{\partial F}{\partial \bar{z}}\, dS, \tag{4}$$

where $z = x + iy, \bar{z} = x - iy$. Now

$$F(x, y) = F\left[\frac{1}{2}(z + \bar{z}), -\frac{1}{2} i (z - \bar{z})\right] = f(z, \bar{z}) \tag{5}$$

say. Therefore from (4) and (5) we finally obtain the theorem which we shall designate as the *area theorem**

$$\oint_C f(z, \bar{z})\, dz = 2i \int_S \frac{\partial f}{\partial \bar{z}}\, dS. \tag{6}$$

The far reaching character of this theorem can be seen from the fact that Cauchy's theorem is included in it as a special case. Indeed if f is a holomorphic function of z inside and on the contour C, we have

$$\oint_C f(z)\, dz = 2i \int_S \frac{\partial f}{\partial \bar{z}}\, dS = 0 \tag{7}$$

since $\partial f / \partial \bar{z} = 0$ for a holomorphic function of z. This is Cauchy's theorem.

We may also note that putting $f = \bar{z}$ in (6) gives the area S as a contour integral namely

$$S = \int_S dS = -\frac{1}{2} i \oint_C \bar{z}\, dz. \tag{8}$$

Another important application is to *integration by parts*. Let u, v be functions of position in the plane. Then

$$2i \int_S \frac{\partial}{\partial \bar{z}} (uv)\, dS = \oint_C uv\, dz$$

* This theorem was first introduced by the author in 1938 as a lemma in 9.51 of the first edition of his *Theoretical Hydrodynamics*. The fundamental nature and importance of the theorem were only perceived later and it was then called the *complex Stokes's theorem*. I am now changing this to the simpler name "area theorem", for the theorem converts an integral over an area into an integral round the contour of the area.

and therefore

$$\int_S v \frac{\partial u}{\partial \bar{z}} \, dS = -\frac{1}{2} i \oint_C uv \, dz - \int_S u \frac{\partial v}{\partial \bar{z}} \, dS \,. \tag{9}$$

3.26. The moment M_3

From 3.2 (7) we have

$$M_3 = \int (y \cdot \widehat{xz} - x \cdot \widehat{yz}) \, dS = -\operatorname{Re} \int iz(\widehat{xz} - i\,\widehat{yz}) \, dS$$

Let $-iM_3'$ denote the imaginary part of the integral on the right. Then we can put

$$\begin{aligned} M_3 + iM_3' &= -\int iz(\widehat{xz} - i\,\widehat{yz}) \, dS \\ &= -i\mu \int z\{\Phi(z) - p\bar{z} - q\bar{z}^2 - r\, z\bar{z}\} \, dS \end{aligned} \tag{1}$$

from 3.2 (5).

We now introduce integrals I, J, K defined as follows

$$\left.\begin{aligned} I &= \int z\bar{z} \, dS = -\frac{1}{4} i \oint_C z\bar{z}^2 \, dz \,, \\ J &= 6i \int z\bar{z}^2 \, dS = \oint_C z\bar{z}^3 \, dz \,, \\ K &= 2i \int z\Phi(z) \, dS = \oint_C z\bar{z}\, \Phi(z) \, dz \,, \end{aligned}\right\} \tag{2}$$

where each has been transformed by the area theorem, 3.25, into an integral round the contour C of a cross-section of the beam. Substituting in (1) we get

$$\frac{M_3 + iM_3'}{\mu} = -\frac{1}{2} K + ipI + \frac{1}{6} qJ - \frac{1}{6} r\bar{J} \,. \tag{3}$$

The moment M_3 is obtained from the real part of the right-hand side, while the real quantity M_3' has no meaning for the present investigation. Adding to (3) its complex conjugate and using 3.2 (6) we get

$$M_3 = -\mu \left\{\frac{1}{4} (K + \bar{K}) + \tau I + \frac{1}{48} (1 - 2\eta) (\beta J + \overline{\beta J})\right\} \tag{4}$$

Observe that I is the polar second moment of area for the cross-section referred to the origin.

We can regard (4) as giving the twist τ when the moment M_3 is given, or as giving the moment M_3 which is required to produce a given twist.

3.27. The position of the load point

The load point is the point $L(x_L, y_L, 0)$ in the antiplane at which the force $(P_1, P_2, 0)$ may be supposed applied.

We shall consider only this force and the moment about the axis OR. Suppose that we first apply an arbitrary couple $-M^0$ in the antiplane. Next let us apply the force $(P_1, P_2, 0)$ at the load point. This loading is statically equivalent to a force $(P_1, P_2, 0)$ applied at the origin together with a couple — M in the antiplane, where

$$M = M^0 + x_L P_2 - y_L P_1 = M^0 + \frac{1}{2} i (z_L \bar{P} - \bar{z}_L P) \tag{1}$$

and

$$P = P_1 + i P_2, \; z_L = x_L + i y_L . \tag{2}$$

We note that in (1) M^0 is independent of the position of the loadpoint L, and of the force P. Therefore

$$\frac{\partial M}{\partial P} = -\frac{1}{2} i \bar{z}_L \text{ or } \bar{z}_L = 2i \frac{\partial M}{\partial P} \tag{3}$$

Thus when the moment M has been calculated we can find the position of the load point by means of (3).

Formulae alternative to (3) are

$$x_L = \frac{\partial M}{\partial P_2}, \qquad y_L = -\frac{\partial M}{\partial P_1} . \tag{4}$$

3.3. Classification of certain antiplane problems

For simplicity of the present explanation let us suppose that the body-field is absent, and that the origin is at the centroid of the antiplane so that

$$x_G = y_G = 0 . \tag{1}$$

Let us further suppose the lateral surface of the beam to be unloaded

$$X_n = Y_n = Z_n = 0 . \tag{2}$$

Then 3.24 (10), (6) give for the pure antiplane problem, $\widehat{zz}_p = 0$,

$$\left.\begin{aligned} C_2 m &= -\frac{P_3}{ES} \\ A_2 H + B_2 A &= -\frac{M_1}{E} \\ A_2 B + B_2 H &= \frac{M_2}{E} \end{aligned}\right\} \tag{3}$$

$$\left.\begin{aligned} C_1 m &= 0 \\ A_1 H + B_1 A &= -\frac{P_2}{E} \\ A_1 B + B_1 H &= -\frac{P_1}{E} \end{aligned}\right\} . \tag{4}$$

There are six *coefficients* (A_1, B_1, C_1), (A_2, B_2, C_2).

The force and moment applied to the free end have *components* (P_1, P_2, P_3), (M_1, M_2, M_3).

We then have the following cases of pure antiplane stress

(i) *Simple extension.* Here all components vanish except P_3 and $P_3 = -ST$. Then of the coefficients, C_2 alone survives and $C_2 m = T/E$. Thus

$$\widehat{zz} = T . \tag{5}$$

This is the case of a bar pulled by tension uniformly distributed over the end.

(ii) *Bending by terminal couples.* Here all components vanish except M_1, M_2 so that all coefficients vanish except

$$A_2 = \frac{HM_1 + AM_2}{E(AB - H^2)}, \quad B_2 = \frac{-BM_1 - HM_2}{E(AB - H^2)}$$

and

$$\widehat{zz} = E(A_2 x + B_2 y) . \tag{6}$$

Thus $\widehat{zz}$ is an odd function of (x, y) changing sign with change of sign of (x, y).

To illustrate take $B_2 = 0$, $H = 0$. Then $\widehat{zz} = EA_2 x$, and applied to the free end of the beam this produces a couple M_2 in the plane of Rx, fig. 3.3 (i).

(iii) *Torsion.* Suppose that all components vanish except $M_3 = -M$. Then all the coefficients vanish and

$$\widehat{zz} = 0 \tag{7}$$

while the moment M is produced by the shears $\widehat{yz}$, $\widehat{zx}$ over the free end so that

$$M = \int_S (x \cdot \widehat{yz} - y \cdot \widehat{xz})\, dS . \tag{8}$$

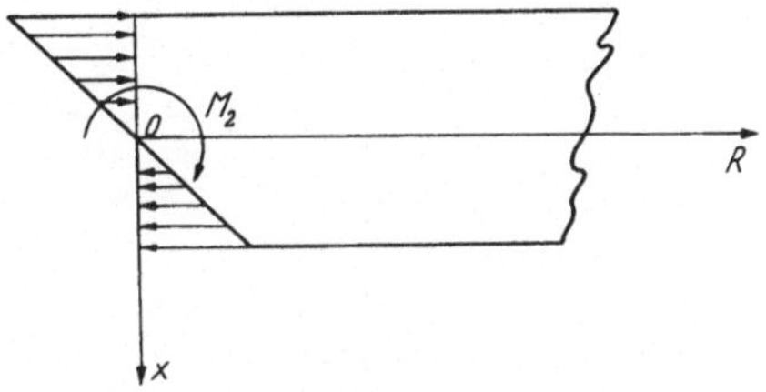

Fig. 3.3 (i)

The couple whose moment is M is called the *twisting couple.* Under its action, with the disposition of fig. 3.3 (ii) (see next page) a generator such as AB becomes a right-handed helical curve AB' on the supposition that A remains fixed.

(iv) *Flexure by terminal loads.* Here all components vanish except P_1 and P_2 so that all coefficients vanish except

$$A_1 = \frac{-AP_1 + HP_2}{E(AB - H^2)}, \quad B_1 = \frac{HP_1 - BP_2}{E(AB - H^2)}$$

and

$$\widehat{zz} = E(A_1 x + B_1 y) R . \tag{9}$$

(v) *Push.* The foregoing cases were based on the condition $Z_n = 0$. This condition is not fulfilled if, for example, we push a circular cylinder through a rough hole. Assuming uniform grip all round, there will be a constant frictional traction along the generators so that we can take $Z_n =$ constant. Therefore, for the pure antiplane part of the problem, on the assumption that all components vanish except $P_3 = Sp_1$, 3.24 (6), (10) give zero for A_1, B_1, A_2, B_2 so that* from 3.2 (11)

$$\widehat{zz}_a = EC_1\, m\, R + EC_2\, m. \tag{10}$$

Push may be regarded a generalization of simple extension.

In the preceding classification we have secured simple isolated cases by reference to the centroid. We do not intend to make this a permanent restriction, rather taking the origin and axes which are mathematically most convenient. Nevertheless it must be pointed out that when we choose a general point as origin, several cases may then occur simultaneously, since the wrench (P_1, P_2, P_3) (M_1, M_2, M_3) has been referred to the axes of reference. Naturally this does not mean that the physical problem depends on our choice of axes of reference, but it does mean that

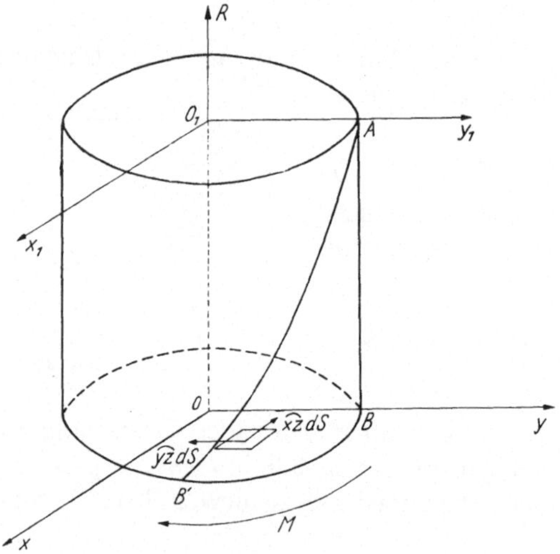

Fig. 3.3 (ii)

without more detailed specification of the wrench the mathematical discussion will intermingle the cases. Thus for example a transverse force applied at a general point will induce torsion as well as flexure, but flexure alone only if the direction of the force is along an axis of symmetry of the section.

* The existence of this case was first noted and discussed by FILON (2) in a paper which also introduced the term "antiplane".

3.4. The equations which give the displacement in pure antiplane stress

For pure antiplane stress we have $\widehat{xx} = \widehat{xy} = \widehat{yy} = 0$ and therefore the stress tensor, 1.4 (5), reduces to

$$\widehat{xz}(\boldsymbol{i};\boldsymbol{k} + \boldsymbol{k};\boldsymbol{i}) + \widehat{yz}(\boldsymbol{j};\boldsymbol{k} + \boldsymbol{k};\boldsymbol{j}) + \widehat{zz}\,(\boldsymbol{k};\boldsymbol{k})\,. \tag{1}$$

Therefore, from Hooke's law in the form

$$\mathsf{S} = \lambda \mathsf{I} \mathsf{D}_I + 2\mu\,\mathsf{D} \tag{2}$$

we have

$$\left.\begin{aligned} 0 &= \lambda(e_{xx} + e_{yy} + e_{zz}) + 2\mu e_{xx} \\ 0 &= \lambda(e_{xx} + e_{yy} + e_{zz}) + 2\mu e_{yy} \end{aligned}\right\} \tag{3}$$

$$0 = 2\,\mu e_{xy} = \mu\left(\frac{\partial v}{\partial x} + \frac{\partial u}{\partial y}\right). \tag{4}$$

From (3), by subtraction, $e_{xx} = e_{yy}$ or

$$\frac{\partial u}{\partial x} = \frac{\partial v}{\partial y}\,,\ \text{while}\ \frac{\partial v}{\partial x} = -\frac{\partial u}{\partial y} \tag{5}$$

from (4). These are the Cauchy-Riemann equations which show that the displacement υ is a function of z but not of $\bar{z}$,

$$\upsilon = u + iv = \upsilon(z, R)\,, \tag{6}$$

since υ must, in general, depend also on R.

Again adding the equations of (3) and observing that $\eta = \lambda/[2(\lambda + \mu)]$ we get

$$e_{xx} + e_{yy} = \frac{-2\eta\widehat{zz}}{E} \tag{7}$$

whence from 3.2 (12), since in pure antiplane stress $\widehat{zz}_p = 0$,

$$\frac{\partial u}{\partial x} + \frac{\partial v}{\partial y} = -2\eta\,\{(A_1 x + B_1 y + C_1 m)\,R + A_2 x + B_2 y + C_2 m\}\,.$$

In this write

$$\left.\begin{aligned} &A_1 + iB_1 = \beta\,, A_2 + iB_2 = \delta \\ &C_1 m \quad = \gamma\,, C_2 m \quad = \theta_0 \end{aligned}\right\} \tag{8}$$

Then

$$\frac{\partial \upsilon}{\partial z} + \frac{\partial \bar{\upsilon}}{\partial \bar{z}} = -\eta\,\{(\bar{\beta} z + \beta \bar{z} + 2\gamma)\,R + \bar{\delta} z + \delta \bar{z} + 2\theta_0\}\,. \tag{9}$$

Again

$$\frac{\widehat{xz} - i\widehat{yz}}{\mu} = 2e_{zx} - 2ie_{zy} = \frac{\partial u}{\partial R} + \frac{\partial w}{\partial x} - i\left(\frac{\partial v}{\partial R} + \frac{\partial w}{\partial y}\right) = \frac{\partial \bar{\upsilon}}{\partial R} + \frac{2\partial w}{\partial z}\,.$$

Therefore using the shear function $\Phi(z)$ of 2.9 (12) we have

$$\frac{\partial \bar{\upsilon}}{\partial R} + 2\frac{\partial w}{\partial z} = \Phi(z) - p\bar{z} - q\bar{z}^2 - r z\bar{z}\,, \tag{10}$$

where, from 3.2 (6),

$$p = i\tau + (1+\eta)(\gamma - \varepsilon m),\ q = \left(\frac{1}{4} + \frac{1}{2}\eta\right)\beta,\ r = \frac{1}{2}\bar{\beta}. \tag{11}$$

Also $e_{zz} = \dfrac{\partial w}{\partial R}$. Therefore

$$\frac{\partial w}{\partial R} = \frac{1}{2}\{(\beta\bar{z} + \bar{\beta}z + 2\gamma)R + (\bar{\delta}z + \delta\bar{z} + 2\theta_0)\}. \tag{12}$$

To find the displacement we have to obtain υ and w in the forms

$$\upsilon = \upsilon(z, R),\ w = w(z, \bar{z}, R)$$

from (9), (10), (12).

3.41. Expressions for the displacement in pure antiplane stress

Differentiate 3.4 (10) with respect to R getting

$$\frac{\partial^2\bar{\upsilon}}{\partial R^2} + 2\frac{\partial}{\partial z}\left(\frac{\partial w}{\partial R}\right) = 0$$

and in this substitute $\partial w/\partial R$ from 3.4 (12). We then get

$$\frac{\partial^2\bar{\upsilon}}{\partial R^2} = -\bar{\beta}R - \bar{\delta} \text{ and therefore } \frac{\partial^2\upsilon}{\partial R^2} = -\beta R - \delta.$$

Integrating twice with respect to R we get

$$\upsilon = -\frac{1}{6}\beta R^3 - \frac{1}{2}\delta R^2 + Rf_1(z) + f_2(z) \tag{1}$$

where $f_1(z), f_2(z)$ are functions of z to be determined. Substitute for υ and $\bar{\upsilon}$ in 3.4 (9) and rearrange to get

$$R[f_1'(z) + \eta\bar{\beta}z + \eta\gamma] + R[\bar{f}_1'(\bar{z}) + \eta\beta\bar{z} + \eta\gamma]$$
$$= -[f_2'(z) + \eta\bar{\delta}z + \eta\theta_0] - [\bar{f}_2'(\bar{z}) + \eta\delta\bar{z} + \eta\theta_0].$$

This equation is of the form $R(A + \bar{A}) = B + \bar{B}$. Since the right hand side is independent of R we must have $A + \bar{A} = 0$, $B + \bar{B} = 0$ and therefore A and B are each pure imaginary. Since A and B are functions of z only we must have $A = ik$, $B = il$ where k and l are real constants. Thus

$$f_1'(z) + \eta\bar{\beta}z + \eta\gamma = ik, \qquad f_2'(z) + \eta\bar{\delta}z + \eta\theta_0 = il$$

and therefore finally

$$f_1(z) = -\frac{1}{2}\eta\bar{\beta}z^2 + (ik - \eta\gamma)z + \alpha_0 \tag{2}$$

$$f_2(z) = -\frac{1}{2}\eta\bar{\delta}z^2 + (il - \theta_0\eta)z + \beta_0 \tag{3}$$

where α_0, β_0 are arbitrary complex constants.

Therefore from (1)

$$u + iv = \upsilon = -\frac{1}{6}\beta R^3 - \frac{1}{2}\delta R^2 + R\left\{-\frac{1}{2}\eta\bar{\beta} z^2 + (ik - \eta\gamma) z + \alpha_0\right\}$$
$$-\frac{1}{2}\eta\bar{\delta} z^2 + (il - \theta_0\eta) z + \beta_0 . \tag{4}$$

To find w we have

$$2dw = 2\frac{\partial w}{\partial z} dz + 2\frac{\partial w}{\partial \bar{z}} d\bar{z} + 2\frac{\partial w}{\partial R} dR .$$

Substitute for $\partial w/\partial z$, $\partial w/\partial \bar{z}$, $\partial w/\partial R$ from 3.4 (10), (12) and rearrange. We then find

$$2dw = d\left\{{}'\Phi(z) + {}'\bar{\Phi}(\bar{z}) + \frac{1}{2}R^2(\beta\bar{z} + \bar{\beta} z + 2\gamma) + R(\delta\bar{z} + \bar{\delta} z + 2\theta_0)\right.$$
$$\left. - \bar{\alpha}_0 z - \alpha_0\bar{z} + [(1+\eta)\varepsilon m - \gamma] z\bar{z} - \frac{1}{4}\beta\bar{z}^2 z - \frac{1}{4}\bar{\beta} z^2\bar{z}\right\}$$
$$+ i(\tau - k)(z\, d\bar{z} - \bar{z}\, dz) .$$

Since dw is an exact differential we must remove the last term by choosing $k = \tau$ and then integration gives

$$2w = {}'\Phi(z) + {}'\bar{\Phi}(\bar{z}) + \frac{1}{2}R^2(\beta\bar{z} + \bar{\beta} z + 2\gamma)$$
$$+ R(\delta\bar{z} + \bar{\delta} z + 2\theta_0) - \alpha_0\bar{z} - \bar{\alpha}_0 z - \frac{1}{4}\bar{\beta} z^2\bar{z} - \frac{1}{4}\beta\bar{z}^2 z \tag{5}$$
$$+ [(1+\eta)\varepsilon m - \gamma] z\bar{z} + 2c_0$$

where c_0 is an arbitrary real constant for w is real. Finally putting $\alpha_0 = \omega_2 - i\omega_1$, $l = \omega_3$, $\beta_0 = a_0 + ib_0$ we have, since $k = \tau$

$$\upsilon = u + iv = -\frac{1}{6}\beta R^3 - \frac{1}{2}\delta R^2 + \left\{-\frac{1}{2}\eta\bar{\beta} z^2 + (i\tau - \eta\gamma) z\right\} R$$
$$-\frac{1}{2}\eta\bar{\delta} z^2 - \eta\theta_0 z + (R\omega_2 - y\omega_3 + a_0) + i(x\omega_3 - R\omega_1 + b_0) \tag{6}$$

$$w = \frac{1}{2}[{}'\Phi(z) + {}'\bar{\Phi}(\bar{z})] + \frac{1}{4}R^2(\beta\bar{z} + \bar{\beta} z + 2\gamma) + \frac{1}{2}R(\delta\bar{z} + \bar{\delta} z + 2\theta_0)$$
$$-\frac{1}{8}\beta\bar{z}^2 z - \frac{1}{8}\bar{\beta} z^2\bar{z} + \frac{1}{2}[(1+\eta)\varepsilon m - \gamma] z\bar{z} + \omega_1 y - \omega_2 x + c_0 . \tag{7}$$

The last terms in each expression correspond to a rigid body movement of rotation $(\omega_1, \omega_2, \omega_3)$ and translation (a_0, b_0, c_0).

3.5. The boundary condition for the pure antiplane problem for isotropic beams

The problem is to determine the shear function $\Phi(z)$ of 3.2 (5). The boundary condition is

$$l\,\widehat{zx} + m\,\widehat{zy} = Z_n \text{ at the lateral surface,} \tag{1}$$

$$l = \frac{\partial y}{\partial s},\; m = -\frac{\partial x}{\partial s}.$$

Now

$$l\,\widehat{zx} + m\,\widehat{zy} = \operatorname{Re}(l + im)\,(\widehat{zx} - i\,\widehat{zy})$$

$$= \frac{1}{2}(l + im)\,(\widehat{zx} - i\,\widehat{zy}) + \frac{1}{2}(l - im)\,(\widehat{zx} + i\,\widehat{zy}).$$

But

$$l + im = \frac{\partial y}{\partial s} - i\frac{\partial x}{\partial s} = -i\frac{\partial z}{\partial s}.$$

Therefore the boundary condition (1) can be written

$$(\widehat{xz} - i\,\widehat{yz})\frac{\partial z}{\partial s} - (\widehat{xz} + i\,\widehat{yz})\frac{\partial \bar z}{\partial s} = 2iZ_n \tag{2}$$

at the lateral surface,

Substitute for $\widehat{xz} - i\,\widehat{yz}$ from 3.2 (5). Then

$$\Phi(z)\frac{\partial z}{\partial s} - \bar\Phi(\bar z)\frac{\partial \bar z}{\partial s} = F(z, \bar z) \text{ at the lateral surface,} \tag{3}$$

$$F(z, \bar z) = \left\{(p\bar z + q\bar z^2 + r\,z\bar z)\frac{\partial z}{\partial s} - (\bar p z + \bar q z^2 + \bar r\,z\bar z)\frac{\partial \bar z}{\partial s}\right\} + 2i\frac{Z_n}{\mu}. \tag{4}$$

Thus (3) and (4) constitute the required boundary condition.

3.51. Solution of the pure antiplane problem for isotropic beams

Let the cross-section of the beam be a simply-connected region L bounded by the contour C, the sense of description of which leaves L on the left and the region R outside C on the right.

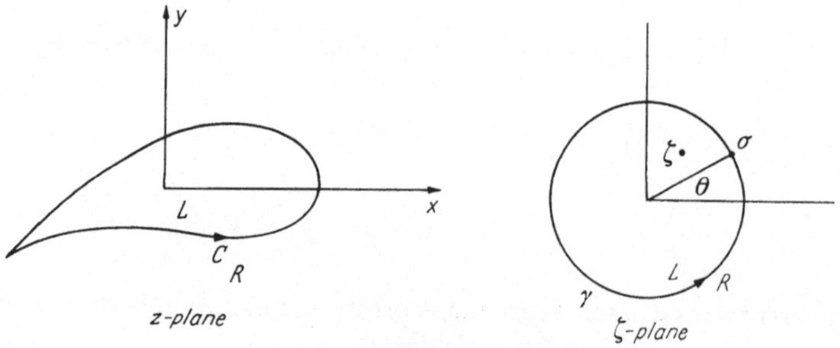

Fig. 3.51

Let

$$z = m(\zeta) \tag{1}$$

map the region L on the interior of the unit circumference γ or $|\sigma| = 1$ in the ζ-plane, σ denoting a general point on γ. Here

$$z = m(\sigma), \bar{z} = \bar{m}(\bar{\sigma}) = \bar{m}\left(\frac{1}{\sigma}\right) \text{ on } \gamma . \tag{2}$$

Let

$$\Phi(z) = \Phi[m(\zeta)] = \omega(\zeta) . \tag{3}$$

Then the boundary condition 3.5 (3) becomes

$$\omega(\sigma)\, m'(\sigma) \frac{d\sigma}{ds} + \bar{\omega}\left(\frac{1}{\sigma}\right) \bar{m}'\left(\frac{1}{\sigma}\right) \frac{1}{\sigma^2} \frac{d\sigma}{ds} = F\left\{m(\sigma), \bar{m}\left(\frac{1}{\sigma}\right)\right\}$$

where

$$F\left\{m(\sigma), \bar{m}\left(\frac{1}{\sigma}\right)\right\} = \left\{p\,\bar{m}\left(\frac{1}{\sigma}\right) + q\left[\bar{m}\left(\frac{1}{\sigma}\right)\right]^2 + r\,m(\sigma)\,\bar{m}\left(\frac{1}{\sigma}\right)\right\} m'(\sigma) \frac{d\sigma}{ds}$$

$$+ \left\{\bar{p}\,m(\sigma) + \bar{q}\,[m(\sigma)]^2 + \bar{r}\,m(\sigma)\,\bar{m}\left(\frac{1}{\sigma}\right)\right\} \bar{m}'\left(\frac{1}{\sigma}\right) \frac{1}{\sigma^2} \frac{d\sigma}{ds} + 2i \frac{Z_n}{\mu} .$$

Here

$$Z_n = Z_n(z, \bar{z}) = Z_n\left[m(\sigma), \bar{m}\left(\frac{1}{\sigma}\right)\right] . \tag{4}$$

Therefore finally

$$\omega(\sigma)\, m'(\sigma) + \bar{\omega}\left(\frac{1}{\sigma}\right) \bar{m}'\left(\frac{1}{\sigma}\right) \frac{1}{\sigma^2} = H(\sigma) + \frac{2iZ_n}{\mu} \frac{ds}{d\sigma} \tag{5}$$

$$\begin{aligned} H(\sigma) = {} & \left\{p\,m\left(\frac{1}{\sigma}\right) + q\left[\bar{m}\left(\frac{1}{\sigma}\right)\right]^2 + r\,m(\sigma)\,\bar{m}\left(\frac{1}{\sigma}\right)\right\} m'(\sigma) \\ & + \left\{\bar{p}\,m(\sigma) + \bar{q}\,[m(\sigma)]^2 + \bar{r}\,m(\sigma)\,\bar{m}\left(\frac{1}{\sigma}\right)\right\} \bar{m}'\left(\frac{1}{\sigma}\right) \frac{1}{\sigma^2} . \end{aligned} \tag{6}$$

Since $\Phi(z)$ is holomorphic in L, so is $\omega(\sigma)$ when σ is in the region L to the left of γ. Therefore $\bar{\omega}(1/\sigma)$ is holomorphic in the region R to the right of γ. Thus when σ is in R, $\frac{1}{\sigma^2} m'\left(\frac{1}{\sigma}\right) \bar{\omega}\left(\frac{1}{\sigma}\right)$ is holomorphic and vanishes at infinity.

Now multiply (5) by $d\sigma/[2\pi i(\sigma - \zeta)]$ where ζ is a point in L and integrate round γ so that

$$\frac{1}{2\pi i} \oint_\gamma \frac{\omega(\sigma)}{\sigma - \zeta} m'(\sigma)\, d\sigma + \frac{1}{2\pi i} \oint_\gamma \frac{\frac{1}{\sigma^2} \bar{\omega}\left(\frac{1}{\sigma}\right) \bar{m}'\left(\frac{1}{\sigma}\right) d\sigma}{\sigma - \zeta}$$

$$= \frac{1}{2\pi i} \oint_\gamma \frac{H(\sigma)\, d\sigma}{\sigma - \zeta} + \frac{2i}{2\pi i \mu} \oint_\gamma \frac{Z_n}{\sigma - \zeta} \frac{ds}{d\sigma} d\sigma .$$

Now apply Cauchy's integral formula [MILNE-THOMSON (6)]. Then the first integral is $\omega(\zeta)\, m'(\zeta)$ and the second integral vanishes. Therefore

$$\Phi(z)\, m'(\zeta) = \omega(\zeta)\, m'(\zeta) = \frac{1}{2\pi i} \oint_{\gamma} \frac{H(\sigma)\, d\sigma}{\sigma - \zeta} + \frac{1}{\mu \pi} \oint_{C} \frac{Z_n\, d s}{\sigma - \zeta} \tag{7}$$

and this equation determines $\Phi(z)$ in terms of the parameter ζ, for the functions $H(\sigma)$, Z_n are known on the boundary.

Notes (i) If the right-hand side of (7) vanishes

$$\Phi(z) \equiv 0 . \tag{8}$$

In particular this will be the case if p, q, r, Z_n all vanish, for example, if there is no body-field, $Z_n = 0$ and A_1, B_1, C_1, τ vanish.

(ii) The vanishing of τ means that no torsion is involved.

(iii) Whenever the mapping function $m(\zeta)$ is rational $H(\sigma)$ is a rational function and therefore

$$I = \frac{1}{2\pi i} \int_{\gamma} \frac{H(\sigma)\, d\sigma}{\sigma - \zeta} \tag{9}$$

can be expressed in terms of elementary functions.

(iv) If we can write

$$H(\sigma) = H_1(\sigma) + H_2(\sigma) \tag{10}$$

where all and only the negative powers of σ occur in $H_2(\sigma)$, then from (9)

$$I = H_1(\zeta) . \tag{11}$$

3.52. Expression of shears in curvilinear coordinates

If

$$\zeta = r e^{i\theta} , \tag{1}$$

the curves $r =$ constant, $\theta =$ constant, are respectively concentric circles about, and rays issuing from $\zeta = 0$. By the mapping

$$z = m(\zeta) \tag{2}$$

these curves map into a net of orthogonal curves in the z-plane on a member of one system of which $r =$ constant, and on a member of the other system $\theta =$ constant. We call (r, θ) curvilinear coordinates, for when r and θ have given values say r_0, θ_0, they fix a point p of polar coordinates (r_0, θ_0) in the ζ-plane, and its map P is fixed in the z-plane as the intersection of the curves $r = r_0$, $\theta = \theta_0$.

The shear at P can be described by either of the vectors $\widehat{xz} - i\,\widehat{yz}$ or $\widehat{rz} - i\,\widehat{\theta z}$, where $\widehat{rz}$ is the component at P in the direction in which r increases i.e. normal to the curve $r =$ constant or tangential to the curve $\theta =$ constant, and $\widehat{\theta z}$ is the component at P in the direction in which θ

increases i.e. normal to the curve $\theta =$ constant or tangential to the curve $r =$ constant. If the normal at P to the curve $r =$ constant makes angle α with the x-axis, we have

$$\widehat{rz} - i\,\widehat{\theta z} = (\widehat{xz} - i\,\widehat{yz})\,e^{i\alpha}. \tag{3}$$

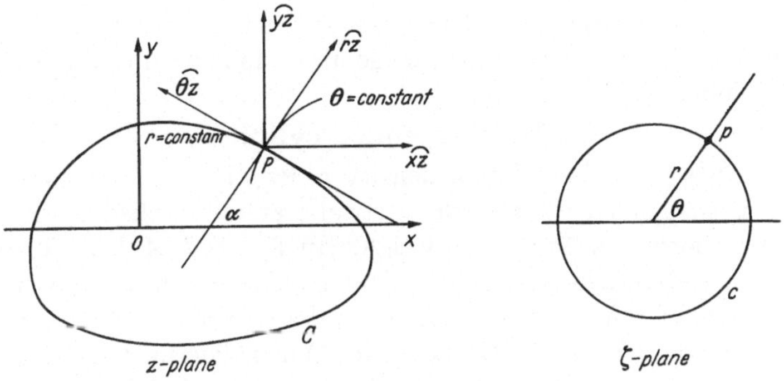

Fig. 3.52

Now if dz is an increment of z at P in the direction of the normal to the curve $r =$ constant, we have

$$dz = |dz|\,e^{i\alpha},\; d\zeta = |d\zeta|\,e^{i\theta}.$$

Therefore

$$e^{i\alpha} = \frac{dz}{|dz|} = \frac{m'(\zeta)\,d\zeta}{|m'(\zeta)\,d\zeta|} = e^{i\theta}\,\frac{m'(\zeta)}{|m'(\zeta)|} = \frac{\zeta}{r}\,\frac{m'(\zeta)}{|m'(\zeta)|}. \tag{4}$$

Therefore

$$\widehat{rz} - i\widehat{\theta z} = \frac{\zeta}{r}\,\frac{m'(\zeta)}{|m'(\zeta)|}\,(\widehat{xz} - i\,\widehat{yz}). \tag{5}$$

The argument applies to any other vector at P. Thus for the displacement

$$u_r - i u_\theta = \frac{\zeta}{r}\,\frac{m'(\zeta)}{|m'(\zeta)|}\,(u - iv), \tag{6}$$

where u_r, u_θ are the components in the directions in which r and θ increase respectively.

3.6. Simple extension

We consider a beam fixed at one end in some prescribed manner and in equilibrium under no forces except a longitudinal pull P applied at the centroid of the free end. We suppose this force P to be the resultant of uniform longitudinal stress T applied at the free end. If S is the area of the cross-section, $P = TS$.

Then from **3.3** (i) all the coefficients vanish except $C_2 m = T/E$. Thus if there is no twisting, $\tau = 0$ and therefore in **3.41** (6), (7) all the constants vanish except $\theta_0 = T/E$. Also from 3.51 we can take $\Phi(z) = 0$. Therefore apart from a rigid body movement

$$u + iv = -\eta z T/E, w = RT/E$$

$$u = \frac{-\eta x T}{E}, v = \frac{-\eta y T}{E}, w = \frac{RT}{E}, \widehat{zz} = T$$

and all other stresses vanish. This problem has been solved as an example of our general methods, but the solution follows directly from the assumption $\widehat{zz} = T$.

3.62. Suspended cylinder

A homogeneous cylinder of density ϱ, length L, and cross-sectional area S is suspended by uniform vertical traction applied to the upper end and deforms under its own weight. In fig. 3.62, $A'B'C'D'$ shows a section of the undeformed cylinder by a plane which contains the line of centroids of the cross-sections. We take this line as R-axis and the centroid O' of the base $B'C'$ as origin. When deformed under its own weight we shall suppose that the line of centroids remains vertical and that the cylinder does not rotate. Without loss of generality we may suppose in addition that the centroid G of the top section is fixed. The condition for this is $\boldsymbol{q} = 0$ at the point $(0, 0, L)$.

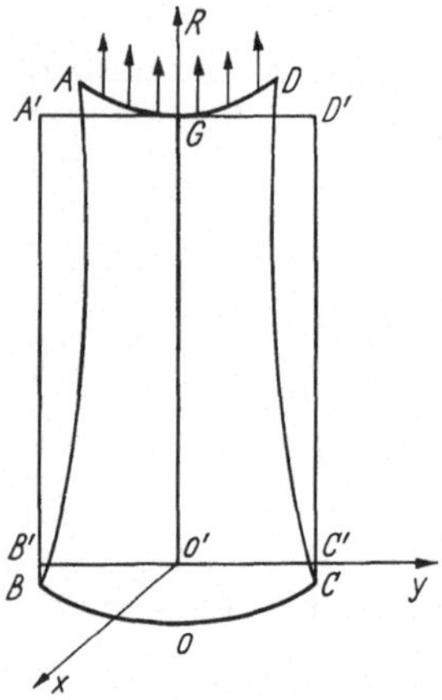

Fig. 3.62

Other conditions of fixing could of course be envisaged to include, for example, arbitrary rotation.

The lateral surface and the bottom face are free of applied force. We shall prove that the problem for infinitesimal strain can be regarded as one of antiplane stress.

On the assumption of antiplane stress we have in the notation of 3.24, $\widehat{zz}_p = 0$, $b_3 = g\varrho$, $k_1 = -g\varrho S/E$, $k_2 = k_3 = 0$, $l_1 = l_2 = l_3 = 0$. Therefore

$$A_1 = B_1 = 0, A_2 = B_2 = C_2 = 0, C_1 m = g\varrho/E, \varepsilon m = g\varrho/E. \quad (1)$$

Therefore in 3.4 (8)

$$\beta = 0, \delta = 0, \gamma = g\varrho/E, \theta_0 = 0. \quad (2)$$

By hypothesis there is no twisting and so $\tau = 0$. Therefore in 3.2 (6) $p = q = r = 0$. Thus using 3.51 (7) $\Phi(z) = 0$ and therefore

$${}'\Phi(z) + {}'\overline{\Phi}(\bar{z}) = \text{constant}$$

and without loss of generality we may take the constant to be zero. Thus it appears that the only non-vanishing stress component is

$$\widehat{zz} = g\varrho R. \quad (3)$$

For the displacement we have from 3.41 (6) and (7)

$$u + iv = \frac{-\eta g\varrho}{E} Rz + (R\omega_2 - y\omega_3 + a_0) + i(x\omega_3 - R\omega_1 + b_0)$$

$$w = \frac{1}{2}\frac{g\varrho}{E}[R^2 + \eta(x^2 + y^2)] + \omega_1 y - \omega_2 x + c_0 .$$

Since there is no rotation $\omega_1 = \omega_2 = \omega_3 = 0$. Since $G(0, 0, L)$ remains fixed $a_0 = 0$, $b_0 = 0$, $c_0 = -\frac{1}{2} g\varrho L^2/E$.

Thus finally

$$u = -\frac{\eta g\varrho}{E} x,\; v = -\frac{\eta g\varrho}{E} y,\; w = \frac{1}{2}\frac{g\varrho}{E}[R^2 - L^2 + \eta(x^2 + y^2)] . \tag{4}$$

Note that points on the axis $x = 0$, $y = 0$ undergo the displacement

$$w = -\frac{1}{2}\frac{g\varrho}{E}(L^2 - R^2)$$

which is negative and is of maximum magnitude at $R = 0$, where its value is $- g\varrho L^2/(2E)$.

The point (x_1, y_1, R_1) is displaced to (x, y, R) where $x = x_1 + u$, $y = y_1 + v$, $R = R_1 + w$ and therefore the cross-section $R_1 =$ constant warps into the paraboloid of revolution.

$$R = R_1 + w = R_1 + \frac{1}{2}\frac{g\varrho}{E}(R_1^2 - L^2) + \frac{1}{2}\frac{g\varrho\eta}{E}(x^2 + y^2) .$$

These paraboloids all derive from

$$R = \frac{1}{2}\frac{g\varrho\eta}{E}(x^2 + y^2)$$

by vertical displacement.

In particular the free end $R_1 = 0$ is deformed into the paraboloid

$$R = \frac{1}{2}\frac{g\varrho}{E}\{\eta(x^2 + y^2) - L^2\} .$$

3.7. Bending by terminal couples

We consider a beam in the absence of a body-field, whose lateral surface is free of applied force.

We suppose the beam to be kept in equilibrium by couples M and $-M$ in axial planes.

Taking the centroid as origin we have from 3.3 (6)

$$\widehat{zz} = E(A_2 x + B_2 y) , \tag{1}$$

where A_2 and B_2 are given by 3.24 (10), (11) in the form

$$A_2 = \frac{HM_1 + AM_2}{E(AB - H^2)} ,\; B_2 = -\frac{BM_1 + HM_2}{E(AB - H^2)} , \tag{2}$$

M_1 and M_2 being the components of the couple M along the axes Ox, Oy respectively.

Since for arbitrary torsion $\widehat{zz} = 0$, a torsion solution could be imposed on (1). Thus if our bending problem involves torsion τ, we can suppose this removed by superposing the torsion solution for $-\tau$. Assuming this to have been done, we have from **3.51** (8), $\Phi(z) = 0$ and therefore

$$\widehat{yz} = \widehat{xz} = 0 \text{ everywhere}. \tag{3}$$

In the notation of 3.4 (8) we have

$$\beta = 0, \gamma = 0, \theta_0 = 0, \delta = A_2 + iB_2 \tag{4}$$

and therefore from **3.41**, if we take the point $(0, 0, L)$ to be fixed and allow no rotation, we have

$$u + iv = -\frac{1}{2}\delta R^2 - \frac{1}{2}\eta\bar{\delta}z^2 + \frac{1}{2}\delta L^2 \tag{5}$$

$$w = \frac{1}{2}R(\delta\bar{z} + \bar{\delta}z) = R(A_2x + B_2y). \tag{6}$$

From (6) we see that longitudinal fibres in the plane

$$A_2x + B_2y = 0 \tag{7}$$

undergo no elongation. This plane is called the *neutral plane*. Since passage through the plane (7) changes the sign of $A_2x + B_2y$, fibres on one side of this plane are elongated and those on the other side are compressed.

The deformation displaces a point (a, b, c) to (x, y, R) where

$$x = a + u, y = b + v, R = c + w.$$

In particular the point $(0, 0, c)$ on the line of centroids $x = 0$, $y = 0$ is displaced to

$$x = -\frac{1}{2}A_2(c^2 - L^2), y = -\frac{1}{2}B_2(c^2 - L^2), R = c. \tag{8}$$

The points (8) lie in the plane

$$A_2y - B_2x = 0 \tag{9}$$

which therefore contains the deformed line of centroids. This plane is called the *plane of bending*.

The plane of bending and the neutral plane are perpendicular.

The *plane of the couple* is

$$M_2y + M_1x = 0. \tag{10}$$

The plane of bending and the plane of the couple are parallel if and only if

$$\frac{M_2}{M_1} = \frac{A_2}{-B_2} = \frac{HM_1 + AM_2}{BM_1 + HM_2},$$

and this can happen, in general, only if the vector of the couple is along a principal dynamical axis of the section i. e. if $H = 0$ *and* either $M_1 = 0$ or $M_2 = 0$, or if $A = B$ and $M_1 = M_2$.

Returning to the line of centroids given by (8) we see that it is the unique parabola formed by the intersection of the plane of bending (9) with either of the parabolic cylinders

$$y = -\frac{1}{2} B_2 (R^2 - L^2) \text{ or } x = -\frac{1}{2} A_2 (R^2 - L^2) .$$

Let ϱ be the radius of curvature of the line of centroids. Since this line is unstretched ($w = 0$) we can put $s = R$, where s is the arc measured from the origin and

$$\frac{1}{\varrho^2} = \left(\frac{d^2 x}{d s^2}\right)^2 + \left(\frac{d^2 y}{d s^2}\right)^2 = A_2^2 + B_2^2 = \delta \bar{\delta} \tag{11}$$

and we determine ϱ from (2) in terms of A, B, H, M_1, M_2. We note that ϱ is constant so that the line of centroids deforms into a circular arc of large radius.

In particular if we suppose Ox to be a principal axis of inertia of the section, i. e. $H = 0$, and if we take $M_1 = 0$, we get

$$\frac{1}{\varrho} = \frac{M_2}{EB} ,$$

the Euler-Bernoulli law of bending, that the moment is proportional to the curvature of the line of centroids. Here B is the principal moment about the axis of the couple M_2. The quantity EB is called *the flexural rigidity*.

This simple relation applies only in the conditions stated. It is a special case of the general law given by (11).

We now prove that particles which lie in a plane which is perpendicular to the line of centroids prior to deformation continue to lie in a plane, perpendicular to the plane of bending, after deformation.

Proof. The point (x_1, y_1, c) in the plane $R = c$ normal to the line of centroids is displaced by the deformation to (x, y, R), where

$$x = x_1 + u, \; y = y_1 + v, \; R = c + w = c(1 + A_2 x_1 + B_2 y_1) . \tag{12}$$

From (5), $u + iv = -\frac{1}{2}\delta c^2 - \frac{1}{2}\eta \delta (x_1^2 - y_1^2 + 2i x_1 y_1) + \frac{1}{2}\delta L^2$. But from (11), $|\delta| = |\bar{\delta}| = 1/\varrho$ which is small. Therefore in (12) we can take $x = x_1$, $y = y_1$ and the points under consideration lie in the plane

$$R = c(1 + A_2 x + B_2 y)$$

which is perpendicular to the plane of bending (9). Q. E. D.

3.8. Circular cylinder pushed into a hole

We consider a long rivet or pile, in the form of a circular cylinder, gripped tightly all round under uniform lateral pressure $\tilde{\omega}$, and driven against friction by uniform pressure p_1 applied at one end, with uniform pressure p_2 applied at the other end.

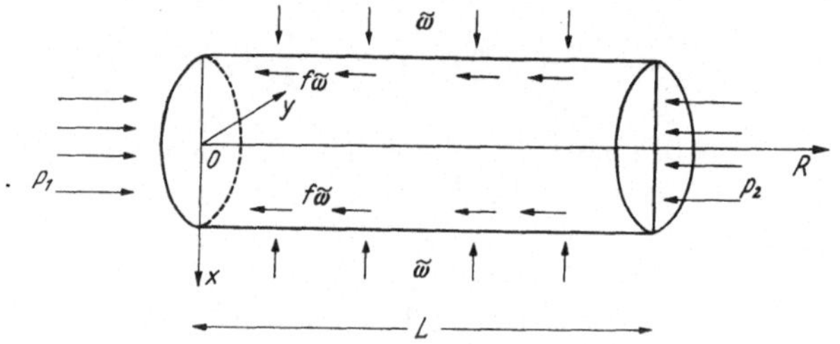

Fig. 3.8

We take as antiplane $R = 0$ the end at which the pressure p_1 is applied. The pressure p_2 is applied at the end $R = L$, where L is the length of the cylinder, and the resultant forces on the ends are $p_1 \pi a^2$, $-p_2 \pi a^2$ where a is the radius. We omit body force and assume that there is no torsion or rotation.

Let f be the coefficient of friction. Then with our usual notation the lateral force (X_n, Y_n, Z_n) is given by

$$X_n = -\tilde{\omega}\cos\theta\,,\ Y_n = -\tilde{\omega}\sin\theta\,,\ Z_n = -f\tilde{\omega}\,. \tag{1}$$

Thus here we have two problems

(i) a plane deformation under the loading $(-\tilde{\omega}\cos\theta, -\tilde{\omega}\sin\theta, 0)$, that is uniform pressure $\tilde{\omega}$ round the lateral surface

(ii) an antiplane problem under the loading $(0, 0, -f\tilde{\omega})$ on the lateral surface and end thrusts $p_1 \pi a^2$, $-p_2 \pi a^2$.

The longitudinal stress component is, 3.2 (10), (11) and 3.3 (10),

$$\widehat{zz} = \widehat{zz}_a + \widehat{zz}_p\,, \tag{2}$$

$$\widehat{zz}_a = E C_1 m R + E C_2 m\,. \tag{3}$$

As regards problem (i) uniform pressure on the boundary produces uniform pressure inside and so

$$\widehat{xx} = -\tilde{\omega}\,,\ \widehat{yy} = -\tilde{\omega}\,,\ \widehat{xy} = 0\,. \tag{4}$$

Therefore

$$\widehat{zz}_p = -2\eta\,\tilde{\omega}$$

and therefore from (2), (3)

$$\widehat{zz} = -2\eta\varpi + E C_2 m + E C_1 m R\,.$$

Now $\widehat{zz} = -p_1$ when $R = 0$ and $\widehat{zz} = -p_2$ when $R = L$. Therefore

$$-p_1 = -2\eta\,\tilde{\omega} + E C_2 m, \quad -p_2 = -2\eta\,\tilde{\omega} + E C_2 m + E C_1 m L .$$

Therefore finally

$$\widehat{zz} = -p_1 + \frac{(p_1 - p_2) R}{L}, \quad E C_1 m = \frac{p_1 - p_2}{L}, \quad E C_2 m = -p_1 + 2\eta\,\tilde{\omega} . \tag{5}$$

Turning now to problem (ii) we have from 3.2 (5), (6)

$$\frac{\widehat{xz} - i\widehat{yz}}{\mu} = \Phi(z) - (1+\eta)\, C_1 m \bar{z} = \Phi(z) - (1+\eta)\,\frac{(p_1 - p_2)\bar{z}}{E L} . \tag{6}$$

Now from the conditions of statical equilibrium, by resolving perpendicularly to the antiplane, we have

$$\pi a^2 (p_1 - p_2) = 2\pi a L f \tilde{\omega} \tag{7}$$

and from 3.5 (3), (4) the boundary condition is

$$\Phi(z)\frac{\partial z}{\partial s} - \bar{\Phi}(\bar{z})\frac{\partial \bar{z}}{\partial s} = (1+\eta)\,\frac{p_1 - p_2}{EL}\left(\bar{z}\frac{\partial z}{\partial s} - z\frac{\partial \bar{z}}{\partial s}\right) + \frac{2 i Z_n}{\mu} . \tag{8}$$

But on the boundary $z = a e^{i\theta}$, $s = a\theta$ and so $\partial z/\partial s = i z/a$. Therefore observing that $E = 2\mu(1+\eta)$ and that, from (7), $(p_1 - p_2)/(2L) = f\tilde{\omega}/a$, the boundary condition (8) becomes

$$z\Phi(z) + \bar{z}\bar{\Phi}(\bar{z}) = 0 .$$

Therefore from 3.51 (8), $\Phi(z) = 0$, so that from (6)

$$\widehat{xz} - i\,\widehat{yz} = -\frac{1}{2}(p_1 - p_2)\,\bar{z}/L$$

or

$$\widehat{xz} = -\frac{1}{2}(p_1 - p_2)\, x/L, \quad \widehat{yz} = -\frac{1}{2}(p_1 - p_2)\, y/L .$$

In terms of polar coordinates (r, θ) these stresses $\widehat{xz}$, $\widehat{yz}$, are equivalent to a radial distribution of shear $\widehat{rz}$ over the cross-section given by

$$\widehat{rz} = -\frac{1}{2}(p_1 - p_2)\frac{r}{L}, \tag{9}$$

while the system (4) is equivalent to $\widehat{rr} = \widehat{\theta\theta} = -\tilde{\omega}$, $\widehat{r\theta} = 0$. Thus the only non-zero stresses are

$$\widehat{zz} = -p_1 + (p_1 - p_2)\,R/L, \quad \widehat{rr} = \widehat{\theta\theta} = -\tilde{\omega}, \quad \widehat{rz} = -\frac{1}{2}(p_1 - p_2)\, r/L. \tag{10}$$

We note that (9) implies a distribution of radial shear applied to the two flat ends of the cylinder which was not present in the original statement of the problem, but is necessary to satisfy the conditions exactly. Nevertheless this distribution (9) is self-equilibrating and therefore we may assume that if it is removed, or replaced by any other self-equilibrating distribution, the effects of the change will, by de Saint-Venant's principle of equipollent loads, become insensible at a distance from the ends.

We may observe that the statical relation (7), used above, implies that when the coefficient of friction is known, there is only one pressure $\tilde{\omega}$ which will adapt itself to a given pressure difference $p_1 - p_2$.

3.81. Displacement for the pushed cylinder

In 3.4 (8) we have $\beta = \delta = 0$,

$$\gamma = C_1 m = \frac{p_1 - p_2}{EL}, \quad \theta_0 = C_2 m = \frac{-p_1 + 2\,\eta\pi}{E} \tag{1}$$

from 3.8 (5)

Therefore from 3.41 (6), (7), apart from a rigid body displacement,

$$u + iv = -\,\eta z (R\gamma + \theta_0)\,, \tag{2}$$

$$w = \frac{1}{2}\gamma(R^2 - z\bar{z}) + R\theta_0\,. \tag{3}$$

In cylindrical coordinates (r, θ, R) the radial displacement u_r and the transverse displacement u_θ in a plane parallel to the antiplane will be given by

$$u_r + iu_\theta = (u + iv)e^{-i\theta} = -\,\eta r(R\gamma + \theta_0)\,.$$

Therefore the transverse displacement u_θ vanishes and

$$u_r = -\,\eta r(\gamma R + \theta_0)\,, \quad w = \frac{1}{2}\gamma(R^2 - r^2) + R\theta_0\,. \tag{4}$$

It follows that longitudinal fibres which lie in a meridional plane before displacement remain in the same plane after displacement.

Also

$$\left(\frac{\partial u_r}{\partial R}\right)_{r=a} = -\,\eta\gamma a = -\,\frac{p_1 - p_2}{E}\,\frac{a}{L}\,\eta$$

which is negative since $p_1 > p_2$. Therefore the lateral expansion of the cylinder is greatest at the driven end ($R = 0$) and decreases as we proceed in the direction of the drive. This result is based on the assumption that $\tilde{\omega}$ is constant.

3.82. Lines of principal stress for the pushed cylinder

Fig. 3.82 (i) shows a general point M in a meridian plane of the pushed cylinder, distant r from the axis OR. By the symmetry one principal direction at M is perpendicular to the meridian plane, and the other two principal directions lie in this plane. Let the tangent to a line of principal stress in the meridian plane at M make the angle χ with OR.

Let S_1, S_2 be the principal stresses at M. Then the fundamental stress combinations

$$\Theta = \widehat{rr} + \widehat{zz}, \qquad \Phi = \widehat{rr} - \widehat{zz} + 2i\widehat{zr}$$

become on rotation of the axes through the angle χ

$$\Theta_1 = S_2 + S_1, \qquad \Phi_1 = S_2 - S_1.$$

But [MILNE-THOMSON (6) 1.24] $\Theta_1 = \Theta$, $\Phi_1 = \Phi e^{2i\chi}$. Therefore

$$2S_1 = \Theta - \Phi e^{2i\chi}, \quad 2S_2 = \Theta + \Phi e^{2i\chi}, \tag{1}$$

$$\Phi e^{2i\chi} = \bar{\Phi} e^{-2i\chi}, \tag{2}$$

since $\Phi_1 = \bar{\Phi}_1$. This last result gives

$$\tan 2\chi = \frac{2\widehat{rz}}{\widehat{zz} - \widehat{rr}} = \frac{-(p_1 - p_2)\dfrac{r}{L}}{-p_1 + \dfrac{p_1 - p_2}{L} R + \tilde{\omega}} = \frac{r}{R_0 - R}, \tag{3}$$

where

$$R_0 = \frac{(p_1 - \tilde{\omega})L}{p_1 - p_2} = \frac{p_1 L}{p_1 - p_2} - \frac{a}{2f}. \tag{4}$$

If J is the point $(0, 0, R_0)$, and if we introduce polar coordinates (ϱ, θ) in the meridian plane with origin at J and OR for initial line, we have

$$\tan 2\chi = \frac{r}{R_0 - R} = \tan(\pi - \theta).$$

Therefore $2\chi = \pi - \theta$ and if φ is the angle between the radius vector ϱ and the tangent to a line of principal stress at M, we have $\varphi = \chi + (\pi - \theta) = 3(\pi - \theta)/2$.
Therefore

$$\frac{\varrho\, d\theta}{d\varrho} = \tan\varphi = \cot\frac{3\theta}{2}$$

is the differential equation of a line of principal stress. Integrating we get for the lines of principal stress

$$\varrho^{3/2}\cos\frac{3\theta}{2} = \text{constant}.$$

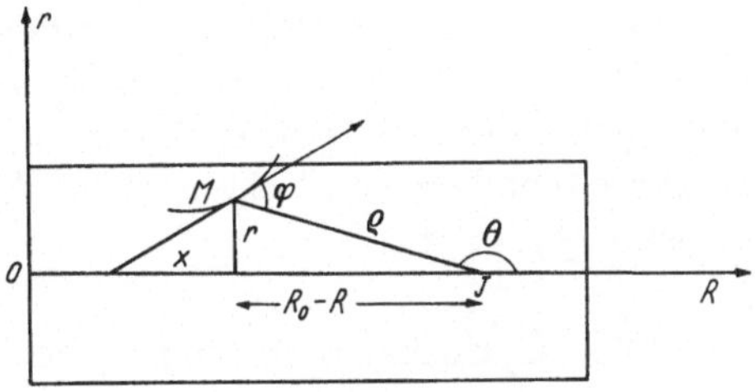

Fig. 3.82 (i)

These lines are shown in fig. 3.82 (ii), which is adapted from [FILON (2)].

As regards size and shape the lines are independent of p_1, p_2, $\tilde{\omega}$. Their position however depends upon the position of the point J.

This point is an isotropic point (1.22) at which the three principal stresses are equal. The point J is inside the cylinder if $0 < R_0 < L$, i. e. from (4) if

$$0 < \frac{p_1 - \tilde{\omega}}{p_1 - p_2} < 1$$

or

$$p_1 > \tilde{\omega} > p_2 .$$

In the case of a peg driven into a long hole we can take $p_2 = 0$ and this condition is always satisfied.

In the case of a pile driven into the ground p_2 will generally be considerable and the point J will be outside the cylinder.

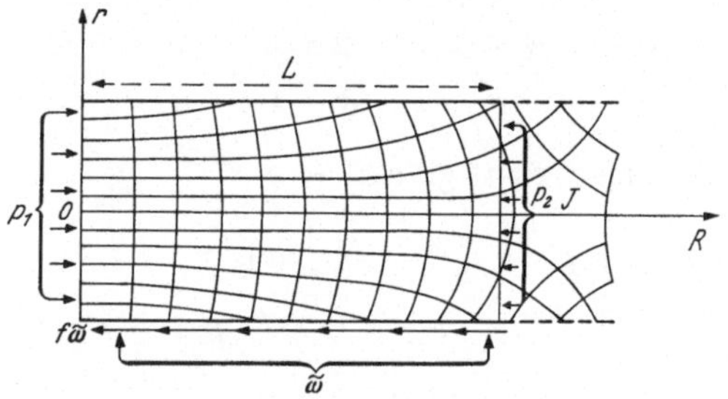

Fig. 3.82 (ii)

By considering the magnitudes of the differences of pairs of principal stresses namely $S_1 + \tilde{\omega}$, $S_2 + \tilde{\omega}$, $S_2 - S_1$, we verify that $S_2 - S_1$ has the greatest magnitude, namely

$$S_2 - S_1 = \Phi_1 = \left[-\tilde{\omega} + p_1 - (p_1 - p_2)\frac{R}{L} \right] \frac{R_0 - R}{\varrho} + (p_1 - p_2)\frac{r^2}{L\varrho} .$$

Using (4) and 3.8 (7) we find that

$$S_2 - S_1 = \frac{2f\tilde{\omega}\varrho}{a} .$$

If we take this greatest stress difference as the criterion for failure, we see that the points where yielding is first to be expected are those for which ϱ, the distance from J, is greatest, that is to say, in general, points on the circumference of the section of the beam at which the driving pressure p_1 is applied.

3.83. Cylindrical roller gripped between parallel planes

A long cylindrical roller, tightly gripped between two rough parallel planes, is forced through against friction by a thrust P applied to the end $R = 0$, the end $R = L$ being free from thrust.

Let N be the normal force per unit length supposed to be uniformly distributed, and let f be the coefficient of friction. The force of friction is fN per unit length. Therefore the lateral surface is subjected to concentrated shears $F = fN$ per unit length along the generators in contact with the planes and so

$$P = 2fNL = 2FL . \tag{1}$$

The problem therefore reduces to

(i) the plane problem of a cylinder gripped between concentrated forces N at the extremities of a diameter AB.

(ii) the antiplane problem of push against concentrated shears fN, fN.

The solution of the plane problem [MILNE-THOMSON (6) 5.27] gives the complex stresses

$$W(z) = \frac{N}{\pi}\left\{-\frac{1}{z-a} + \frac{1}{z+a} - \frac{1}{a}\right\}, \tag{2}$$

$$w(z) = \frac{N}{\pi}\left\{\frac{1}{z-a} - \frac{1}{z+a} - \frac{a}{(z-a)^2} - \frac{a}{(z+a)^2}\right\} \tag{3}$$

whence we can find $\widehat{xx}$, $\widehat{xy}$, $\widehat{yy}$. These yield a self-equilibrating system of tractions over the plane ends.

For $\widehat{zz}_p$ we have

$$\widehat{zz}_p = \eta(\widehat{xx} + \widehat{yy}) = \eta[W(z) + \overline{W}(\bar{z})].$$

Hence if $z+a = r_1 e^{i\theta_1}$, $z-a = r_2 e^{i(\pi-\theta_2)}$, (4)

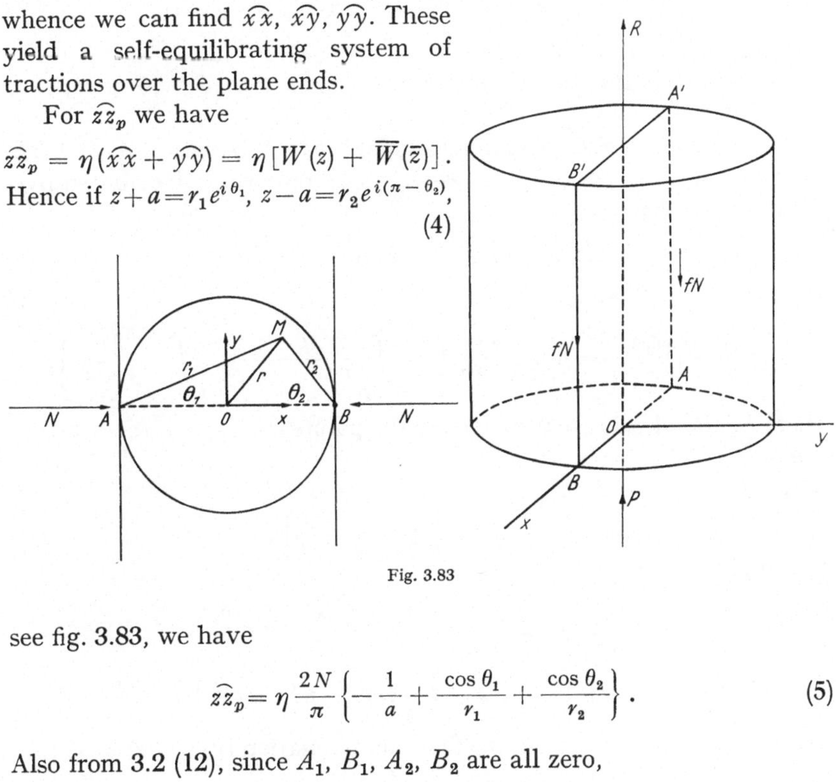

Fig. 3.83

see fig. 3.83, we have

$$\widehat{zz}_p = \eta \frac{2N}{\pi}\left\{-\frac{1}{a} + \frac{\cos\theta_1}{r_1} + \frac{\cos\theta_2}{r_2}\right\}. \tag{5}$$

Also from 3.2 (12), since A_1, B_1, A_2, B_2 are all zero,

$$\widehat{zz} = \widehat{zz}_p + EC_2 m + EC_1 mR.$$

But $\widehat{zz} = -\dfrac{P}{\pi a^2}$ when $R = 0$ and $\widehat{zz} = 0$ when $R = L$. Therefore

$$\widehat{zz} = -\frac{P}{\pi a^2}\left(1 - \frac{R}{L}\right). \tag{6}$$

Therefore

$$E C_1 m = \frac{P}{\pi a^2 L}, \qquad E C_2 m = -\frac{P}{\pi a^2} - \widehat{zz}_p\,. \tag{7}$$

Now consider the antiplane problem. On the boundary we have $\partial z/\partial s = iz/a$, and therefore from **3.5** (3)

$$z\Phi(z) + \bar{z}\bar{\Phi}(\bar{z}) = (1+\eta)\frac{P}{E\pi a^2 L}\,2z\bar{z} + \frac{2aZ_n}{\mu}$$

$$= \frac{P}{\mu\pi L} + \frac{2aZ_n}{\mu}\,.$$

Observe that on the boundary $\bar{z} = a^2/z$ and therefore by the method of **3.51**

$$\Phi(z) = \frac{1}{2\pi i}\oint \frac{P\,dt}{\pi\mu L t(t-z)} + \frac{1}{2\pi i}\oint \frac{2aZ_n dt}{\mu t(t-z)}\,, \tag{8}$$

wherein the first integral vanishes.

To evaluate the second integral we note that $Z_n = 0$ except on AA', BB'.

Consider narrow strips of infinitesimal breadth 2ε with AA' and BB' as centre lines. Suppose the stress Z_n to be uniformly distributed over these strips. Then

$$2(2\varepsilon L\cdot Z_n) = -2fNL = -P$$

and so $Z_n = -P/(4\varepsilon L)$ (9)

$$\frac{1}{2\pi i}\oint \frac{2aZ_n dt}{\mu t(t-z)} = \lim_{\varepsilon\to 0}\frac{iaP}{4L\mu\pi}\left[\frac{1}{\varepsilon}\int_{B_1B_2}\frac{dt}{t(t-z)} + \frac{1}{\varepsilon}\int_{A_1A_2}\frac{dt}{t(t-z)}\right],$$

where B_1, B_2, A_1, A_2 are respectively the points $a - i\varepsilon$, $a + i\varepsilon$, $-a + i\varepsilon$, $-a - i\varepsilon$.

Now

$$\frac{1}{t(t-z)} = \frac{1}{z}\left[\frac{1}{t-z} - \frac{1}{t}\right].$$

Therefore

$$\int_{B_1B_2}\frac{dt}{t(t-z)} = \frac{1}{z}\left[\ln(t-z) - \ln t\right]_{B_1}^{B_2}$$

$$= \frac{1}{z}\{\ln(a - z + i\varepsilon) - \ln(a - z - i\varepsilon) - \ln(a + i\varepsilon) + \ln(a - i\varepsilon)\}.$$

Use the logarithmic series in the first approximation $\ln(1+x) = x$, then

$$\int_{B_1B_2}\frac{1}{t(t-z)} = 2i\varepsilon\left(\frac{1}{a-z} - \frac{1}{a}\right).$$

Treating the second integral similarly we get

$$\Phi(z) = \frac{-aP}{2\pi\mu L z}\left(\frac{1}{a-z} + \frac{1}{a+z} - \frac{2}{a}\right) = \frac{P}{2\pi\mu L}\left(\frac{1}{z-a} + \frac{1}{z+a}\right). \tag{10}$$

Now from (7) and 3.2 (6) $p = P/(2\mu\pi a^2 L)$ and therefore

$$\widehat{xz} - i\,\widehat{yz} = \mu\,\Phi(z) - \mu\,p\bar{z} = \frac{P}{2\pi L}\left\{\frac{1}{z-a} + \frac{1}{z+a} - \frac{\bar{z}}{a^2}\right\}. \qquad (11)$$

The stresses are therefore all determined from (2), (3), (6), (11).

EXAMPLES III

1. Calculate the integrals I, J, K of 3.26 (2) in the case of (i) a circle (ii) a cardioid (iii) one loop of the lemniscate of Bernoulli, $r^2 = 2c^2 \cos 2\theta$.

2. The faces of a rectangular bar are the planes

$$x = \pm a, \; y = \pm b, \qquad R = \pm c\,.$$

A small square, whose side is of length l, is drawn on the face $y = b$, the centre of the square being at the point (x_0, b, R_0). The bar is now stretched by tension T applied uniformly to the ends $R = \pm c$. Find the shape into which the square is deformed when one of its sides before deformation makes the angle θ with the x-axis.

Examine in particular the cases $\theta = 0$, (ii) $\theta = \pi/2$, (iii) $\theta = 0$, $R_0 = \frac{1}{2} c$.

3. In the case of the cylinder suspended under gravity show that a generator, vertical before deformation, becomes an arc of a parabola after deformation.

4. In the case of a cylinder suspended under gravity prove that

$$e_{xx} = -\eta \frac{g\varrho R}{E} = e_{yy}, \; e_{zz} = \frac{g\varrho R}{E},$$

$$e_{yz} = e_{zx} = e_{xy} = 0\,.$$

Hence show that the only non-zero stress component is

$$\widehat{zz} = g\varrho R/E\,.$$

5. In the case of a suspended cylinder of weight W show that the elongation coincides with that of the same cylinder in the absence of gravity extended by the resultant force $\frac{1}{2} W$ applied by a uniform traction over the free end.

6. In the case of a beam whose antiplane is $R = 0$, bent by couples (3.7), find the strain coefficients $e_{xx}, e_{yy}, e_{zz}, e_{yz}, e_{zx}, e_{xy}$ and verify that all stress components except $\widehat{zz}$ vanish.

7. A beam whose cross-section is a rectangle $ABCD$ for which $AB = 2a$, $BC = 2b$ is bent by couples M and $-M$ applied at the end faces about principal axes which pass through the centroids and are parallel to AB. Calculate the deformed shape of a typical cross-section.

8. A beam whose cross-section is a square is bent by terminal couples M_1, M_2 applied about perpendicular axes, through the centre of the

square, in the plane of an end. Show that, apart from a rigid body movement, the displacement is of the same magnitude as it would be if the couples were applied about the diagonals of the square.

9. When a bar is bent by terminal couples M, $-M$ applied about principal axes through the centroid in the planes of the ends, show that the potential energy of deformation of the part of the bar between two normal sections at distance b apart is $\frac{1}{2}Mb/\varrho$, where ϱ is the radius of curvature of the bent line of centroids.

Give a physical interpretation of this result.

10. When a bar is bent by terminal couples applied about a principal axis through the centroid of the ends, show that the neutral plane is deformed into an anticlastic surface whose principal curvatures are in the ratio $1/\eta$, where η is Poisson's ratio.

11. A rectangular plate is bounded by the three pairs of planes $x = \pm a$, $y = \pm b$, $R = \pm c$. The plate is held bent by couples M_1, $-M_1$ applied about the lines $x = a$, $R = 0$ and $x = -a$, $R = 0$, and also by couples M_2, $-M_2$ applied about the lines $y = b$, $R = 0$ and $y = -b$, $R = 0$. Draw a diagram to show the system. By superposing solutions, show that the displacements, apart from a rigid body movement, can be written in the form

$$u = (A - \eta B)\, xR, \quad v = (B - \eta A)\, yR\,,$$

$$w = -\frac{1}{2}(A - \eta B)\, x^2 - \frac{1}{2}(B - \eta A)\, y^2 - \frac{1}{2}\eta(A + B) R^2\,,$$

where A and B are constants.

12. In Ex. 11 prove that the principal curvatures of the deformed plane $R = h$ are

$$\frac{1}{\varrho_1} = \eta B - A, \quad \frac{1}{\varrho_2} = \eta A - B$$

and hence prove that the potential energy of the bent plate per unit area is

$$\frac{1}{z}\frac{Eh^3}{1-\eta^2}\left[\left(\frac{1}{\varrho_1} + \frac{1}{\varrho_2}\right)^2 - 2(1-\eta)\frac{1}{\varrho_1\varrho_2}\right].$$

13. For a cylinder pushed through a hole use the notation of 3.82 to prove that the principal stresses in the meridian plane are

$$\frac{f\varpi}{a}\left[R + R_0 - \frac{2p_1L}{p_1 - p_2} + \varrho\right], \quad \frac{f\varpi}{a}\left[R + R_0 - \frac{2p_1L}{p_1 - p_2} - \varrho\right],$$

and hence show that the maximum difference of pairs of principal stress is given by the difference of the principal stresses in the meridian plane.

14. Exhibit graphically the magnitude of the greater of the principal stresses in the meridian plane along a generator as a function of R using the formulae of Ex. 13.

15. In the case of a cylinder gripped between parallel planes, 3.83, discuss the distribution of principal stress. Trace the lines of shearing stress over a cross-section and evaluate the displacement, in both cases paying special attention to the neighbourhood of the lines of grip.

16. A circular cylinder of radius a is pushed by pressure uniformly applied over one end through a rough hole which grips the cylinder with a pressure $\tilde{\omega}(1 - \cos n\theta)$, $n > 1$. Prove that the stresses are

$$\widehat{rr} = \frac{1}{2}\tilde{\omega}\left\{n\left(\frac{r}{a}\right)^{n-2} - (n-2)\left(\frac{r}{a}\right)^{n}\right\}\cos n\theta\,,$$

$$\widehat{\theta\theta} = \frac{1}{2}\tilde{\omega}\left\{-n\left(\frac{r}{a}\right)^{n-2} + (n+2)\left(\frac{r}{a}\right)^{n}\right\}\cos n\theta\,,$$

$$\widehat{r\theta} = \frac{1}{2}n\tilde{\omega}\left\{\left(\frac{r}{a}\right)^{n} - \left(\frac{r}{a}\right)^{n-2}\right\}\sin n\theta\,,$$

$$\widehat{zz} = -2f\tilde{\omega}\left(\frac{R}{a}\right) + 2\eta\tilde{\omega}\left(\frac{r}{a}\right)^{n}\cos n\theta\,,$$

where f is the coefficient of friction (FILON).

17. In the problem of a cylinder gripped between parallel planes (3.83) find the stress components $\widehat{xx}$, $\widehat{xy}$, $\widehat{yy}$, and verify that they correspond to a boundary unloaded except at two points.

18. In Ex. 17 show that the stress system $\widehat{xx}$, $\widehat{xy}$, $\widehat{yy}$ is, in the notation of fig. 3.83, equivalent to all round tension $N/(\pi a)$ together with radial distributions of pressure $\widehat{rr} = -2N\cos\theta_1/\pi a_1$ radiating from A and $\widehat{r_2 r_2} = -2N\cos\theta_2/\pi r^2$ radiating from B.

19. In the problem if a cylinder gripped between parallel planes (3.83) find explicitly the shears $\widehat{xz}$, $\widehat{yz}$ and show that they give no load at the boundary except at two points.

20. In Ex. 18 show that the resultant of the stress

$$zz_p = \eta(\widehat{r_1 r_1} + \widehat{r_2 r_2})$$

is an axial thrust $4\eta N a$ acting through the centre of the cross-section.

21. Examine the displacement components when a cylinder is gripped between two planes and is pushed through against friction.

22. Let $P(x, y)$, $Q(x, y)$ be analytic functions in the region inside and upon the contour C. Prove by means of the area theorem (3.25) that

$$\int_C (P\,dx + Q\,dy) = \int_S \left(\frac{\partial Q}{\partial x} - \frac{\partial P}{\partial y}\right) dS\,,$$

$$\int_C (P\,dy - Q\,dx) = \int_S \left(\frac{\partial P}{\partial x} + \frac{\partial Q}{\partial y}\right) dS\,.$$

Chapter IV

The torsion of isotropic beams

In this chapter the torsion problem for isotropic cylindrical or prismatic beams is considered. Methods of approaching the problem are examined and the general solution is obtained whenever the cross-section is simply connected and can be mapped on a circle. Applications of the principles of virtual displacements and virtual stresses to give approximate solutions are illustrated.

4.1. The torsion problem

In the classification of 3.3 the pure torsion problem arises when

$$\widehat{zz} = 0; \text{ and } M_3 \neq 0, Z_n = 0. \tag{1}$$

Therefore from 3.2 (5)

$$\frac{\widehat{xz} - i\widehat{yz}}{\mu} = \Phi(z) - i\tau\bar{z} \tag{2}$$

where τ is the angle of twist per unit length (see 4.22).

This equation contains the whole theory of the pure torsion of an isotropic beam twisted by an axial couple applied to the free end, in the absence of body force.

The torsion problem consists in finding the function $\Phi(z)$, for when this function is found all the stresses and displacements can be expressed in terms of $\Phi(z)$ as will appear from the following pages.

It is mathematically convenient to introduce real valued conjugate functions ϕ, ψ and the *complex torsion potential* $w(z)$ defined by

$$\Phi(z) = \tau w(z) = \tau(\phi + i\psi) \tag{3}$$

and then (2) can be replaced by

$$\frac{\widehat{xz} - i\widehat{yz}}{\mu\tau} = \frac{dw(z)}{dz} - i\bar{z}, \tag{4}$$

whence we have at once

$$\left.\begin{aligned} \widehat{xz} &= \mu\tau\left(\frac{\partial\psi}{\partial y} - y\right) = \mu\tau\left(\frac{\partial\phi}{\partial x} - y\right) \\ \widehat{yz} &= \mu\tau\left(-\frac{\partial\psi}{\partial x} + x\right) = \mu\tau\left(\frac{\partial\phi}{\partial y} + x\right) \end{aligned}\right\}. \tag{5}$$

We also note that

$$2i\psi = w(z) - \bar{w}(\bar{z})$$

so that

$$2i\frac{\partial\psi}{\partial z} = \frac{dw(z)}{dz}$$

and therefore from (4)

$$\frac{\widehat{xz} - i\widehat{yz}}{\mu\tau} = i\left(2\frac{\partial\psi}{\partial z} - \bar{z}\right). \tag{6}$$

Moreover ϕ and ψ are harmonic functions since they are the real and imaginary parts of a holomorphic function of z and therefore

$$\nabla^2\phi = 0, \ \nabla^2\psi = 0 . \tag{7}$$

We now define the real-valued *torsion shear stress function* Ω by

$$\Omega = \psi - \frac{1}{2} z\bar{z} \tag{8}$$

from which, using (6), we see that

$$\frac{\widehat{xz} - i\widehat{yz}}{\mu\tau} = 2i\frac{\partial\Omega}{\partial z} \tag{9}$$

so that

$$\widehat{xz} = \mu\tau\frac{\partial\Omega}{\partial y}, \qquad \widehat{yz} = -\mu\tau\frac{\partial\Omega}{\partial x} . \tag{10}$$

4.2. Lines of shearing stress

Def. A line of shearing stress is a line such that its tangent at every point is in the direction of the shearing stress vector at that point.

The lines of shearing stress are furnished by the curves

$$\Omega = c \tag{1}$$

for different values of the constant c.

Proof: Let L be a line of shearing stress. By definition the component of the stress vector along the normal, direction cosines $\partial y/\partial s$, $-\partial x/\partial s$, is zero where ds is an element of arc of L. Thus

$$\widehat{xz}\frac{\partial y}{\partial s} - \widehat{yz}\frac{\partial x}{\partial s} = 0 \tag{2}$$

or from 4.1 (10) $\frac{\partial\Omega}{\partial s} = 0$.

Therefore integrating along L

$$\Omega = \text{constant} . \qquad \text{Q.E.D.}$$

We can now determine the boundary condition. If C is the contour of a cross-section, since by hypothesis the boundary is unloaded, condition (2) holds on C and therefore

$$\Omega = \text{constant on } C \tag{3}$$

and this is the boundary condition.

Now from 4.1 (8) and (7)

$$\nabla^2\Omega = \nabla^2\psi - 2 = -2 . \tag{4}$$

Therefore the shear stress function Ω satisfies (4) everywhere in the material and the condition (3) on the boundary. These conditions suffice to determine Ω save for an irrelevant added constant.

An equivalent set of conditions is obtained from (3) and (4), using 4.1 (7), (8) namely

$$\nabla^2 \psi = 0 \text{ throughout the material} \tag{5}$$

$$\psi - \frac{1}{2}(x^2 + y^2) = \text{constant on the boundary}\,. \tag{6}$$

In the above investigation we have relied solely on the statement in 3.3 of the torsion problem as one of antiplane stress. No hypothesis whatever has been made concerning the displacement.

We can use (3) and 4.1 (9) to show that the resultant force due to shear on a cross-section is zero. For the resultant is

$$\int (\widehat{xz} - i\widehat{yz})\, dS = 2i\mu\tau \int \frac{\partial \Omega}{\partial z}\, dS = -\mu\tau \oint \Omega\, d\bar{z} = 0\,,$$

on using the area theorem, 3.25, and the fact that Ω is constant on the boundary C. Thus it is verified that the action of the shear on the cross-section can reduce only to a couple.

4.22. The displacement

In the notation of 3.4 (8) we have $\widehat{zz} = 0$ so that

$$\beta = \gamma = \delta = \theta_0 = 0$$

and therefore from 3.41 (6) and (7) the displacement (u, v, w) is given by

$$u + iv = i\tau z R + (R\omega_2 - y\omega_3 + a_0) + i(x\omega_3 - R\omega_1 + b_0)\,, \tag{1}$$

$$w = \frac{1}{2}\tau[w(z) + \bar{w}(\bar{z})] + (\omega_1 y - \omega_2 x + c_0)\,, \tag{2}$$

where the last terms correspond to a rigid body movement of translation (a_0, b_0, c_0) and rotation $(\omega_1, \omega_2, \omega_3)$.

Thus, apart from a rigid body movement of the material as a whole, the components of the displacement are

$$(-\tau y R, \tau x R, \tau\phi)\,. \tag{3}$$

If we wish u, v to vanish at the fixed end, we can take this rigid body movement to consist solely of the rotation $(0, 0, -\tau L)$ and then we get for the displacement

$$[\tau y(L - R), -\tau x(L - R), \tau\phi]\,. \tag{4}$$

Thus if we consider points in a cross-section the radius vector to the point (x, y) is rotated through the angle τR in the plane of the cross-section, while the point is moved through the distance $\tau\phi = \tau\phi(x, y)$

out of the plane. Thus, unless ϕ is constant, the planes of the cross-sections are *warped* by the deformation.

It appears from the above that τR is an angle and so, as stated in 4.1, τ is the angle of twist per unit length.

4.24. Hydrodynamic analogies

First analogy. The functions $w(z)$, ϕ, ψ are respectively mathematically identical with the complex potential, the velocity potential and the stream function of the steady irrotational motion of an inviscid liquid occupying a vessel of the same shape as the prism (or cylinder), which rotates about an axis parallel to the generators with unit angular velocity [Milne-Thomson (5)].

Let (u, v) be the components* of the fluid velocity. Then

$$u - iv = -\frac{dw(z)}{dz} = -\frac{\widehat{xz} - i\widehat{yz}}{\mu\tau} - i\bar{z} \tag{1}$$

from 4.1 (4). Therefore

$$-\frac{\widehat{xz}}{\mu\tau} = u + y, \qquad -\frac{\widehat{yz}}{\mu\tau} = v - x. \tag{2}$$

This interprets the shears in terms of the velocity $(u + y,\ v - x)$ of a fluid particle *relative* to the prism.

Observe that in this analogy the vessel rotates and carries the x- and y-axes with it, but (u, v) are the components of absolute velocity of a fluid particle resolved parallel to the instantaneous positions of the axes.

It follows from (2) that the lines of shearing stress in the twisted prism, coincide with the paths of the particles of the fluid relative to the rotating prism in the analogy.

The first analogy presents the advantage that inferences can be based on the well explored field of irrotational motion.

Second analogy. The function Ω of 4.1 (8) is mathematically identical with the stream function of the steady rotational motion of an inviscid liquid, which occupies a fixed vessel of the same shape as the prism, circulating with uniform vorticity -2.

This follows from 4.1 (8), (7) since the vorticity is

$$\nabla^2 \Omega = \nabla^2 \psi - \nabla^2 \left(\frac{1}{2} z\bar{z}\right) = -2.$$

In this analogy the x- and y-axes remain fixed and if (u, v) are the velocity components of the fluid motion, 4.1 (10) shows that

$$-\frac{\widehat{xz}}{\mu\tau} = u, \qquad -\frac{\widehat{yz}}{\mu\tau} = v. \tag{3}$$

* There need be no confusion between the velocity components in the analogy and the displacements in the torsion problem.

From this it follows that the lines of shearing stress coincide with the paths of the fluid particles in this analogy.

The second hydrodynamical analogy leads to a conclusion of practical importance. Suppose a shaft transmitting a couple to contain a cylindrical hole of circular cross-section whose axis is parallel to that of the shaft. We shall suppose the diameter of the cavity to be small compared with that of the shaft and the distance of the cavity from the surface to be great compared with its diameter. The problem is then nearly the same as that of liquid streaming past a circular cylinder. Now we know [MILNE-THOMSON (5)] that the maximum speed of the liquid occurs on the boundary of the cylinder and is twice that of the main stream. The inference is that in the torsion problem the maximum shear in the neighbourhood of the cavity may be about twice that which would exist if there were no cavity.

Third analogy. The function Ω of 4.1 (8) is mathematically identical with the fluid speed of the laminar motion of a viscous liquid, of viscosity λ, which flows under a pressure gradient of magnitude 2λ, through a pipe whose cross-section is the same as that of the prism [MILNE-THOMSON (5)].

In this analogy the lines of shearing stress, $\Omega =$ constant, coincide with the lines in a cross-section of the pipe on which the fluid speed has the same magnitude.

4.26. Maximum shearing stress

Let

$$\sigma^2 = (\widehat{xz})^2 + (\widehat{yz})^2 = 4\mu^2\tau^2 \frac{\partial \Omega}{\partial z} \frac{\partial \Omega}{\partial \bar{z}} \tag{1}$$

give the square of the resultant shearing stress. We shall prove that σ^2 can have no maximum value at any point *inside* the material, and consequently that the maximum values of σ^2, if any, must occur on the boundary.

Proof. Since

$$\nabla^2 = 4 \frac{\partial^2}{\partial z\, \partial \bar{z}} \text{ and } \nabla^2 \Omega = -2, \tag{2}$$

we have

$$\nabla^2(\sigma^2) = 16\mu^2\tau^2 \frac{\partial^2}{\partial z\, \partial \bar{z}} \left(\frac{\partial \Omega}{\partial z} \cdot \frac{\partial \Omega}{\partial \bar{z}} \right) = 16\mu^2\tau^2 \left(\frac{1}{4} + \frac{\partial^2 \Omega}{\partial z^2} \cdot \frac{\partial^2 \Omega}{\partial \bar{z}^2} \right) > 0 \tag{3}$$

since the last term is the product of two conjugate complex numbers and is therefore never negative.

Now if σ^2 has a maximum value at a point P inside the material, we can find a curve γ surrounding P and wholly inside the material such that the outward normal derivative $\partial(\sigma^2)/\partial n$ is negative at every point of γ and therefore

$$\oint_\gamma \frac{\partial(\sigma^2)}{\partial n}\, ds < 0. \tag{4}$$

But by Gauss's theorem [MILNE-THOMSON (5)]

$$\oint_\gamma \frac{\partial(\sigma^2)}{\partial n}\, ds = \int_S \nabla^2(\sigma^2)\, dS > 0 \tag{5}$$

from (3). Since (4) and (5) furnish a contradiction the theorem is proved.

The above argument applies to the maximum value of any *sub-harmonic function* ϕ i.e. a function such that $\nabla^2\phi > 0$ inside the region of its definition.

4.28. Change of axis of twist

In the foregoing we have assumed that the beam is twisted about an axis which is parallel to the generators and passes through the point $z = 0$ i.e. $(0, 0, 0)$ in the antiplane $R = 0$. Let us suppose that the same twist is applied (using the language of plane geometry) not about the point $z = 0$ but the point $z = c$. By the first hydrodynamic analogy the problem is the same as if the rotating prism had its axis of rotation transferred in the same way. But rotation with unit angular velocity about $z = c$ is the same as rotation with unit angular velocity about $z = 0$, together with a velocity $-i\bar{c}$ of the point $z = 0$. Therefore for the rotating fluid, if suffix 1 refers to parallel axes $x_1 y_1$ through $z = c$, we have

$$w_1(z_1) = w(z) - i\bar{c}z, \qquad z = z_1 + c. \tag{1}$$

Therefore from 4.1 (4)

$$\frac{\widehat{x_1 z_1} - i\widehat{y_1 z_1}}{\mu\tau} = \frac{d w_1(z_1)}{d z_1} - i\bar{z}_1 = \frac{d w(z)}{d z} - i\bar{z} = \frac{\widehat{xz} - i\widehat{yz}}{\mu\tau}. \tag{2}$$

Therefore the stress is the same whether we twist about $z = 0$ or $z = c$.

To find the effect on the displacement we have from 4.22

$$u_1 + iv_1 = i\tau z_1 R = i\tau(z - c)R = u + iv - i\tau c R, \tag{3}$$

$$\begin{aligned}(w)_1 &= \frac{1}{2}\tau[w_1(z_1) + \bar{w}_1(\bar{z}_1)] = \frac{1}{2}\tau[w(z) + \bar{w}(\bar{z}) - i\bar{c}z + ic\bar{z}] \\ &= w - \frac{1}{2}i\tau(\bar{c}z - c\bar{z}).\end{aligned} \tag{4}$$

Put

$$c = \alpha + i\beta. \tag{5}$$

Then (u_1, v_1, w_1) is obtained by adding to (u, v, w) the rigid body movement due to the rotation $(\tau\alpha, \tau\beta, 0)$ about the origin.

To sum up; a translation of the axis of twist has no effect on the stress distribution but alters the displacement by a rigid body movement.

4.3. The twisting moment

From 4.1 (4) we have

$$\widehat{xz} - i\,\widehat{yz} = \mu\tau\left[\frac{dw(z)}{dz} - i\bar{z}\right]$$

and therefore from 3.3 (8) the twisting moment M is

$$\begin{aligned} M &= \int_S (x\cdot\widehat{yz} - y\cdot\widehat{xz})\,dS = \mathrm{Re}\int_S iz(\widehat{xz} - i\cdot\widehat{yz})\,dS \\ &= \mu\tau\int_S iz\left[\frac{dw(z)}{dz} - i\bar{z}\right]dS \end{aligned} \tag{1}$$

since, as will appear from (3) below, this integral is real.
Now by the area theorem, 3.25,

$$\begin{aligned} \int_S iz\frac{dw(z)}{dz}\,dS &= \frac{1}{2}\oint_C z\bar{z}\frac{dw(z)}{dz}\,dz \\ &= -\frac{1}{2}i\oint_C [w(z) - \bar{w}(\bar{z}) + \text{constant}]\frac{dw(z)}{dz}\,dz \end{aligned}$$

since on the boundary

$$w(z) - \bar{w}(\bar{z}) = 2i\,\psi = i\,z\bar{z} + \text{constant}\,. \tag{2}$$

Also by Cauchy's theorem, since $w(z)$ is holomorphic over the cross-section

$$\oint_C [w(z) + \text{constant}]\frac{dw(z)}{dz}\,dz = 0\,.$$

Therefore

$$\int_S iz\frac{dw(z)}{dz}\,dS = \frac{1}{2}i\oint_C \bar{w}(\bar{z})\frac{dw(z)}{dz}\,dz = -\int_S \frac{d\bar{w}(\bar{z})}{d\bar{z}}\frac{dw(z)}{dz}\,dS\,, \tag{3}$$

on using the area theorem again.
Also

$$T = \frac{1}{2}\int_S \frac{dw(z)}{dz}\frac{d\bar{w}(\bar{z})}{d\bar{z}}\,dS \tag{4}$$

is real and non-negative since the integrand is the product of complex conjugate numbers.

It follows from (3) and (4) that

$$\int_S iz\frac{dw(z)}{dz}\,dS = -\int_S i\bar{z}\frac{d\bar{w}(\bar{z})}{d\bar{z}}\,dS = -2T\,.$$

Thus the left-hand integral of (3) is real and therefore the integral of (1) is also real. Taking the conjugate of the sum of the first and last integrals of (3), we have

$$0 = \mu\tau \int_S \frac{d\overline{w}(\bar z)}{d\bar z}\left[\frac{d w(z)}{dz} - i\bar z\right] dS\,.$$

Adding this to (1) we get

$$\frac{M}{\mu\tau} = \int_S \left[\frac{d w(z)}{dz} - i\bar z\right]\left[\frac{d\overline{w}(\bar z)}{d\bar z} + i z\right] dS \tag{5}$$

and since the integrand is the product of conjugate complex numbers we see that

$$\frac{M}{\mu\tau} > 0\,. \tag{6}$$

This is a strict inequality for the integral (5) can never vanish unless

$$\frac{d w(z)}{dz} \equiv i\bar z$$

which is impossible for the left side is independent of $\bar z$.

Returning to (1) we note that

$$\int z\bar z\, dS = \int (x^2 + y^2)\, dS = I \tag{7}$$

the polar second moment of area with respect to the origin. Therefore (1) and (3) give

$$M = \mu\tau(I - 2T) \tag{8}$$

and from (6) it follows that

$$I - 2T > 0\,. \tag{9}$$

To calculate $I - 2T$ we use the area theorem, 3.25. Thus

$$\begin{aligned} I - 2T &= \int_S \frac{\partial}{\partial \bar z}\left(\frac{1}{2} z\bar z^2\right) dS - \int_S \frac{\partial}{\partial \bar z}\left(\overline{w}(\bar z)\frac{d w(z)}{dz}\right) dS \\ &= -\frac{1}{4} i \oint_C z\bar z^2 dz + \frac{1}{2} i \oint_C \overline{w}(\bar z)\, d(w(z)) \qquad (10) \\ &= -\frac{1}{4} i \oint_C z\bar z^2\, dz - \frac{1}{2}\oint_C w(z)\, d(z\bar z)\,, \end{aligned}$$

on using (2) and Cauchy's theorem.

Since on the contour C, $\bar z$ is a function of z deducible from the equation $f(x, y) = 0$ of C, the integrals can be evaluated by Cauchy's residue theorem or any other suitable means.

4.32. The twisting moment in terms of Ω

From 4.3 the twisting moment is given by the real integrals.

$$M = \int_S (x \cdot \widehat{yz} - y \cdot \widehat{zx})\, dS = -2\mu\tau \int_S z \frac{\partial \Omega}{\partial z}\, dS$$

$$= -2\mu\tau \int \left[\frac{\partial}{\partial z}(\Omega z) - \Omega\right] dS \tag{1}$$

$$= -2\mu\tau \left\{\frac{1}{2} i \oint_C \Omega z\, d\bar{z} - \int \Omega\, dS\right\},$$

using the area theorem, 3.25.

To interpret this suppose that the cross-section is bounded by curves C_1 and C_2 of which C_2 lies entirely within C_1.
Then

$$\frac{1}{2} i \oint_C \Omega z\, d\bar{z} = \frac{1}{2} i \oint_{C_1} \Omega z\, d\bar{z} - \frac{1}{2} i \oint_{C_2} \Omega z\, d\bar{z}$$

$$= \frac{1}{2} i \Omega_1 \oint_{C_1} z\, d\bar{z} - \frac{1}{2} i \Omega_2 \oint_{C_2} z\, d\bar{z} \tag{2}$$

where Ω_1, Ω_2 are the constant values assumed by Ω on C_1 and C_2 respectively. Also from the area theorem, the areas enclosed by C_1 and C_2 are respectively

$$A_1 = \frac{1}{2} i \oint_{C_1} z\, d\bar{z}, \quad A_2 = \frac{1}{2} i \oint_{C_2} z\, d\bar{z}. \tag{3}$$

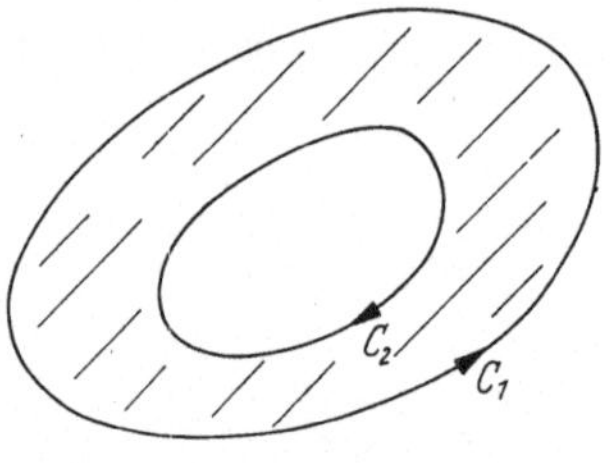

Fig. 4.32

Therefore

$$M = 2\mu\tau \left\{- A_1 \Omega_1 + A_2 \Omega_2 + \int_S \Omega\, dS\right\} \tag{4}$$

where the integral is extended to the region S between C_1 and C_2.

In the case of a simply connected cross-section we have $A_2 = 0$ and therefore

$$M = 2\mu\tau \left(- A_1 \Omega_1 + \int_S \Omega\, dS\right) = 2\mu\tau \int_S (\Omega - \Omega_1)\, dS. \tag{5}$$

If we write

$$\Omega_0 = \Omega - \Omega_1 \tag{6}$$

so that Ω_0 is that form of Ω which vanishes on the boundary, we get

$$M = 2\mu\tau \int \Omega_0\, dS. \tag{7}$$

4.34. Torsional rigidity

If we write

$$M = \tau D, \tag{1}$$

the quantity D is called the *torsional rigidity*, since for a given value of M a large value of D corresponds to a small twist τ.

From 4.3 (5), (8) we have

$$D = \frac{M}{\tau} = \mu(I - 2T) = \mu \int \left[\frac{dw(z)}{dz} - i\bar{z}\right]\left[\frac{d\bar{w}(\bar{z})}{d\bar{z}} + iz\right] dS \tag{2}$$

so that, as implied by 4.3 (5)

$$D > 0. \tag{3}$$

This is a strict inequality, for the integral in (2) can not vanish.

We shall now prove that the torsional rigidity is independent of the actual position of the axis of twist.

Let us as in 4.28 transfer the axis of twist (using the language of plane geometry) from $z = 0$ to $z = c$. Then if suffix 1 refers to the origin $z = c$ with coordinates (x_1, y_1) we have as in 4.28 (1)

$$w_1(z_1) = w(z) - i\bar{c}z, \qquad z = z_1 + c.$$

Then from (2)

$$\begin{aligned} D_1 &= \mu \int \left[\frac{dw_1(z_1)}{dz_1} - i\bar{z}_1\right]\left[\frac{d\bar{w}_1(\bar{z}_1)}{d\bar{z}_1} + iz_1\right] dS \\ &= \mu \int \left[\frac{dw(z)}{dz} - i(\bar{z}_1 + \bar{c})\right]\left[\frac{d\bar{w}(\bar{z})}{d\bar{z}} + i(z_1 + c)\right] dS = D \end{aligned}$$

since $z_1 + c = z$. Q.E.D.

It follows that for a given couple M, the torsional rigidity D is invariant. Therefore from (2) $\mu(I - 2T)$ is invariant. We note that in the first hydrodynamic analogy

$$\begin{aligned} \frac{dw(z)}{dz} - i\bar{z} &= -u + iv - i(x - iy) = (-u - y) - i(-v + x) \\ &= \left(\frac{\partial\phi}{\partial x} - y\right) - i\left(\frac{\partial\phi}{\partial y} + x\right) = \left(\frac{\partial\psi}{\partial y} - y\right) + i\left(\frac{\partial\psi}{\partial x} - x\right) \end{aligned}$$

and therefore from (2)

$$\frac{D}{\mu} = \int_S \left[\left(\frac{\partial\phi}{\partial x} - y\right)^2 + \left(\frac{\partial\phi}{\partial y} + x\right)^2\right] dS \tag{4}$$

$$= \int_S \left[\left(\frac{\partial\psi}{\partial x} - x\right)^2 + \left(\frac{\partial\psi}{\partial y} - y\right)^2\right] dS. \tag{5}$$

We also note that, from 4.32 (7), in the case of a simply connected cross-section

$$\frac{D}{\mu} = \int_S 2\Omega_0 \, dS. \tag{6}$$

4.36. Maximum torsional rigidity

We have proved that

$$D = \mu (I - 2T) > 0, \tag{1}$$

where

$$T = \frac{1}{2} \int_S \frac{dw(z)}{dz} \cdot \frac{d\overline{w}(\overline{z})}{d\overline{z}} \, dS \geqq 0. \tag{2}$$

Now in the first hydrodynamic analogy, T is the kinetic energy, per unit thickness, of the liquid, of unit density, which fills the rotating prism.

These observations furnish simple proofs of the following theorems concerning the maximum of the torsional rigidity.

Theorem 4.36 (1). Consider cylindrical beams of the same material, whose cross-sections are simply connected and have the same polar second moment I about the centroid of the cross-section. The beam for which the torsional rigidity is greatest is the circular cylindrical beam.

Proof. For a circular cylinder, containing liquid, rotating about its axis of figure, $w(z) =$ constant, [MILNE-THOMSON (5) 9.71] and therefore $T = 0$. In this case $D = \mu I$, and from (1), (2) this is the greatest value which D can assume. Q.E.D.

If a is the radius of the circle, $I = \pi a^4/2$, and therefore a is determined.

Theorem 4.36 (2). Consider cylindrical beams of the same material whose cross-sections are doubly-connected and have the same polar second moment I about the centroid of the cross-section. The beam for which the torsional rigidity is greatest is the concentric circular cylindrical tube.

The proof is the same as for Theorem 4.36 (1) since $T = 0$ in the hydrodynamic analogy when rotation takes place about the axis of figure of the concentric circular tube.

Let a be the outer and b the inner radius. Then

$$\frac{1}{2} \pi (a^4 - b^4) = I \tag{3}$$

so that in this case the assignment of I determines only the difference of the fourth powers of the radii. All tubes whose radii satisfy (3) will have the same torsional rigidity.

4.37. A minimum property of the torsional rigidity

Theorem 4.37.

Let

$$J = \int_S \left(\frac{dW(z)}{dz} - i\overline{z} \right) \left(\frac{d\overline{W}(\overline{z})}{d\overline{z}} + iz \right) dS \tag{1}$$

taken over the area of the cross-section S of a beam, where $W(z)$ is holomorphic over the cross-section. Let $w(z)$ give the solution of the torsion problem for the beam, so that the torsional rigidity is given by

$$\frac{D}{\mu} = \int_S \left(\frac{d w(z)}{d z} - i\bar{z}\right)\left(\frac{d\bar{w}(\bar{z})}{d\bar{z}} + i z\right) d S .$$

Then

$$\frac{D}{\mu} \leqq J , \tag{2}$$

with equality only if

$$W'(z) \equiv w'(z) . \tag{3}$$

Proof. Let

$$W(z) = w(z) + w_1(z) , \tag{4}$$

where $w_1(z)$ is holomorphic over the cross-section. Then

$$\begin{aligned} J &= \int_S \left(\frac{d w(z)}{d z} - i\bar{z} + \frac{d w_1(z)}{d z}\right)\left(\frac{d\bar{w}(\bar{z})}{d z} + i z + \frac{d\bar{w}_1(\bar{z})}{d\bar{z}}\right) d S \\ &= \frac{D}{\mu} + \int_S \frac{d w_1(z)}{d z}\,\frac{d\bar{w}_1(\bar{z})}{d\bar{z}}\, d S + H \end{aligned} \tag{5}$$

where

$$H = \int_S \left[\frac{d\bar{w}_1(\bar{z})}{d\bar{z}}\left(\frac{d w(z)}{d z} - i\bar{z}\right) + \frac{d w_1(z)}{d z}\left(\frac{d\bar{w}(\bar{z})}{d\bar{z}} + i z\right)\right] d S$$

and therefore using the area theorem, 3.25,

$$2iH = \oint_C \frac{d w_1(z)}{d z}\,(\bar{w}(\bar{z}) + i z\bar{z})\, d z - \oint_C \frac{d\bar{w}_1(\bar{z})}{d\bar{z}}\,(w(z) - i z\bar{z}))\, d\bar{z} . \tag{6}$$

Now on the boundary C

$$i z\bar{z} = 2 i \psi + \text{constant} = w(z) - \bar{w}(\bar{z}) + \text{constant}$$

and therefore

$$\bar{w}(\bar{z}) + i z\bar{z} = w(z) + \text{constant}$$

$$w(z) - i z\bar{z} = \bar{w}(\bar{z}) + \text{constant}$$

and therefore in (6) the integrals vanish by Cauchy's theorem so that $H = 0$. Therefore (5) becomes

$$J = \frac{D}{\mu} + 2 T_1 , \tag{7}$$

where

$$2 T_1 = \int \frac{d w_1(z)}{d z}\,\frac{d\bar{w}_1(\bar{z})}{d\bar{z}}\, d S \geqq 0$$

since the integrand is the product of conjugate complex quantities. Thus

$$J \geqq D/\mu \text{ with equality only if } d w_1(z)/d z = 0$$

which proves the theorem.

It follows from this theorem that if we substitute an approximate solution $W(z)$ for the exact solution $w(z)$ of the torsion problem, the torsional rigidity calculated from the approximate solution will be greater than the exact value.

4.38. A maximum property of the torsional rigidity

We consider a beam whose cross-section S is simply connected and is bounded by a contour C. Let Ω_0 be the torsion shear stress function 4.1 (8) which vanishes on C so that, 4.32 (7), the twisting moment is

$$M = \mu\tau \int_S 2\,\Omega_0\, dS\,. \tag{1}$$

We now prove the following theorem.

Theorem 4.38. Let $F = F(z, \bar{z})$ be a real valued function analytic on the cross-section S of a beam and vanishing on the contour C;

$$F \equiv 0 \quad \text{on } C\,. \tag{2}$$

Let

$$K = \int_S \left\{4F - 4\frac{\partial F}{\partial z}\frac{\partial F}{\partial \bar{z}}\right\} dS \tag{3}$$

and let D, where

$$\frac{D}{\mu} = \int_S 2\,\Omega_0\, dS\,, \tag{4}$$

be the torsional rigidity.

Then

$$\frac{D}{\mu} \geqq K \tag{5}$$

with equality only if

$$F \equiv \Omega_0\,. \tag{6}$$

Proof. Write

$$F = \Omega_0 + f\,. \tag{7}$$

From (2), since $\Omega_0 = 0$ on C, we have

$$f \equiv 0 \quad \text{on } C\,. \tag{8}$$

Substitute for F in (3). Then the integrand becomes

$$4\,\Omega_0 + 4f - 2\frac{\partial \Omega_0}{\partial z}\left(\frac{\partial \Omega_0}{\partial \bar{z}} + 2\frac{\partial f}{\partial \bar{z}}\right) - 2\frac{\partial \Omega_0}{\partial \bar{z}}\left(\frac{\partial \Omega_0}{\partial z} + 2\frac{\partial f}{\partial z}\right) - 4\frac{\partial f}{\partial z}\frac{\partial f}{\partial \bar{z}}$$

$$= 4\,\Omega_0 + 4f - 4\frac{\partial f}{\partial z}\frac{\partial f}{\partial \bar{z}} - 2\frac{\partial}{\partial \bar{z}}\left\{(\Omega_0 + 2f)\frac{\partial \Omega_0}{\partial z}\right\}$$

$$- 2\frac{\partial}{\partial z}\left\{(\Omega_0 + 2f)\frac{\partial \Omega_0}{\partial \bar{z}}\right\} + (\Omega_0 + 2f)\frac{4\partial^2 \Omega_0}{\partial z\,\partial \bar{z}}$$

$$= 2\,\Omega_0 - 4\frac{\partial f}{\partial z}\frac{\partial f}{\partial \bar{z}} - 2\frac{\partial}{\partial \bar{z}}\left\{(\Omega_0 + 2f)\frac{\partial \Omega_0}{\partial z}\right\} - 2\frac{\partial}{\partial z}\left\{(\Omega_0 + 2f)\frac{\partial \Omega_0}{\partial \bar{z}}\right\}$$

since, by 4.2 (4), $4\partial^2 \Omega_0/\partial z\,\partial \bar{z} = -2$.

Putting this form of the integrand in (3) and using the area theorem we get

$$K=\int\limits_S 2\Omega_0\,dS-4\int\limits_S \frac{\partial f}{\partial z}\frac{\partial f}{\partial \bar z}\,dS+i\oint\limits_C(\Omega_0+2f)\frac{\partial\Omega_0}{\partial z}\,dz$$
$$-i\oint\limits_{\bar C}(\Omega_0+2f)\frac{\partial\Omega_0}{\partial\bar z}\,d\bar z. \tag{9}$$

But $\Omega_0+2f=0$ on C and so the last two integrals vanish. Also

$$H=4\int\limits_S \frac{\partial f}{\partial z}\frac{\partial f}{\partial \bar z}\,dS$$

is never negative, for the integrand is the product of conjugate complex numbers. Therefore from (4) and (9)

$$\frac{D}{\mu}=K+H$$

and since $H\geqq 0$ and vanishes only when $f\equiv 0$ the theorem is proved.

It follows from this theorem that if we substitute an approximate torsion shear function F, obeying (2), for the exact function Ω_0, the torsional rigidity calculated from the approximation will be less than the actual value.

Combining theorems 4.37 and 4.38 we have

$$K\leqq\frac{D}{\mu}\leqq J$$

and so we can always find upper and lower bounds for D/μ.

These inequalities are best possible since they become equalities when the exact solutions are used.

4.39. Potential energy

From 1.82 (12) and 1.98 (5) the strain-energy function is

$$\frac{1}{2\mu}\left[(\widehat{xz})^2+(\widehat{yz})^2\right] \tag{1}$$

and therefore from 4.1 (4) the potential energy per unit length of cylinder is

$$V=\frac{1}{2}\mu\tau^2\int\left[\frac{dw(z)}{dz}-i\bar z\right]\left[\frac{d\bar w(\bar z)}{d\bar z}+iz\right]dS \tag{2}$$
$$=\frac{1}{2}M\tau=\frac{1}{2}D\tau^2 \tag{3}$$

from 4.3 (5), 4.34 (1).

Thus

$$M=\frac{\partial V}{\partial\tau}. \tag{4}$$

4.4. Solution by conformal mapping

The general solution for the antiplane problem in the case of an isotropic beam whose simply connected cross-section C can be mapped on the region within the unit circumference γ, $|\zeta| = 1$, by $z = m(\zeta)$ is given in 3.51.

In the case of pure torsion a great simplification becomes possible since 3.51 (6) reduces to

$$H(\sigma) = i\tau\left\{\overline{m}\left(\frac{1}{\sigma}\right) m'(\sigma) - \frac{1}{\sigma^2} m(\sigma)\, \overline{m}'\left(\frac{1}{\sigma}\right)\right\} = i\tau \frac{d}{d\sigma}\left\{m(\sigma)\, \overline{m}\left(\frac{1}{\sigma}\right)\right\} \quad (2)$$

If therefore we write

$$'\Phi(z) = {}'\Phi[m(\zeta)] = \tau\omega(\zeta)\,, \quad (3)$$

we have from 3.51 (7)

$$\Phi(z)\, m'(\zeta) = \tau\omega'(\zeta) = i\tau \cdot \frac{1}{2\pi i} \oint_\gamma \frac{\frac{d}{d\sigma}\left[m(\sigma)\, \overline{m}\left(\frac{1}{\sigma}\right)\right] d\sigma}{\sigma - \zeta}$$

$$= i\tau \cdot \frac{1}{2\pi i}\left[\frac{m(\sigma)\, \overline{m}\left(\frac{1}{\sigma}\right)}{\sigma - \zeta}\right]_\gamma + i\tau \cdot \frac{1}{2\pi i} \oint_\gamma \frac{m(\sigma)\, \overline{m}\left(\frac{1}{\sigma}\right) d\sigma}{(\sigma - \zeta)^2}$$

on integration by parts. The first term on the right vanishes and therefore finally

$$\Phi(z)\, m'(\zeta) = \tau\omega'(\zeta) = i\tau \frac{\partial}{\partial\zeta}\left[\frac{1}{2\pi i} \oint_\gamma \frac{m(\sigma)\, \overline{m}\left(\frac{1}{\sigma}\right) d\sigma}{\sigma - \zeta}\right]. \quad (4)$$

The advantage of this formulation is that it leads to the calculation of a simpler integral, namely that in the square brackets of (4). Thus in the notation of 4.1 (3) we have from (4)

$$w(z) = \omega(\zeta) = i \cdot \frac{1}{2\pi i} \oint_\gamma \frac{m(\sigma)\, \overline{m}\left(\frac{1}{\sigma}\right) d\sigma}{\sigma - \zeta}\,. \quad (5)$$

When we can write $m(\sigma)\, \overline{m}(1/\sigma)$ in the form

$$m(\sigma)\, \overline{m}\left(\frac{1}{\sigma}\right) = M_1(\sigma) + M_2(\sigma)\,, \quad (6)$$

where all and only the negative powers of σ occur in $M_2(\sigma)$, we have by Cauchy's formula

$$w(z) = \omega(\zeta) = \frac{i}{2\pi i} \oint_\gamma \frac{m(\sigma)\, \overline{m}\left(\frac{1}{\sigma}\right) d\sigma}{\sigma - \zeta} = iM_1(\zeta) \quad (7)$$

and therefore, from (4),

$$\Phi(z) = \frac{\tau\omega'(\zeta)}{m'(\zeta)} = i\tau\frac{M_1'(\zeta)}{m'(\zeta)}, \qquad z = m(\zeta) \,. \tag{8}$$

In this result we may regard ζ as a parameter. From 4.1 (2) we then have

$$\frac{\widehat{xz} - i\,\widehat{yz}}{\mu\tau} = i\left\{\frac{M_1'(\zeta)}{m'(\zeta)} - \overline{m}(\overline{\zeta})\right\}. \tag{9}$$

4.42. Cross-section a circle

Let the beam be in the form of a circular cylinder of radius a. With the origin at the centre, the circumference of a cross-section is $|z| = a$ which is mapped on the unit circumference by

$$z = a\zeta = m(\zeta) \,.$$

Therefore 4.4 (6)

$$M_1(\sigma) + M_2(\sigma) = a\sigma \cdot \frac{a}{\sigma} = a^2 \,.$$

Therefore $M_1(\zeta) = a^2$ and from 4.4 (8), (9)

$$\Phi(z) = 0$$

$$\frac{\widehat{xz} - i\,\widehat{yz}}{\mu\tau} = -\, i\overline{z}$$

$$\widehat{xz} = -\,\mu\tau y, \qquad \widehat{yz} = \mu\tau x \,,$$

From 4.22, apart from a rigid body movement of the material as a whole the displacement is given by

$$u = -\,\tau y R, \qquad v = \tau x R, \qquad w = 0 \,.$$

Thus in this case plane cross-sections remain plane and parallel to the antiplane, when the circular cylinder is twisted about its axis.

If, however, we twist about say the point $(\alpha, \beta, 0)$ the point (x, y, h) which lies in the plane $R = h$ moves, by 4.28, to

$$x - \tau h(y - \beta),\ y + \tau h(x - \alpha),\ h + \tau(\alpha y - \beta x) \,.$$

Thus the particles of a cross-section after twisting still lie in a plane but the plane is no longer parallel to its original position. This property is characteristic of a *circular* cylinder only.

4.44. Cross-section a cardioid

Let the cross-section be the cardioid $r = 2c(1 + \cos\theta)$.
Its periphery C is mapped on the unit circumference $|\zeta| = 1$ by

$$z = m(\zeta) = c(1 + \zeta)^2 \,. \tag{1}$$

Therefore in the notation of 4.4 (6)

$$M_1(\sigma) + M_2(\sigma) = c^2(1+\sigma)^2\left(1+\frac{1}{\sigma}\right)^2 = \frac{c^2(1+\sigma)^4}{\sigma^2},$$

$$M_1(\sigma) = c^2(\sigma^2 + 4\sigma + 6)$$

so that

$$\Phi(z) = i\tau\frac{c^2(2\zeta+4)}{2c(1+\zeta)} = i\tau c\frac{\zeta+2}{\zeta+1}$$

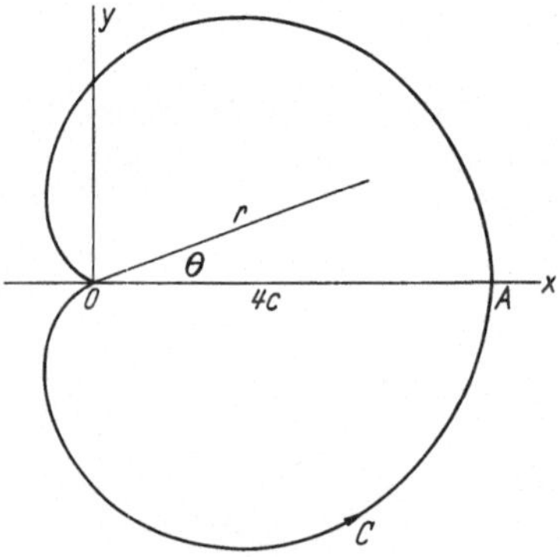

Fig. 4.44

and therefore from 4.4 (9)

$$\frac{\widehat{xz} - i\widehat{yz}}{\mu\tau} = ic\left\{\frac{\zeta+2}{\zeta+1} - (1+\bar{\zeta})^2\right\},$$

which with (1) solves the problem. In the present case we can actually eliminate ζ to give

$$\frac{\widehat{xy} - i\widehat{yz}}{\mu\tau} = ic\left\{\sqrt{\frac{c}{z}} + 1 - \frac{\bar{z}}{c}\right\}$$

and we notice that the stress becomes infinite at the cusp, $z = 0$, a circumstance characteristic of a reentrant point. In such cases, infinite stress being a physical impossibility, there is a region round the cusp in which plastic yielding occurs.

4.46. Cross-section one loop of Bernoulli's lemniscate

The equation of the curve in question may be taken in the form

$$r^2 = 2c^2\cos 2\theta \tag{1}$$

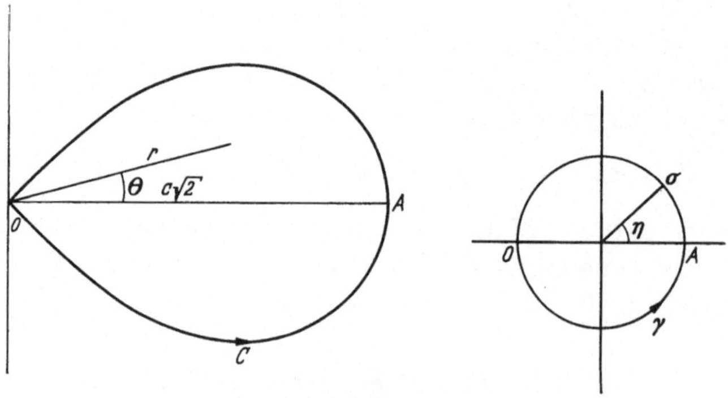

Fig. 4.46 (i)

which is mapped on the unit circumference γ, $|\zeta| = 1$, by

$$z = c(1+\sigma)^{1/2}, \quad \sigma = e^{i\eta}, \tag{2}$$

where the branch of the double valued function $(1+\sigma)^{1/2}$ is chosen to be that which reduces to $+1$ when $\sigma = 0$.
The mapping function $m(\sigma) = c(1+\sigma)^{1/2}$ is no longer rational and the method of 4.4 (6) fails.

We must therefore evaluate

$$\frac{\omega(\zeta)}{i} = \frac{1}{2\pi i} \oint\limits_{\gamma} \frac{m(\sigma)\,\overline{m}\left(\frac{1}{\sigma}\right) d\sigma}{\sigma - \zeta} = \frac{1}{2\pi i} c^2 \oint\limits_{\gamma} \frac{(1+\sigma)\, d\sigma}{\sigma^{1/2}(\sigma-\zeta)}. \tag{3}$$

To evaluate $\omega(\zeta)$ consider the contour shown in fig. 4.46 (ii) which differs from the interior of the unit circle γ by removing the interior

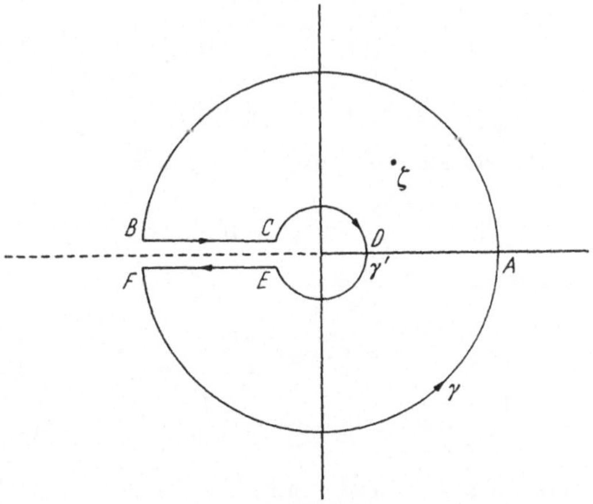

Fig. 4.46 (ii)

of the loop $BCDEF$ whose boundary consists of two lines BC, FE one on each side of the real axis and the circumference CDE or γ' of radius r. Inside this contour the integrand is holomorphic except at the pole ζ and therefore by Cauchy's formula

$$\frac{1}{2\pi i} \int\limits_{ABCDEF} \frac{(1+\sigma)\, d\sigma}{\sigma^{1/2}(\sigma-\zeta)} = \frac{1+\zeta}{\zeta^{1/2}}. \tag{4}$$

Let

$$u = \sigma^{-1/2}(1+\sigma).$$

Then (4) states that

$$\int\limits_{FAB} \frac{u}{\sigma-\zeta} d\sigma + \int\limits_{BC} \frac{u}{\sigma-\zeta} d\sigma + \int\limits_{CDE} \frac{u}{\sigma-\zeta} d\sigma + \int\limits_{EF} \frac{u}{\sigma-\zeta} d\sigma = 2\pi i \frac{1+\zeta}{\zeta^{1/2}}.$$

Now let the distance between BC and EF, and the radius of $\gamma' \to 0$. Then we get

$$\frac{1}{ic^2}\omega(\zeta) + \frac{1}{2\pi i} \int\limits_{BC} \frac{(1+\sigma)\, d\sigma}{\sigma^{1/2}(\sigma-\zeta)} + \frac{1}{2\pi i} \int\limits_{EF} \frac{(1+\sigma)\, d\sigma}{\sigma^{1/2}(\sigma-\zeta)} = \frac{1+\zeta}{\zeta^{1/2}}, \tag{5}$$

since the integral round $\gamma' \to 0$ when $r \to 0$, and BC, EF are now supposed to lie in the real axis.

Let $x (> 0)$ denote distance from the centre of γ, Then on BC, $\sigma = x e^{i\pi}$, $\sigma^{1/2} = i x^{1/2}$ while on EF, $\sigma = x e^{-i\pi}$ and $\sigma^{1/2} = -i x^{1/2}$. Then

$$\frac{1}{i c^2} \omega(\zeta) = \frac{1+\zeta}{\zeta^{1/2}} - \frac{1}{\pi} \int_0^1 \frac{(1-x)\,dx}{x^{1/2}(x+\zeta)}.$$

Put $x = t^2$ and we get

$$\frac{1}{i c^2} \omega(\zeta) = \frac{1+\zeta}{\zeta^{1/2}} - \frac{2}{\pi} \int_0^1 \left\{-1 + \frac{\zeta+1}{t^2+\zeta}\right\} dt$$

$$= \frac{1+\zeta}{\zeta^{1/2}} - \frac{2}{\pi} \left\{-1 - \frac{\zeta+1}{\zeta^{1/2}} \left[\tan^{-1} \zeta^{1/2} - \frac{\pi}{2}\right]\right\}$$

and so $\omega(\zeta) = \dfrac{2ic^2}{\pi} \dfrac{\zeta+1}{\zeta^{1/2}} \tan^{-1} \zeta^{1/2} + \text{constant}$. (6)

The constant is irrelevant and thus from 4.4 (5)

$$w(z) = \frac{2ic^2}{\pi} \frac{\zeta+1}{\zeta^{1/2}} \tan^{-1} \zeta^{1/2}. \tag{7}$$

For the stress we have

$$\frac{\widehat{xz} - i\widehat{yz}}{\mu\tau} = \frac{dw(z)}{dz} - i\bar{z} = \frac{dw(z)}{d\zeta} \cdot \frac{1}{m'(\zeta)} - i\,\overline{m}(\bar{\zeta})$$
$$= \frac{2ic^2}{\pi} \cdot \left[\frac{1}{2}\left(\frac{1}{\zeta^{1/2}} - \frac{1}{\zeta^{3/2}}\right) \tan^{-1} \zeta^{1/2} + \frac{1}{2\zeta}\right] \frac{2(1+\zeta)^{1/2}}{c} - ic(1+\bar{\zeta})^{1/2}. \tag{8}$$

Now at $z = 0$ which is a salient point $1 + \zeta = 0$ so that $\zeta = \bar{\zeta} = -1$ and the stress vanishes, as it should.

Also by Gregory's series

$$\tan^{-1} \zeta^{1/2} = \zeta^{1/2} \left(1 - \frac{\zeta}{3} + \frac{\zeta^2}{5} - \cdots\right)$$

and the expansion of (8) in ascending powers of ζ near $\zeta = 0$ is of the form $A + B\zeta + \cdots$ and contains no negative powers of ζ. Thus there is no singularity at $\zeta = 0$.

4.48. Mapping on a semicircle

From the circle $|\zeta| < 1$ in the ζ-plane, the real axis cuts off a semicircle in the upper half-plane bounded by the curved arc δ and the diameter ε. Let the mapping function

$$z = m(\zeta) \tag{1}$$

map the closed curve C in the z-plane on to $\delta + \varepsilon$ by

$$\begin{aligned} z &= m(\sigma), \quad \sigma = e^{i\theta} \text{ on } \delta \\ z &= m(\xi), \quad -1 \leq \xi \leq 1 \text{ on } \varepsilon. \end{aligned} \tag{2}$$

Let the function $f(z)$ be holomorphic within C and write

$$f(z) = f[m(\zeta)] = F(\zeta) \tag{3}$$

so that $F(\zeta)$ is holomorphic in the semicircle.

Let the real part of $F(\zeta)$ be subject to the boundary conditions

$$\mathrm{Re}[F(\zeta)] = U(\sigma) \text{ on } \delta\,, \tag{4}$$

$$\mathrm{Re}[F(\zeta)] = V(\xi) \text{ on } \varepsilon\,. \tag{5}$$

It can be shown [DEUTSCH] that the boundary value problem is solved by

$$\begin{aligned} f(z) = F(\zeta) &= \frac{1}{\pi i}\int_{\delta} U(\sigma)\left(\frac{1}{\sigma-\zeta}+\frac{1}{\sigma-\dfrac{1}{\zeta}}\right)d\sigma \\ &+ \frac{1}{\pi i}\int_{-1}^{1} V(\xi)\left(\frac{1}{\xi-\zeta}+\frac{1}{\xi-\dfrac{1}{\zeta}}\right)d\xi + \text{constant}\,. \end{aligned} \tag{6}$$

Consider for example the torsion problem for a beam whose cross-section is the semicircle of radius a in the upper half-plane, centre at the origin.

The mapping function is then

$$z = a\zeta \tag{7}$$

and from 4.2 (6) the boundary conditions are, if $w(z) = W(\zeta)$,

$$\mathrm{Re}\left[\frac{W(\zeta)}{i}\right] = m(\sigma)\,\overline{m}\left(\frac{1}{\sigma}\right) = \frac{1}{2}a^2 \text{ on } \delta$$

$$\mathrm{Re}\left[\frac{W(\zeta)}{i}\right] = m(\xi)\,\overline{m}(\xi) = \frac{1}{2}a^2\xi^2 \text{ on } \varepsilon$$

whence we obtain from (6)

$$w(z) = W(\zeta) = \frac{1}{2}i a^2\zeta^2 + \frac{a^2}{\pi}\left(\zeta+\frac{1}{\zeta}\right) - \frac{a^2}{2\pi}\left(\zeta-\frac{1}{\zeta}\right)^2 \ln\frac{1+\zeta}{1-\zeta}\,. \tag{8}$$

The torsional rigidity is easily calculated to give [STEVENSON]

$$D = \mu a^4\left[\frac{\pi}{2}-\frac{4}{\pi}\right].$$

When the mapping function for a beam is known and the cross-section has an axis of symmetry the above method will yield the torsion of a half-beam obtained by dividing the original beam along the axial plane through the axis of symmetry. The same mapping function will serve both cases. See fig. 5.3.

4.5. The $z\bar{z}$ method

From 4.3 (2) the boundary condition can be written

$$2i\psi = w(z) - \overline{w}(\bar{z}) = i\,z\bar{z} + \text{constant on } C\,. \tag{1}$$

Suppose we can express the equation of the boundary in the form

$$z\bar{z} = f(z) + \bar{f}(\bar{z}) + \text{constant}\,, \tag{2}$$

where $f(z)$ is holomorphic inside C.

Then (1) will be satisfied if we take

$$w(z) = i f(z) + \text{constant}\,, \tag{3}$$

and this solves the problem.

Further suppose that the boundary C is defined as the curve $\xi = \alpha$ in the net

$$z = g(\zeta), \qquad \zeta = \xi + i\eta\,. \tag{4}$$

If we can express the product $z\bar{z}$ on the boundary in the form

$$z\bar{z} = h(\zeta) + \bar{h}(\bar{\zeta}) + \text{constant}, \qquad \xi = \alpha\,, \tag{5}$$

then (1) will be satisfied by

$$w(z) = i h(\zeta) + \text{constant}\,, \tag{6}$$

which with (4) determines the solution in terms of the parameter ζ.

In the above it is essential that $w(z)$, and therefore $f(z)$ or $h(\zeta)$ as the case may be, should be holomorphic in the interior of the boundary C of the cross-section.

We can, if preferred, find the torsion shear function $\Omega = \Omega(z, \bar{z})$ by noting that from 4.1 (8)

$$\Omega = \psi - \frac{1}{2} z\bar{z} = -\frac{i}{2}[w(z) - \bar{w}(\bar{z})] - \frac{1}{2} z\bar{z}\,.$$

Thus (3) leads to

$$\Omega = \frac{1}{2}[f(z) + \bar{f}(\bar{z}) - z\bar{z}] + \text{constant}\,, \tag{7}$$

while (6) leads to

$$\Omega = \frac{1}{2}[h(\zeta) + \bar{h}(\bar{\zeta}) - g(\zeta)\bar{g}(\bar{\zeta})] + \text{constant}\,, \tag{8}$$

which with (4) determines the solution in terms of the parameter ζ.

We now show some examples of the application of this method.

4.52. Cross-section an ellipse

We have identically

$$4\left(\frac{x^2}{a^2} + \frac{y^2}{b^2} - 1\right) = \frac{(z+\bar{z})^2}{a^2} - \frac{(z-\bar{z})^2}{b^2} - 4 = z^2\left(\frac{1}{a^2} - \frac{1}{b^2}\right)$$
$$+ \bar{z}^2\left(\frac{1}{a^2} - \frac{1}{b^2}\right) + 2z\bar{z}\left(\frac{1}{a^2} + \frac{1}{b^2}\right) - 4\,.$$

Thus the equation of the ellipse

$$\frac{x^2}{a^2} + \frac{y^2}{b^2} - 1 = 0 \tag{1}$$

can be expressed in the form

$$z\bar{z} = \frac{1}{2} z^2 \frac{a^2 - b^2}{a^2 + b^2} + \frac{1}{2} \bar{z}^2 \frac{a^2 - b^2}{a^2 + b^2} + \text{constant}$$

and therefore

$$w(z) = \frac{1}{2} i \frac{a^2 - b^2}{a^2 + b^2} z^2 \tag{2}$$

$$\Omega = \frac{1}{4} \frac{a^2 - b^2}{a^2 + b^2} (z^2 + \bar{z}^2) - \frac{1}{2} z\bar{z} + \text{constant}$$

$$= -\frac{a^2 b^2}{a^2 + b^2} \left(\frac{x^2}{a^2} + \frac{y^2}{b^2} - 1 \right), \tag{3}$$

where we have determined the constant term to make $\Omega = 0$ on the boundary.

The lines of stress are $\Omega = c$ a constant or

$$\frac{x^2}{a^2} + \frac{y^2}{b^2} = k^2 \quad \text{a constant}.$$

Thus the lines of stress are homothetic ellipses of semiaxes ka, kb.

It follows that either (2) or (3) also solves the torsion problem for the case where the cross-section of the beam is the region comprised between (1) and an interior homothetic ellipse, an elliptic tube of this cross-section, for the boundaries are then lines of stress, fig. 4.52.

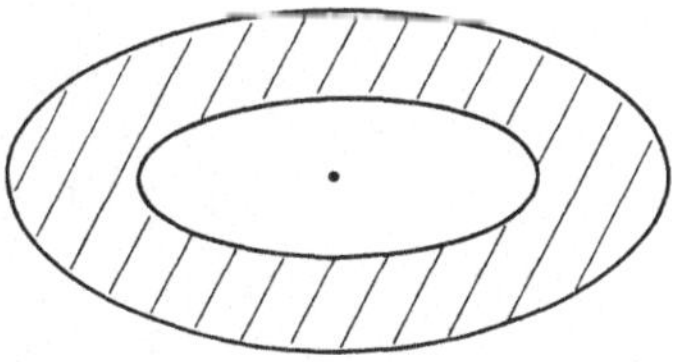

Fig. 4.52

4.53. Cross-section an equilateral triangle

The lines

$$x - a = 0, \; x - y\sqrt{3} + 2a = 0, \; x + y\sqrt{3} + 2a = 0$$

are the sides of an equilateral triangle of altitude $3a$ with the origin at the centroid and one side parallel to the y-axis. The equation of the boundary is

$$F(x, y) = (x - a)(x - y\sqrt{3} + 2a)(x + y\sqrt{3} + 2a) = 0$$

or, writing $z = x + iy$,

$$\frac{1}{2}(z^3 + \bar{z}^3) + 3az\bar{z} - 4a^3 = 0$$

and therefore

$$w(z) = -\frac{iz^3}{6a}.$$

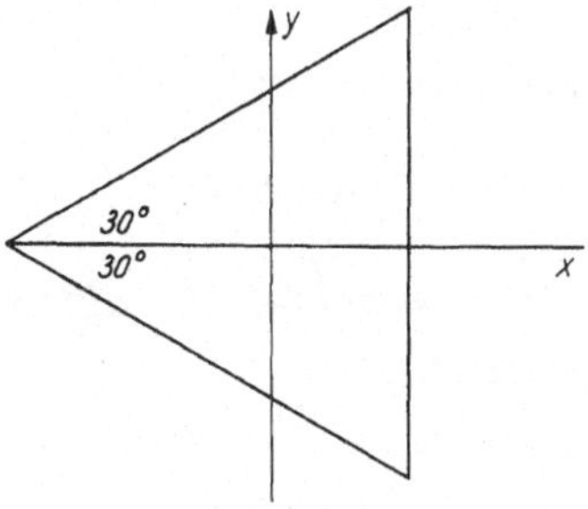

Fig. 4.53

4.54. Cross-section a lune

Let the cross-section be the lune comprised between the circles $x^2 + y^2 - b^2 = 0$, $x^2 + y^2 - 2ax = 0$. Multiplying these equations we get

$$z\bar{z} = a(z + \bar{z}) + b^2 - ab^2\left(\frac{1}{z} + \frac{1}{\bar{z}}\right)$$

so that

$$w(z) = ia\left(z - \frac{b^2}{z}\right).$$

This may be looked upon as giving the torsion of a circular shaft which has a circular groove or notch, whose cross-section is a circular arc.

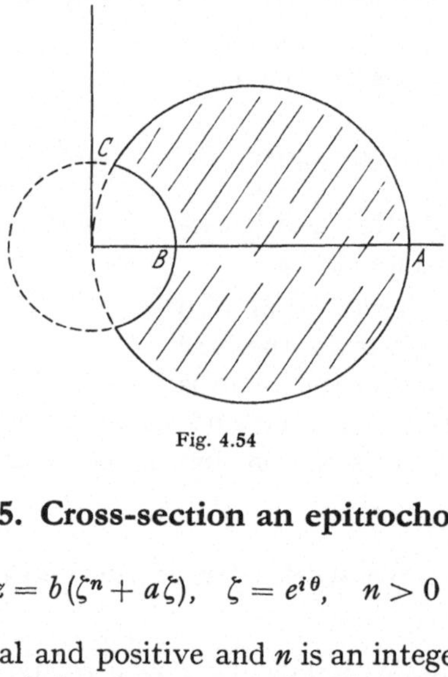

Fig. 4.54

4.55. Cross-section an epitrochoid

The curve

$$z = b(\zeta^n + a\zeta), \quad \zeta = e^{i\theta}, \quad n > 0 \tag{1}$$

where a and b are real and positive and n is an integer, is an *epitrochoid*. When $n = 1$ it is a circle, when $n = 2$ the kidney-shaped elliptic limaçon. The curve is mapped by (1) on the circumference of the unit circle. Now

$$\bar{\zeta} = e^{-i\theta} = \frac{1}{\zeta}.$$

Therefore $\bar{z} = b\left(\frac{1}{\zeta^n} + \frac{a}{\zeta}\right)$

and so on the curve

$$z\bar{z} = b^2\left(1 + a^2 + a\zeta^{n-1} + \frac{a}{\zeta^{n-1}}\right) = b^2(1 + a^2 + a\zeta^{n-1} + a\bar{\zeta}^{n-1}).$$

Thus

$$w(z) = iab^2\zeta^{n-1}, \qquad z = b(\zeta^n + a\zeta) \tag{2}$$

solve the problem of torsion.

4.56. Cross-section Booth's lemniscate

This curve is the inverse of an ellipse with respect to its centre and has for equation

$$z = \frac{k\zeta}{\zeta^2 + a^2}, \qquad \zeta = e^{i\theta} \tag{1}$$

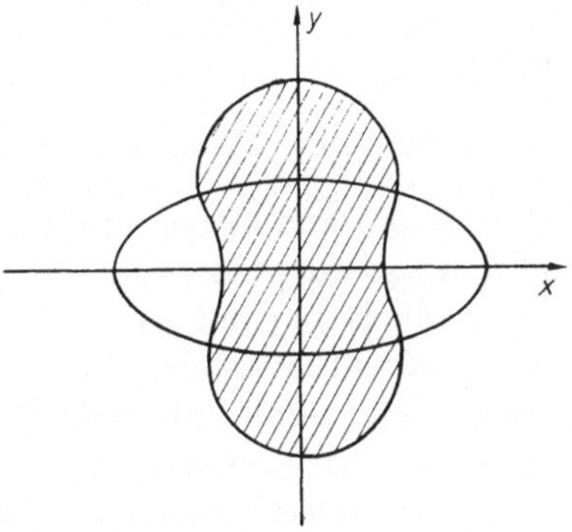

Fig. 4.56

whereby it is mapped on the unit circumference. Then $\bar{\zeta} = e^{-i\theta} = 1/\zeta$, $\zeta\bar{\zeta} = 1$ and therefore on the periphery

$$z\bar{z} = \frac{k^2\zeta\bar{\zeta}}{(\zeta^2 + a^2)(\bar{\zeta}^2 + a^2)} = \frac{k^2\zeta^2}{(\zeta^2 + a^2)(1 + a^2\zeta^2)} = \frac{-k^2a^2}{(\zeta^2 + a^2)(1 - a^4)} + \frac{k^2}{1 - a^4}\,\frac{1}{1 + a^2\zeta^2}.$$

Now

$$\frac{1}{1 + a^2\zeta^2} = \frac{\bar{\zeta}^2}{\bar{\zeta}^2 + a^2\zeta^2\bar{\zeta}^2} = \frac{\bar{\zeta}^2}{\bar{\zeta}^2 + a^2} = 1 - \frac{a^2}{\bar{\zeta}^2 + a^2}.$$

Therefore

$$z\bar{z} = \frac{-k^2a^2}{(\zeta^2 + a^2)(1 - a^4)} - \frac{k^2a^2}{(\bar{\zeta}^2 + a^2)(1 - a^4)} + k^2.$$

Therefore the problem is solved by

$$w(z) = \frac{i a^2 k^2}{(a^4 - 1)(\zeta^2 + a^2)}, \qquad z = \frac{k\zeta}{\zeta^2 + a^2}.$$

4.6. Boundary conditions

With the usual axes of reference, the basic condition for pure antiplane problems in which the lateral surface of the beam is free of applied force is

$$l\,\widehat{xz} + m\,\widehat{yz} = 0 \text{ on } C, \text{ the contour of a cross section,} \tag{1}$$

where $(l, m, 0)$ such that

$$l = \frac{dy}{ds}, \qquad m = -\frac{dx}{ds} \tag{2}$$

are the direction cosines of the outwardly drawn normal to C.

In terms of the conjugate harmonic functions ϕ, ψ of 4.1 (3) we then have from 4.1 (5)

$$l\frac{\partial\phi}{\partial x}+m\frac{\partial\phi}{\partial y}-ly+mx=0 \text{ on } C$$

that is

$$\frac{\partial\phi}{\partial n}=ly-mx \text{ on } C \tag{3}$$

or alternatively as in 4.2 (6)

$$\psi=\frac{1}{2}(x^2+y^2)+\text{constant on } C\,. \tag{4}$$

The direct procedure of 4.4 and 4.5 naturally are derived from the above boundary conditions, so adapted as to render tentative methods unnecessary. But these methods may not always be convenient either on account of mathematical difficulties in their application or because the region occupied by the material is multiply connected.

In such cases we must fall back on such procedure as the boundary conditions (1), (3) or (4) may suggest for each particular problem. This is illustrated in the two sections which follow.

4.61. Rectangular cross-section

Let the cross-section of the beam be the rectangle bounded by $x=\pm a$, $y=\pm b$, where we shall assume that $b\geqq a$.

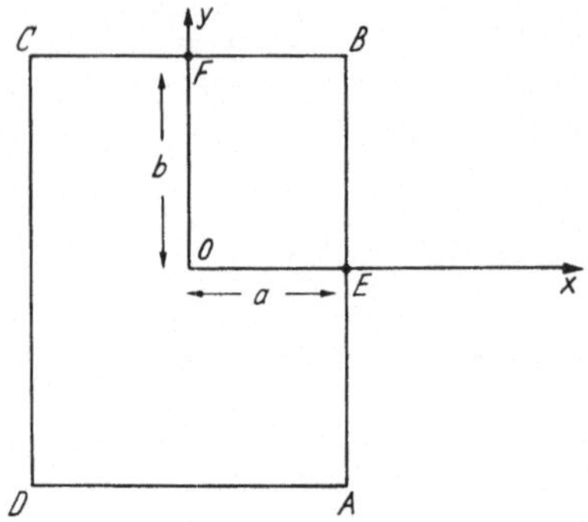

Fig. 4.61

The boundary conditions 4.6 (1) are

$$\widehat{xz}=0 \text{ on } AB,\, CD \text{ i.e. on } x=\pm a$$

$$\widehat{yz}=0 \text{ on } BC,\, DA \text{ i.e. on } y=\pm b\,.$$

Therefore from 4.1 (4)

$$\frac{dw(z)}{dz}+\frac{d\overline{w}(\overline{z})}{d\overline{z}}-2y=0$$

on $z=\pm a+iy$

$$-\frac{dw(z)}{dz}+\frac{d\overline{w}(\overline{z})}{d\overline{z}}+2ix=0$$

on $z=x\pm ib$.

Differentiate these with respect to y and x respectively. Then

$$w''(z)-\overline{w}''(\overline{z})=-2i \text{ on } z=\pm a+iy \tag{1}$$

$$w''(z)-\overline{w}''(\overline{z})=2i \text{ on } z=x\pm ib \tag{2}$$

and we have to find $w(z)$ to satisfy these boundary conditions. Let n be an integer. Then when $z=\pm a+iy$

$$i\cos\frac{(2n+1)\pi z}{2a}=i\cos\left[\pm\frac{(2n+1)\pi}{2}+\frac{(2n+1)i\pi y}{2a}\right]$$

$$=\pm\sinh(2n+1)\frac{\pi y}{2a}$$

which is *real*. This is the crucial step in the procedure. Therefore condition (1) will be satisfied by taking

$$w''(z) = -i + i \sum_{n=0}^{\infty} \lambda_n \cos(\mu_n z) \tag{3}$$

where

$$\mu_n = \frac{(2n+1)\pi}{2a} \tag{4}$$

and the λ_n are arbitrary real constants, which we shall now choose so as to satisfy (2).

Substitute (3) in (2). Then we must have, on $z = x \pm ib$,

$$\begin{aligned} 2i &= -2i + 2i \sum_{n=0}^{\infty} \lambda_n \cos\frac{\mu_n(z+\bar{z})}{2} \cos\frac{\mu_n(z-\bar{z})}{2} \\ 2 &= \sum_{n=0}^{\infty} \lambda_n \cosh(\mu_n b) \cos(\mu_n x)\,. \end{aligned} \tag{5}$$

Multiply this equation by $\cos(\mu_n x)$ and integrate from $x = -a$ to $x = a$. All the trigonometrical integrals on the right vanish except that involving $\cos^2(\mu_n x)$ and we get

$$\lambda_n = \frac{4(-1)^n}{a\,\mu_n \cosh(\mu_n b)}\,. \tag{6}$$

Therefore

$$w''(z) = -i + 4i \sum_{n=0}^{\infty} \frac{(-1)^n \cos(\mu_n z)}{a\,\mu_n \cosh(\mu_n b)}\,.$$

Integrating we get

$$w'(z) = A - iz + 4i \sum_{n=0}^{\infty} \frac{(-1)^n \sin(\mu_n z)}{a\,\mu_n^2 \cosh(\mu_n b)}\,. \tag{7}$$

Here A is an integration constant which is in fact a measure of the shearing stress at the centre of the rectangle. Assuming the stress at the centre to be zero and noting that $\widehat{xz} - i\widehat{yz} = \mu\tau(w'(z) - i\bar{z})$ we have

$$\frac{\widehat{xz} - i\widehat{yz}}{\mu\tau} = -2ix + \frac{16ia}{\pi^2} \sum_{n=0}^{\infty} \frac{(-1)^n \sin\frac{(2n+1)\pi z}{2a}}{(2n+1)^2 \cosh\frac{(2n+1)\pi b}{2a}} \tag{8}$$

and $\widehat{xz}$ and $\widehat{yz}$ are found by equating the real and imaginary parts.

$$\frac{\widehat{xz}}{\mu\tau} = -\frac{16a}{\pi^2} \sum_{n=0}^{\infty} \frac{(-1)^n \cos(\mu_n x)\sinh(\mu_n y)}{(2n+1)^2 \cosh(\mu_n b)}$$

$$\frac{\widehat{yz}}{\mu\tau} = 2x - \frac{16a}{\pi^2} \sum_{n=0}^{\infty} \frac{(-1)^n \sin(\mu_n x)\cosh(\mu_n y)}{(2n+1)^2 \cosh(\mu_n b)}\,.$$

At the midpoint E of the longer side AB we have $\widehat{xz}_E = 0$ and

$$\frac{\widehat{yz}_E}{\mu\tau} = 2a - \frac{16a}{\pi^2} \sum_{n=0}^{\infty} \frac{\operatorname{sech}(\mu_n b)}{(2n+1)^2}$$

while at the midpoint F of the shorter side BC, $\widehat{yz}_F = 0$ and

$$\frac{\widehat{xz}_F}{\mu\tau} = -\frac{16a}{\pi^2}\sum_{n=0}^{\infty}\frac{(-1)^n\tanh(\mu_n b)}{(2n+1)^2}.$$

The maximum stress occurs at the centre of the longer side. Integrating (7), with $A = 0$, and omitting an irrelevant added constant we have

$$w(z) = -\frac{1}{2}iz^2 - 4i\sum_{n=0}^{\infty}\frac{(-1)^n\cos(\mu_n z)}{a\,\mu_n^3\cosh(\mu_n b)}. \qquad (9)$$

This series converges strongly on account of the factor $1/(2n+1)^3$ and the fact that $\cosh(\mu_n b)$ increases rapidly with n. Thus we have the approximation

$$w(z) = -\frac{1}{2}iz^2 - \frac{32ia^2}{\pi^3}\cos\left(\frac{\pi z}{2a}\right)\Big/\cosh\left(\frac{\pi b}{2a}\right).$$

4.63. Confocal elliptic ring

Let the material occupy the ring bounded by the confocal ellipses

$$\xi = \alpha,\ \xi = \beta,\ \alpha > \beta \text{ in the net } z = c\cosh\zeta,\ \zeta = \xi + i\eta. \qquad (1)$$

The boundary conditions to be satisfied are

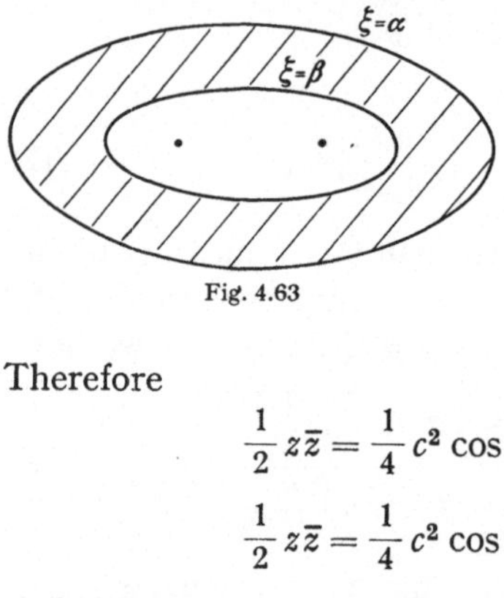

Fig. 4.63

$$\psi - \frac{1}{2}z\bar{z} = \text{constant on } \xi = \alpha$$
$$\text{and on } \xi = \beta.$$

Now

$$\frac{1}{2}z\bar{z} = \frac{1}{2}c^2\cosh\zeta\cosh\bar{\zeta}$$
$$= \frac{1}{4}c^2\left[\cosh(\zeta - \bar{\zeta}) + \cosh(\zeta + \bar{\zeta})\right].$$

Therefore

$$\frac{1}{2}z\bar{z} = \frac{1}{4}c^2\cos 2\eta + \frac{1}{4}c^2\cosh 2\alpha \text{ on } \xi = \alpha$$

$$\frac{1}{2}z\bar{z} = \frac{1}{4}c^2\cos 2\eta + \frac{1}{4}c^2\cosh 2\beta \text{ on } \xi = \beta$$

and therefore

$$\psi = \frac{1}{4}c^2\cos 2\eta + \text{constant on both } \xi = \alpha \text{ and } \xi = \beta.$$

Now

$$\operatorname{Re}\frac{\cosh(2\zeta - \alpha - \beta)}{\cosh(\alpha - \beta)} = \cos 2\eta \text{ when } \xi = \alpha \text{ and when } \xi = \beta.$$

This is the crucial step in the procedure. Therefore

$$\psi = \operatorname{Re}\frac{1}{4}c^2\frac{\cosh(2\zeta - \alpha - \beta)}{\cosh(\alpha - \beta)} + \text{constant}$$

and therefore we can take

$$w(z) = \frac{i c^2}{4} \frac{\cosh(2\zeta - \alpha - \beta)}{\cosh(\alpha - \beta)} \tag{2}$$

and (1) and (2) constitute the solution.

For the torsional rigidity we use 4.3 (10). We find that on the boundary $\xi = \alpha$

$$\overline{w}(\overline{z})\, dw(z) = \frac{1}{16} \frac{c^4}{\cosh^2(\alpha - \beta)} \{\sinh(2\alpha - 2\beta) + i \sin 4\eta\}\, i\, d\eta$$

and on $\xi = \beta$

$$\overline{w}(\overline{z})\, dw(z) = \frac{1}{16} \frac{c^4}{\cosh^2(\alpha - \beta)} \{-\sinh(2\alpha - 2\beta) + i \sin 4\eta\}\, i\, d\eta\, .$$

Since

$$2T = -\frac{1}{2} i \oint_{\xi = \alpha} \overline{w}(\overline{z})\, dw(z) - \frac{1}{2} i \oint_{\xi = \beta} \overline{w}(\overline{z})\, dw(z)$$

we find after reduction that

$$D = \frac{1}{4} \mu \pi c^4 \left\{ \frac{1}{4} (\sinh 2\alpha - \sinh 2\beta) - \tanh(\alpha - \beta) \right\}. \tag{3}$$

4.7. A uniqueness theorem

Consider the region τ within a closed surface Σ and let $\boldsymbol{n}$ be the unit outward normal at the element $d\Sigma$ of Σ. Then if S is a symmetric 2-tensor, and $\boldsymbol{u}$ is an arbitrary (differentiable) vector field, we have

$$\nabla \cdot (\mathsf{S} \cdot \boldsymbol{u}) = \boldsymbol{u} \cdot (\nabla \cdot \mathsf{S}) + (\mathsf{S} \cdot \nabla) \cdot \boldsymbol{u}\, .$$

Integrate through τ and apply Gauss's Theorem 5.46.

Then

$$\int_{\Sigma} \boldsymbol{n} \cdot (\mathsf{S} \cdot \boldsymbol{u})\, d\Sigma - \int_{\tau} \boldsymbol{u} \cdot (\nabla \cdot \mathsf{S})\, d\tau = \int_{\tau} (\mathsf{S} \cdot \nabla) \cdot \boldsymbol{u}\, d\tau\, . \tag{1}$$

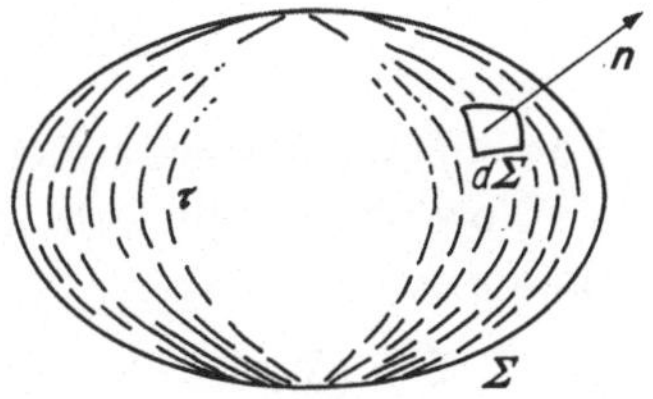

Fig. 4.7

This is a purely mathematical identity, so far without physical content.

Now suppose that S is the stress tensor in a body (not necessarily elastic) which occupies the region τ and is in equilibrium under body-force $\boldsymbol{g}$ per unit mass, surface force $\boldsymbol{F}_n$ per unit area, and displacement $\boldsymbol{q}$. The general conditions of equilibrium are,

$$\nabla \cdot \mathsf{S} + \varrho \boldsymbol{g} = 0 \text{ throughout the material} \tag{2}$$

$$\boldsymbol{n} \cdot \mathsf{S} = \boldsymbol{F}_n \text{ on } \Sigma\, . \tag{3}$$

We shall call (3) the *statical boundary condition.*

Then (1) becomes

$$\int_{\Sigma} \boldsymbol{F}_n \cdot \boldsymbol{u}\, d\Sigma + \int_{\tau} \varrho \boldsymbol{g} \cdot \boldsymbol{u}\, d\tau = \int_{\tau} \mathsf{S} \cdot\cdot (\nabla; \boldsymbol{u})\, d\tau . \tag{4}$$

Observe that (4) holds for an *arbitrary* vector field $\boldsymbol{u}$.

We now prove the following theorem.

Uniqueness theorem. The stress tensor which yields the solution of the problem of equilibrium of a body which obeys Hooke's law, when at every point of the boundary either the surface stress $\boldsymbol{F}_n$ or the displacement $\boldsymbol{q}$ is prescribed, is unique. The displacement $\boldsymbol{q}$ is determinate save for a rigid body movement.

Proof. If possible let there be two distinct solutions with stress tensor, surface force, and displacement $\mathsf{S}^{(1)}$, $\boldsymbol{F}_n^{(1)}$, $\boldsymbol{q}_n^{(1)}$ and $\mathsf{S}^{(2)}$, $\boldsymbol{F}_n^{(2)}$, $\boldsymbol{q}^{(2)}$ respectively.

These hypothetical systems must be subject to the conditions

$$\boldsymbol{F}_n^{(1)} - \boldsymbol{F}_n^{(2)} = 0 \text{ at a point where the surface force is given} \tag{5}$$

$$\boldsymbol{q}^{(1)} - \boldsymbol{q}^{(2)} = 0 \text{ at a point where the displacement is given .} \tag{6}$$

Then from (4)

$$\int_{\Sigma} \boldsymbol{F}_n^{(1)} \cdot \boldsymbol{u}\, d\Sigma + \int_{\tau} \varrho \boldsymbol{g} \cdot \boldsymbol{u}\, d\tau = \int_{\tau} \mathsf{S}^{(1)} \cdot\cdot (\nabla; \boldsymbol{u})\, d\tau$$

$$\int_{\Sigma} \boldsymbol{F}_n^{(2)} \cdot \boldsymbol{u}\, d\Sigma + \int_{\tau} \varrho \boldsymbol{g} \cdot \boldsymbol{u}\, d\tau = \int_{\tau} \mathsf{S}^{(2)} \cdot\cdot (\nabla; \boldsymbol{u})\, d\tau .$$

Subtract and then write $\boldsymbol{u} = \boldsymbol{q}^{(1)} - \boldsymbol{q}^{(2)}$. Then

$$\begin{aligned} &\int_{\tau} (\mathsf{S}^{(1)} - \mathsf{S}^{(2)}) \cdot\cdot [\nabla; (\boldsymbol{q}^{(1)} - \boldsymbol{q}^{(2)})]\, d\tau \\ &= \int_{\Sigma} (\boldsymbol{F}_n^{(1)} - \boldsymbol{F}_n^{(2)}) \cdot (\boldsymbol{q}^{(1)} - \boldsymbol{q}^{(2)})\, d\Sigma = 0 \end{aligned} \tag{7}$$

from (5) and (6) since either $\boldsymbol{F}_n$ or $\boldsymbol{q}$ is given at every point of the boundary.

Now observe that the integrand in the left hand member of (7) is of the form

$$\mathsf{S} \cdot\cdot (\nabla; \boldsymbol{q}) = \mathsf{S} \cdot\cdot \frac{1}{2} (\nabla; \boldsymbol{q} + \boldsymbol{q}; \nabla) + \mathsf{S} \cdot\cdot \frac{1}{2} (\nabla; \boldsymbol{q} - \boldsymbol{q}; \nabla)$$

and the last product of a symmetric tensor S and an antisymmetric tensor $\frac{1}{2} (\nabla; \boldsymbol{q} - \boldsymbol{q}; \nabla)$ vanishes, 1.14. Therefore

$$\mathsf{S} \cdot\cdot (\nabla; \boldsymbol{q}) = \mathsf{S} \cdot\cdot \mathsf{D} \tag{8}$$

which is a strain energy function and therefore never negative.

Therefore the integral on the left of (7) can vanish if and only if

$$\mathsf{S}^{(1)} = \mathsf{S}^{(2)} ,$$

which proves the uniqueness of the stress tensor.

If $\mathsf{D}^{(1)}$, $\mathsf{D}^{(2)}$ are the deformation tensors in the two hypothetical solutions, by Hooke's law

$$\mathsf{D}^{(1)} = \mathsf{K}_{(4)} \cdot\cdot\, \mathsf{S}^{(1)}, \; \mathsf{D}^{(2)} = \mathsf{K}_{(4)} \cdot\cdot\, \mathsf{S}^{(2)}$$

and since $\mathsf{S}^{(1)} = \mathsf{S}^{(2)}$, we have

$$\mathsf{D}^{(1)} - \mathsf{D}^{(2)} = 0$$

or

$$\nabla;\, \boldsymbol{q}_d + \boldsymbol{q}_d;\, \nabla = 0, \qquad \boldsymbol{q}_d = \boldsymbol{q}^{(1)} - \boldsymbol{q}^{(2)} .$$

Now it is easy to prove that the only solution of this equation is of the form

$$\boldsymbol{q}_d = \boldsymbol{a} + \boldsymbol{\omega}_\wedge \boldsymbol{r} ,$$

where $\boldsymbol{a}$ is an arbitrary constant translation, $\boldsymbol{\omega}$ is an arbitrary constant small rotation and $\boldsymbol{r}$ is the position vector. Thus $\boldsymbol{q}^{(1)}$ and $\boldsymbol{q}^{(2)}$ differ by an arbitrary rigid body movement. Q.E.D.

4.71. The principle of virtual displacements

In 4.7 we proved the identity

$$\int_\Sigma \boldsymbol{F}_n \cdot \boldsymbol{u}\, d\Sigma + \int_\tau \varrho \boldsymbol{g} \cdot \boldsymbol{u}\, d\tau = \int_\tau \mathsf{S} \cdot\cdot\, (\nabla;\, \boldsymbol{u})\, d\tau \tag{1}$$

where $\boldsymbol{u}$ is an *arbitrary* vector field, Σ is the surface of the body τ, S is the stress tensor and $\boldsymbol{F}_n$, $\boldsymbol{g}$ are the surface and body forces. Let us give the body a virtual displacement, for which the body remains continuous and for which prescribed displacements are not varied. For example, at a fixed point, $\delta \boldsymbol{q} = 0$.

Let us write

$$\mathsf{D} = \frac{1}{2} (\nabla;\, \boldsymbol{q} + \boldsymbol{q};\, \nabla) \tag{2}$$

so that D coincides with the strain tensor when the deformation is infinitesimal. Then we put

$$\delta U = \mathsf{S} \cdot\cdot\, (\nabla;\, \delta \boldsymbol{q}) = \mathsf{S} \cdot\cdot\, \frac{1}{2} (\nabla; \delta \boldsymbol{q} + \delta \boldsymbol{q};\, \nabla) = \mathsf{S} \cdot\cdot\, \delta \mathsf{D} \tag{3}$$

as in 4.7 (8). Here δU is the virtual work of the stresses in the virtual displacement $\delta \boldsymbol{q}$(1.54). Thus if we define the *variation of the potential energy* by

$$\delta V = \int_\tau \delta U\, d\tau \tag{4}$$

and put $\boldsymbol{u} = \delta \boldsymbol{q}$ in (1), which is permissible since $\boldsymbol{u}$ is arbitrary, we have

$$\int_\Sigma \boldsymbol{F}_n \cdot \delta \boldsymbol{q}\, d\Sigma + \int_\tau \varrho \boldsymbol{g} \cdot \delta \boldsymbol{q}\, d\tau = \delta V . \tag{5}$$

This equation is known as *Lagrange's variational equation.* It embodies the *principle of virtual displacements* or the *principle of virtual work,* namely that the work done by the forces in a virtual displacement is equal to the variation of the potential energy.

When there exists a function U which corresponds with the variation defined by (3), we have the *potential energy* V defined by

$$V = \int_\tau U \, d\tau . \tag{6}$$

Conversely the principle of virtual work implies the equilibrium equations 4.7 (2) and the statical boundary condition 4.7 (3). For assuming (5) to hold, put $\boldsymbol{u} = \delta \boldsymbol{q}$ in 4.7 (1) and subtract the result from (5). We then have

$$\int_\Sigma (\boldsymbol{F}_n - \boldsymbol{n} \cdot \mathsf{S}) \cdot \delta \boldsymbol{q} \, d\Sigma + \int_\tau (\varrho \boldsymbol{g} + \nabla \cdot \mathsf{S}) \cdot \delta \boldsymbol{q} \, d\tau = 0 . \tag{7}$$

Since the field $\delta \boldsymbol{q}$ is completely arbitrary, put $\delta \boldsymbol{q} = 0$ in τ but $\delta \boldsymbol{q} \neq 0$ on Σ. Then (7) becomes

$$\int_\Sigma (\boldsymbol{F}_n - \boldsymbol{n} \cdot \mathsf{S}) \cdot \delta \boldsymbol{q} \, d\Sigma = 0$$

and since the $\delta \boldsymbol{q}$ are arbitrary the integrand is zero which yields equation 4.7 (3). Similarly putting $\delta \boldsymbol{q} = 0$ on Σ but not in τ we get 4.7 (2). Thus the equations

$$\nabla \cdot \mathsf{S} + \boldsymbol{g} \varrho = 0, \qquad \boldsymbol{F}_n = \boldsymbol{n} \cdot \mathsf{S} \tag{8}$$

follow from the principle of virtual displacements.

It is particularly important to notice that the boundary condition $\boldsymbol{F}_n = \boldsymbol{n} \cdot \mathsf{S}$, which we have called the statical boundary condition, is a *consequence* of the principle of virtual displacements. Therefore when we apply the principle of virtual displacements in the form (5), *it is not necessary to satisfy the statical boundary condition beforehand,* for it is satisfied automatically. The same remark applies to the equations of equilibrium $\nabla \cdot \mathsf{S} + \boldsymbol{g} \varrho = 0$.

In (1) and throughout if we write $\boldsymbol{g} - \boldsymbol{a}$ instead of $\boldsymbol{g}$, where $\boldsymbol{a}$ is the acceleration, nothing in the argument is altered and (5) then applies to systems not in equilibrium.

The principle of virtual work is a general principle of mechanics, here stated in terms of continuous material.

The principle can also be stated as follows:

The displacements $\boldsymbol{q}$ in the actual state of equilibrium differ from all other possible displacements, which do not violate the constraints, in that they make stationary the expression

$$V - \int_\tau \varrho \boldsymbol{g} \cdot \boldsymbol{q} \, d\tau - \int_\Sigma \boldsymbol{F}_n \cdot \boldsymbol{q} \, d\Sigma . \tag{9}$$

Stable situations arise only when the stationary value is a minimum.

Thus Lagrange's variational equation (3) can be written

$$\delta \left\{ V - \int_\tau \varrho \boldsymbol{g} \cdot \boldsymbol{q} \, d\tau - \int_\Sigma \boldsymbol{F}_n \cdot \boldsymbol{q} \, d\Sigma \right\} = 0 . \tag{10}$$

Here δ is an operator which varies $\boldsymbol{q}$ alone.

4.72. Application of the principle of virtual displacements to torsion

From the equations for the displacement $\boldsymbol{q}$, 4.22 (3),

$$\boldsymbol{q} = -\boldsymbol{i} \tau y R + \boldsymbol{j} \tau x R + \boldsymbol{k} \tau \phi$$

we see that, in a virtual displacement (in which τ remains constant),

$$\delta \boldsymbol{q} = \boldsymbol{k} \tau \delta \phi$$

or

$$\delta u = 0, \qquad \delta v = 0, \qquad \delta w = \tau \delta \phi . \tag{1}$$

The whole variation then centres on the function ϕ and the variation $\delta \phi$ is completely arbitrary at each point at which $\boldsymbol{q}$ is not prescribed.

Since there are no forces on the lateral surface and no work can be done at the fixed end of the beam, the only forces to take into account are those due to the stress $(\widehat{xz}, \widehat{yz}, 0)$ at the free end; and their virtual work is $\int (\widehat{xz} \, \delta u + \widehat{yz} \, \delta v) \, d\Sigma = 0$ from (1). Therefore, since $\boldsymbol{g} = 0$ by hypothesis, Lagrange's variational equation 4.71 (5) reduces to

$$\delta V = 0 . \tag{2}$$

But from 4.39, $V = \frac{1}{2} D \tau^2$, so that (2) is equivalent to

$$\delta D = 0 , \tag{3}$$

where D is the torsional rigidity.

Therefore from 4.34 (4)

$$\delta \left(\frac{D}{\mu} \right) = \delta \int_S \left[\left(\frac{\partial \phi}{\partial x} - y \right)^2 + \left(\frac{\partial \phi}{\partial y} + x \right)^2 \right] dS = 0 . \tag{4}$$

Thus, since $\delta \dfrac{\partial \phi}{\partial x} = \dfrac{\partial (\delta \phi)}{\partial x}$, $\delta \dfrac{\partial \phi}{\partial y} = \dfrac{\partial (\delta \phi)}{\partial y}$,

$$\int_S \left[\left(\frac{\partial \phi}{\partial x} - y \right) \frac{\partial (\delta \phi)}{\partial x} + \left(\frac{\partial \phi}{\partial y} + x \right) \frac{\partial (\delta \phi)}{\partial y} \right] dS = 0$$

and integrating by parts we get

$$\oint_C \left(\frac{\partial \phi}{\partial n} - l y + m x\right) \delta \phi \, d s - \int_S \delta \phi \, \nabla^2 \phi \, d S = 0 \tag{5}$$

where $l = \partial x/\partial s$, $m = -\partial y/\partial s$ are direction-cosines of the outward normal to C, and $\partial \phi/\partial n$ is the derivative along the normal.

Putting (*a*) $\delta \phi = 0$ on the contour but not inside it (*b*) $\delta \phi = 0$ inside the contour but not on it, we get

$$\nabla^2 \phi = 0 \text{ in the material}\,, \tag{6}$$

$$\frac{\partial \phi}{\partial n} - l x + m y = 0 \text{ on the boundary}\,. \tag{7}$$

Equation (7) is the statical boundary condition. In applying the principle of virtual displacements, as we noticed in 4.71, it is not necessary to satisfy the statical boundary condition beforehand. It is automatically satisfied the more exactly our choice of ϕ approaches the exact value. The same remark applies to equation (6). It is not necessary to choose a harmonic function for ϕ.

4.73. Elliptic cross-section

We seek an approximate solution using the principle of virtual displacements. Let the cross-section be the ellipse

$$\frac{x^2}{a^2} + \frac{y^2}{b^2} = 1\,. \tag{1}$$

To apply the principle of virtual displacements in the form 4.72 (4), assume the approximation ϕ_a given by

$$\phi_a = A\, x y\,, \tag{2}$$

where A is a constant. Then the corresponding approximation to the torsional rigidity is

$$\frac{D_a}{\mu} = \int \{y^2 (A-1)^2 + x^2 (A+1)^2\}\, d S$$

$$= \frac{b^2}{4} \pi a b (A-1)^2 + \frac{a^2}{4} \pi a b (A+1)^2,$$

and $\delta D_a = 0$ if $\partial D_a/\partial A = 0$ i.e. if $A = -(a^2 - b^2)/(a^2 + b^2)$ so that

$$\phi_a = -\frac{a^2 - b^2}{a^2 + b^2}\, x y\,.$$

Comparing this with 4.52 we see that ϕ_a concides with the exact solution.

Had we taken $\phi_a = B(x^2 - y^2)$ we should have obtained, by the same steps, $B = 0$. This indicates that an approximation of this form is unsuitable.

4.74. Rectangular cross-section

We seek an approximate solution using the principle of virtual displacements. For the rectangle of fig. 4.61 take the approximation

$$\phi_a = A\,x\,y + B\,x\,y^3 + C\,x^3 y \tag{1}$$

(which is not a solution of Laplace's equation). Then

$$\begin{aligned}\frac{D_a}{\mu} &= \int_S \{((A-1)\,y + B\,y^3 + 3\,C\,x^2 y)^2 + ((A+1)\,x + 3\,B\,x\,y^2 + C\,x^3)^2\}\,dS\\ &= \frac{4}{3}\,a\,b^3(A-1)^2 + 4\,a\,b^5\left(\frac{b^2}{7} + \frac{3\,a^2}{5}\right)B^2 + 4\,a^5 b\left(\frac{a^2}{7} + \frac{3\,b^2}{5}\right)C^2\\ &+ \frac{4}{3}\,a^3 b\,(A+1)^2 + \frac{8}{5}\,a\,b^5(A-1)\,B + \frac{8}{3}\,a^3 b^3(A+1)\,B\\ &+ \frac{8}{5}\,a^5 b\,(A+1)\,C + \frac{8}{3}\,a^3 b^3(A-1)\,C + \frac{8}{5}\,a^3 b^3(a^2+b^2)\,B\,C\,.\end{aligned} \tag{2}$$

To minimize this we have

$$0 = \delta D_a = \frac{\partial D_a}{\partial A}\,\delta A + \frac{\partial D_a}{\partial B}\,\delta B + \frac{\partial D_a}{\partial C}\,\delta C$$

and so

$$\frac{\partial D_a}{\partial A} = 0,\ \frac{\partial D_a}{\partial B} = 0,\ \frac{\partial D_a}{\partial C} = 0.$$

This gives three *linear* equations for A, B, C whose values we obtain, after reduction, as

$$\left.\begin{aligned} E\,A &= -\,7\,(a^6 - b^6) - 135\,a^2 b^2 (a^2 - b^2)\\ 3E\,B &= -\,7\,a^2(3\,a^2 + 35\,b^2)\\ 3E\,C &= 7\,b^2(35\,a^2 + 3\,b^2)\\ E &= 7\,(a^6 + b^6) + 107\,a^2 b^2 (a^2 + b^2)\end{aligned}\right\} \tag{3}$$

We find D_a from (2) in the form

$$\frac{D_a}{16\,a^3 b\,\mu} = \frac{t^2}{45\,(1+t^2)} \cdot \frac{105\,(1+t^4) + 1\,234\,t^2}{7\,(1+t^4) + 100\,t^2} = k_1(t)\,, \tag{4}$$

say, in terms of the ratio $b/a = t$.

Again from (1) we have for the approximate value $\widehat{yz}_a$ of the shear $\widehat{yz}$

$$\frac{\widehat{yz}_a}{\mu\tau} = A\,x + 3\,B\,x\,y^2 + C\,x^3 + x$$

and this assumes its maximum value at $(a, 0)$ so that

$$\left(\frac{\widehat{yz}_a}{2\,a\,\mu\,\tau}\right)_{\max} = \frac{1}{2}\,(1 + A + C\,a^2) = k_2(t)\,, \tag{5}$$

say.

The following tables, due to Leibenzon, show the comparison between some approximate values of $k_1(t)$, $k_2(t)$ from the above formulae and the corresponding values given by the accurate solution (4.61).

	t	1	2	4	∞
approximate	$k_1(t)$	0.1409	0.2317	0.2875	$\frac{1}{3}$
exact	$k_1(t)$	0.1406	0.229	0.281	$\frac{1}{3}$

	t	1	2	4	∞
approximate	$k_2(t)$	0.6944	0.9828	0.9990	1
exact	$k_2(t)$	0.675	0.930	0.997	1.000

4.75. Parabolic cross-section

We seek an approximate solution using the principle of virtual displacements.

Let the cross-section be the segment of the parabola

$$\frac{2y}{c} = \sqrt{\frac{x}{h}}\,. \tag{1}$$

cut off by the line $x = h$. Taking the approximation

$$\phi_a = A\,xy + By$$

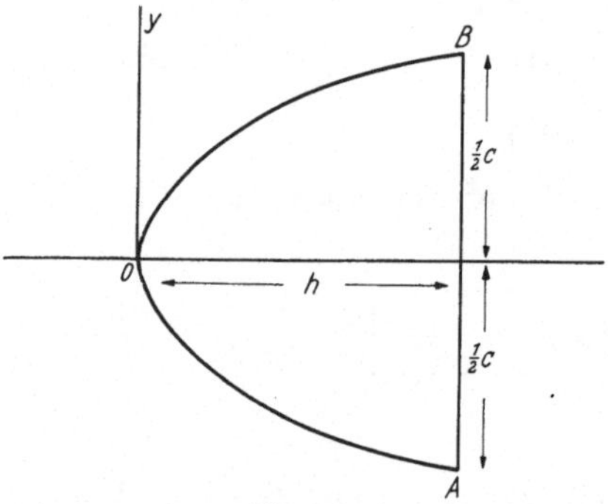

Fig. 4.75

we have from 4.34 (4)

$$\begin{aligned}\frac{D_a}{\mu} &= \int_S \{(A-1)^2 y^2 \\ &\quad + [(A+1)\,x + B]^2\}\,dS \\ &= (A-1)^2\,\frac{c^3 h}{30} \\ &\quad + (A+1)^2\,\frac{2ch^3}{7} \\ &\quad + (A+1)\,B\,\frac{4ch^2}{5} \\ &\quad + B^2\,\frac{2ch}{3}\,.\end{aligned} \tag{2}$$

To minimize this we have

$$\frac{\partial D_a}{\partial A} = 0, \quad \frac{\partial D_a}{\partial B} = 0$$

which give two linear equations whose solution is

$$A = \frac{35c^2 - 48h^2}{35c^2 + 48h^2}\,, \quad B = \frac{-42c^2 h}{35c^2 + 48h^2}\,. \tag{3}$$

Substitute these in (2) putting $c = h$. Then the estimate of D_a is $0.0770\,\mu h^4$ and from 4.37 this is an overestimate so that

$$D < 0.0770\,\mu h^4 . \tag{4}$$

4.77. The membrane analogy

Consider a rigid flat plate in which there is a hole bounded by the simple closed curve C. Let a uniform elastic membrane fixed to the plate all along C cover the hole and be stretched to a uniform tension T, so that the membrane is in the form of a plane curve bounded by C. Take this plane as the (x, y) plane.

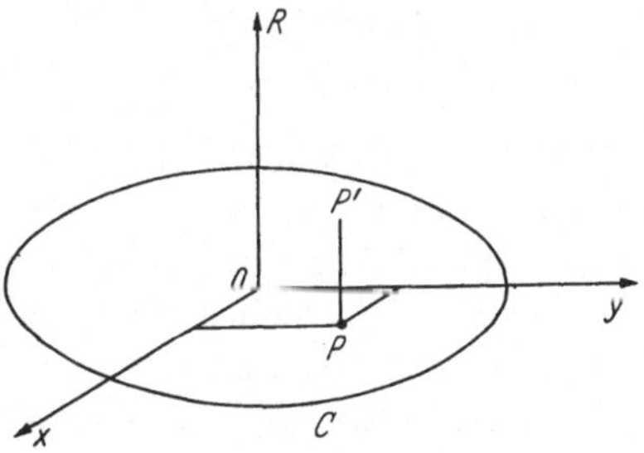

Fig. 4.77

Let us now subject the membrane to a uniform pressure q per unit area on the side for which R is negative. This will displace a point $P(x, y, 0)$ of the membrane to the position $P'(x + u, y + v, w)$, so that the membrane will be deformed from a plane to a curved surface but with the same edge curve C. This will evoke stresses other than T in the membrane but if $w = w(x, y)$ is taken to be small these stresses will be small in comparison with T. Correspondingly u, v will be negligible compared with w and we shall have

$$e_{yz} = \frac{1}{2}\frac{\partial w}{\partial y}, \qquad e_{zx} = \frac{1}{2}\frac{\partial w}{\partial x} .$$

Therefore on the basis of 1.82 (11) the potential energy of deformation is

$$V = \frac{1}{2} T \int\limits_S \left[\left(\frac{\partial w}{\partial x}\right)^2 + \left(\frac{\partial w}{\partial y}\right)^2\right] dS \tag{1}$$

over the plane area S enclosed by the contour C.

Therefore the equation of virtual work will be

$$-\,\delta V + \int q\,\delta w\, dS = 0 \tag{2}$$

where δw is an arbitrary variation consistent with $w = 0$ on C i.e.

$$\delta w = 0 \text{ on } C . \tag{3}$$

Therefore from (1) and (2)

$$\int\limits_S q\,\delta w\, dS - T \int\limits_S \left(\frac{\partial w}{\partial x}\frac{\partial(\delta w)}{\partial x} + \frac{\partial w}{\partial y}\frac{\partial(\delta w)}{\partial y}\right) dS = 0 .$$

Integrating by parts and observing the boundary condition (3) we get

$$\int_S (q + T\,\nabla^2 w)\,\delta w\,dS = 0$$

and since δw is arbitrary

$$\nabla^2 w + \frac{q}{T} = 0. \tag{4}$$

In this equation write

$$w = \frac{q}{2T}\,\Omega. \tag{5}$$

Then it becomes

$$\nabla^2 \Omega = -2$$

which is identical with 4.2 (4). Thus measurement of w on the membrane will yield numbers proportional to the torsion stress function Ω of the torsion problem for the cross-section whose contour is C. Also the level lines of the deformed membrane will coincide with the lines of shearing stress and the volume enclosed by the membrane and the plane of C will measure the torsional rigidity

$$D = 2\mu \int_S \Omega\,dS.$$

This is the *membrane analogy*. See e.g. [SOKOLNIKOFF].

4.8. The principle of virtual stresses

We start from the stress equilibrium equation and boundary condition 4.7 (2), (3) namely

$$\nabla \cdot \mathsf{S} + \boldsymbol{g}\varrho = 0 \text{ throughout the material} \tag{1}$$

$$\boldsymbol{F}_n = \boldsymbol{n} \cdot \mathsf{S} \text{ on the boundary.} \tag{2}$$

We make a variation in the stress tensor S so that it becomes

$$\mathsf{S} + \delta\mathsf{S} \tag{3}$$

where $\delta\mathsf{S}$ is *statically possible* that is to say does not violate the constraints. Therefore from (1)

$$\nabla \cdot (\mathsf{S} + \delta\mathsf{S}) + \boldsymbol{g}\varrho = 0$$

and therefore, using (1),

$$\nabla \cdot (\delta\mathsf{S}) = 0. \tag{4}$$

In order that (3) may be statically possible it is necessary also to produce a change in the existing system of surface forces $\boldsymbol{F}_n$, which satisfy (2), to

$$\boldsymbol{F}_n + \delta\boldsymbol{F}_n. \tag{5}$$

Therefore, from (2), we must have $\boldsymbol{F}_n + \delta \boldsymbol{F}_n = \boldsymbol{n} \cdot (\mathsf{S} + \delta \mathsf{S})$ or

$$\delta \boldsymbol{F}_n = \boldsymbol{n} \cdot \delta \mathsf{S} . \tag{6}$$

Now with D given by 4.71 (2), define

$$\delta U^* = \delta \mathsf{S} \cdot\cdot \nabla ; \boldsymbol{q} = \delta \mathsf{S} \cdot\cdot \mathsf{D} , \tag{7}$$

the work of the virtual stress $\delta \mathsf{S}$, and

$$\delta V^* = \int_\tau \delta U^* d\tau , \tag{8}$$

the *variation of the complementary potential energy*. (See (10) below). In the identity 4.7 (1) write $\delta \mathsf{S}$ for S, $\boldsymbol{q}$ for $\boldsymbol{u}$ and use (4) and (6). Then we have

$$\delta V^* = \int_\Sigma \delta \boldsymbol{F}_n \cdot \boldsymbol{q} \, d\Sigma . \tag{9}$$

This is *Castigliano's variational equation*. Here $\boldsymbol{q} \cdot \delta \boldsymbol{F}_n$ is the work imagined done by the virtual stress $\delta \boldsymbol{F}_n$ when its point of application moves through the actual displacement $\boldsymbol{q}$.

When there exists a function U^* corresponding with the variation defined by (7), we call

$$V^* = \int_\tau U^* d\tau \tag{10}$$

the *complementary potential energy*.

In this case Castigliano's variational equation can be expressed in the form that

$$\delta \left[V^* - \int_\Sigma \boldsymbol{F}_n \cdot \boldsymbol{q} \, d\Sigma \right] = 0 \tag{11}$$

i.e. that the *complementary energy*

$$V^* - \int_\Sigma \boldsymbol{F}_n \cdot \boldsymbol{q} \, d\Sigma \tag{12}$$

is stationary and in fact for stability is a minimum.

From the definition of δU^* in (7) and δU in 4.71 (3) we have

$$\delta U + \delta U^* = \delta (\mathsf{S} \cdot\cdot \mathsf{D})$$

which explains the use of the adjective "complementary".

This distinction disappears in the case of *infinitesimal elastic deformation*, for from 1.82 (4) we then have

$$\delta U + \delta U^* = 2 \delta U \tag{13}$$

so that $\delta U = \delta U^*$ and therefore $V = V^*$.

Thus in the elastic case, (12) states the *principle of minimum complementary energy* that the complementary energy

$$V - \int_{\Sigma} \boldsymbol{F}_n \cdot \boldsymbol{q} \, d\Sigma \tag{14}$$

is a minimum.

The principle can also be stated as follows.

For a body which obeys Hooke's law the actual stresses which keep the body in equilibrium are distinguished from all other possible stresses, consistent with the constraints, in that they render the complementary energy a minimum.

If the surface forces are all prescribed we have $\delta \boldsymbol{F}_n = 0$ and therefore (14) states *Castigliano's theorem of minimum strain energy* namely that when a body which obeys Hooke's law is kept in equilibrium by prescribed surface forces, the potential energy that is the total strain energy is a minimum.

4.81. Application of the principle of virtual stresses to torsion

We have to minimize the complementary energy

$$\int_{\tau} U \, d\tau - \int_{\Sigma} \boldsymbol{F}_n \cdot \boldsymbol{q} \, d\Sigma = -\frac{1}{2} \mu \tau^2 K L \,, \tag{1}$$

say, where L is the length of the beam. Here the volume integral is taken throughout the volume occupied by the material and the surface integral is taken over the bounding surface.

Since the only non-vanishing stress components are $\widehat{x\bar{z}}$ and $\widehat{y\bar{z}}$, in the case of isotropic material we have from 1.82 (12) and 4.1 (10)

$$U = \frac{1}{2} \mu \tau^2 \left[\left(\frac{\partial \Omega}{\partial x} \right)^2 + \left(\frac{\partial \Omega}{\partial y} \right)^2 \right] = 2 \mu \tau^2 \frac{\partial \Omega}{\partial z} \frac{\partial \Omega}{\partial \bar{z}} \tag{2}$$

and therefore

$$\int_{\tau} U \, d\tau = 2 \mu \tau^2 L \int_{S} \frac{\partial \Omega}{\partial z} \frac{\partial \Omega}{\partial \bar{z}} \, dS \tag{3}$$

over the area of the cross-section. Now consider $\int_{\Sigma} \boldsymbol{F}_n \cdot \boldsymbol{q} \, d\Sigma$. On the lateral surface of the beam $\boldsymbol{F}_n = 0$ so that $\boldsymbol{F}_n \cdot \boldsymbol{q} = 0$.

On the fixed end $R = L$, we have from 4.22 (4) $\boldsymbol{q} = \tau \phi \boldsymbol{k}$, $\boldsymbol{n} = \boldsymbol{k}$, $\boldsymbol{F}_n = \boldsymbol{n} \cdot \mathsf{S} = \boldsymbol{k} \cdot \mathsf{S} = \boldsymbol{i} \, \widehat{x\bar{z}} + \boldsymbol{j} \, \widehat{y\bar{z}}$ and therefore $\boldsymbol{F}_n \cdot \boldsymbol{q} = 0$.

On the free end $R = 0$, we have from 4.22 (4)

$$\begin{aligned} &\boldsymbol{q} = \boldsymbol{i} \tau y L - \boldsymbol{j} \tau x L + \boldsymbol{k} \tau \phi, \qquad \boldsymbol{n} = -\boldsymbol{k}, \\ &\boldsymbol{F}_n = -\boldsymbol{k} \cdot \mathsf{S} = -\boldsymbol{i} \, \widehat{x\bar{z}} - \boldsymbol{j} \, \widehat{y\bar{z}} \,. \text{ Therefore} \\ &\boldsymbol{F}_n \cdot \boldsymbol{q} = -L \mu \tau^2 \left(y \frac{\partial \Omega}{\partial y} + x \frac{\partial \Omega}{\partial x} \right) = -L \mu \tau^2 \left(z \frac{\partial \Omega}{\partial z} + \bar{z} \frac{\partial \Omega}{\partial \bar{z}} \right). \end{aligned} \tag{4}$$

Therefore from (1) and (2)

$$K = \int_S \left[-4 \frac{\partial \Omega}{\partial z} \frac{\partial \Omega}{\partial \bar{z}} - 2z \frac{\partial \Omega}{\partial z} - 2\bar{z} \frac{\partial \Omega}{\partial \bar{z}} \right] dS \tag{5}$$

taken over a cross-section S.

Now

$$2\bar{z} \frac{\partial \Omega}{\partial \bar{z}} + 2z \frac{\partial \Omega}{\partial z} = 2 \frac{\partial}{\partial \bar{z}} (\Omega \bar{z}) + 2 \frac{\partial}{\partial z} (\Omega z) - 4\Omega .$$

Therefore by the area theorem, 3.25 ,

$$\begin{aligned} \int_S \left(2\bar{z} \frac{\partial \Omega}{\partial \bar{z}} + 2z \frac{\partial \Omega}{\partial z}\right) dS &= -i \oint_C \Omega \bar{z}\, dz + i \oint_C \Omega z\, d\bar{z} \\ &\quad - 4 \int_S \Omega\, dS = -4 \int_S \Omega\, dS \end{aligned} \tag{6}$$

if we take $\Omega = 0$ on C.

Therefore

$$K = \int_S \left(4\Omega - 4 \frac{\partial \Omega}{\partial z} \frac{\partial \Omega}{\partial \bar{z}}\right) dS \tag{7}$$

and to apply the principle of virtual stresses we choose a function Ω_a which depends on one or more parameters $A, B, \ldots$, and which satisfies the condition

$$\Omega_a = O \text{ on the boundary} \tag{8}$$

We substitute Ω_a for Ω in (7) and then choose the parameters so that $\delta K = 0$. This leads to a set of equations

$$\frac{\partial K}{\partial A} = 0, \qquad \frac{\partial K}{\partial B} = 0, \ldots , \tag{9}$$

from which we determine the parameters $A, B, \ldots$, and so the approximation Ω_a to the torsion shear function.
Moreover if D_a is the value of μK for the function Ω_a we see from 4.38 that, for the torsional rigidity D,

$$D > D_a .$$

Thus D_a furnishes an approximation, to the torsional rigidity, which is too small.

4.83. Composite profiles

Consider a curve C_1 whose equation is

$$\frac{y}{a} = \psi\left(\frac{x}{c}\right) = \psi(t), \qquad t = \frac{x}{c} \tag{1}$$

where

$$\psi(0) = 0, \ \psi(1) = 0 . \tag{2}$$

We shall suppose this curve as far as the part between the points (0, 0) and $(c, 0)$ is concerned to be of the general shape shown in fig. 4.83 (i) that is to say it has a single turning point M_1 the point (x_0, h), where $0 < x_0 < c$, given by $\psi'(t) = 0$.

If we take a second curve C_2 given by

$$\frac{y}{b} = \psi\left(\frac{x}{c}\right) = \psi(t), \qquad t = \frac{x}{c}, \tag{3}$$

the two curves C_1, C_2 will form a "crescent" or an "oval" according as ab is positive or negative.

These forms we shall term *composite profiles*. A useful and comprehensive family of composite profiles has been introduced by Leibenzon under the name of *aerofoil profiles*. These are defined by

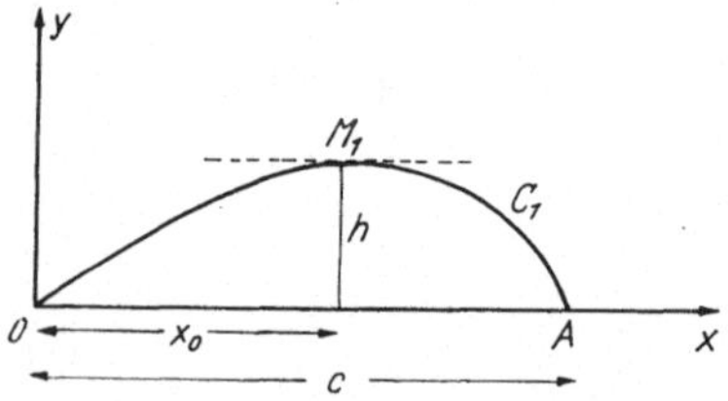

Fig. 4.83 (i)

$$\psi(t) = t^m (1 - t^p)^q, \tag{4}$$

where m, p, q are non-negative numbers.

The curve C_1 corresponding to $m = \frac{1}{2}$, $p = 1$, $q = \frac{1}{2}$ is a semiellipse, while $m = p = q = 1$ gives a parabola.

The points M_1, M_2 of maximum camber are given by

$$t_0 = \frac{x_0}{c} = \left(\frac{m}{m + pq}\right)^{1/p}. \tag{5}$$

These are points at which the shearing stress has a stationary value.

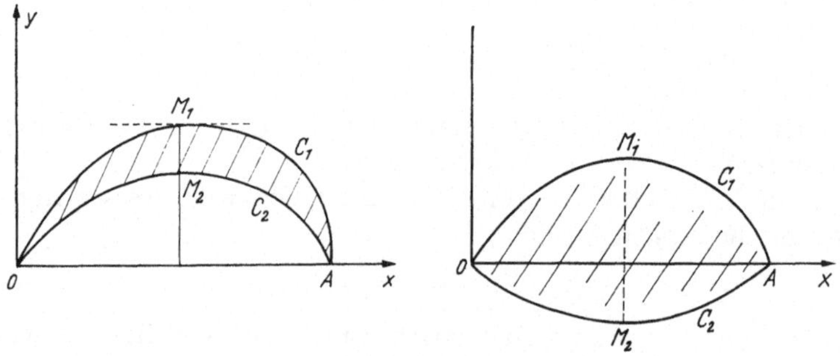

Fig. 4.83 (ii)

We seek a torsion shear function Ω determined by

$$\Omega(x, y) = [y - a\,\psi]\,[y - b\,\psi]\,\omega(x, y) \tag{6}$$

which satisfies the boundary condition

$$\Omega = 0 \text{ on } C_1 \text{ and } C_2 \tag{7}$$

As a *first approximation* take

$$\omega(x, y) = A, \text{ a constant,}$$

so that the first approximation to Ω is Ω_a where

$$\Omega_a = A(y - a\psi)(y - b\psi) = AF(x, y) \tag{8}$$

say.

Therefore by the principle of virtual stresses we have to choose A so that the integral

$$K = \int_S \left(4\Omega_a - 4\frac{\partial\Omega_a}{\partial z}\frac{\partial\Omega_a}{\partial \bar{z}}\right) dS = \int_S \left[4\Omega_a - \left(\frac{\partial\Omega_a}{\partial x}\right)^2 - \left(\frac{\partial\Omega_a}{\partial y}\right)^2\right] dS$$
$$= \int_S \left\{4AF - A^2\left[\left(\frac{\partial F}{\partial x}\right)^2 + \left(\frac{\partial F}{\partial y}\right)^2\right]\right\} dS.$$

is stationary. Therefore $\partial K/\partial A = 0$ and so

$$\int_S 4F\,dS - 2A\int_S \left[\left(\frac{\partial F}{\partial x}\right)^2 + \left(\frac{\partial F}{\partial y}\right)^2\right] dS = 0$$

or

$$A = \frac{2\int_S F\,dS}{\int_S \left[\left(\frac{\partial F}{\partial x}\right)^2 + \left(\frac{\partial F}{\partial y}\right)^2\right] dS}.$$

$$= \frac{2\int_0^c dx \int_{b\psi}^{a\psi} F\,dy}{\int_0^c dx \int_{b\psi}^{a\psi} \left[\left(\frac{\partial F}{\partial x}\right)^2 + \left(\frac{\partial F}{\partial y}\right)^2\right] dy} \tag{9}$$

Now $F = (y - a\psi)(y - b\psi) = y^2 - (a+b)\psi y + ab\psi^2$. Remembering that $\psi = \psi(t) = \psi(x/c)$, we have

$$\int_{b\psi}^{a\psi} F\,dy = \left[\frac{y^3}{3} - (a+b)\psi\frac{y^2}{2} + ab\psi^2 y\right]_{b\psi}^{a\psi} = -\frac{1}{6}(a-b)^3\psi^3. \tag{10}$$

$$\int_{b\psi}^{a\psi} \left[\left(\frac{\partial F}{\partial x}\right)^2 + \left(\frac{\partial F}{\partial y}\right)^2\right] dy$$
$$= \int_{b\psi}^{a\psi} \left[\left(-\frac{a+b}{c}y\psi' + \frac{2ab}{c}\psi\psi'\right)^2 + (2y - (a+b)\psi)^2\right] dy$$
$$= \frac{1}{3}(a-b)^3\psi^3 + \frac{(a-b)^3}{3c^2}\psi^3\psi'^2(a^2 - ab + b^2).$$

Changing the variable from x to t we have from (9)

$$A = \frac{-\int_0^1 \psi^3 dt}{\int_0^1 \psi^3 dt + \frac{a^2 - ab + b^2}{c^2} \int_0^1 \psi^3 \psi'^2 dt} = \frac{-1}{1 + \frac{a^2 - ab + b^2}{c^2} \varrho},$$

where

$$\varrho = \int_0^1 \psi^3 \psi'^2 dt \Big/ \int_0^1 \psi^3 dt . \tag{11}$$

To calculate the approximation D_a to the torsional rigidity we have from (7) and 4.34 (6)

$$\frac{D_a}{\mu} = \int 2\Omega_a dS = 2A \int F\, dS = -\frac{1}{3}(a-b)^3 c A \int_0^1 \psi^3 dt$$

$$= \frac{\frac{1}{3}(a-b)^3 c \int_0^1 \psi^3 dt}{1 + \frac{a^2 - ab + b^2}{c^2} \varrho} \qquad \text{from (10)}$$

4.85. Cross-section a parabolic crescent

We seek an approximate solution by the principle of virtual stresses. Put $m = p = q = 1$ in 4.83 (4). We get

$$\psi(t) = t(1-t) \tag{1}$$

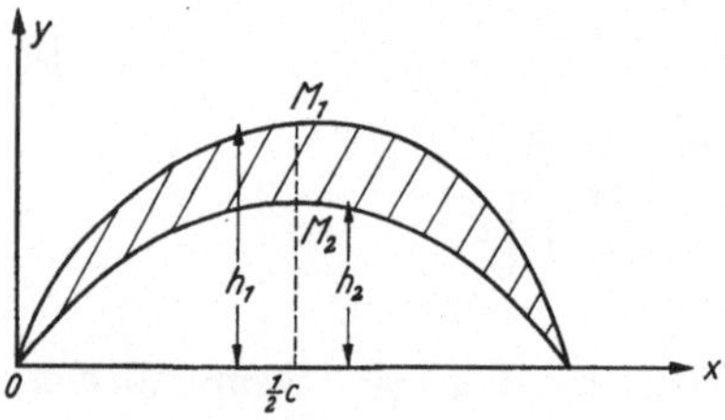

Fig. 4.85

and therefore

$$\frac{y}{a} = \frac{x}{c}\left(1 - \frac{x}{c}\right), \qquad \frac{y}{b} = \frac{x}{c}\left(1 - \frac{x}{c}\right) \tag{2}$$

give the boundaries of a parabolic crescent in which the ordinates of M_1, M_2, the points of maximum camber, are

$$h_1 = \frac{a}{4}, \quad h_2 = \frac{b}{4} . \tag{3}$$

Since

$$\int_0^1 t^m (1-t)^n dt = \frac{\Gamma(m+1)\,\Gamma(n+1)}{\Gamma(m+n+2)}$$

[MILNE-THOMSON (4)] we have

$$\int_0^1 \psi^3 dt = \frac{1}{140}, \quad \int_0^1 \psi^3 \psi'^2 dt = \int_0^1 3t^2(1-t)^3 dt - \int_0^1 3t^3(1-t)^2 dt = \frac{1}{1260}$$

and therefore 4.83 (11) gives

$$\varrho = \frac{1}{9}$$

and therefore

$$\frac{D_a}{\mu} = \frac{\frac{1}{3}(a-b)^3 c \cdot \frac{1}{140}}{1 + \frac{a^2 - ab + b^2}{9c^2}} = \frac{16}{105} \frac{(h_1 - h_2)^3 c}{1 + \frac{16}{9} \frac{h_1^2 - h_1 h_2 + h_2^2}{c^2}}.$$

If in this we put $h_1 = h$, $h_2 = 0$, we get the parabolic segment discussed, with different axes, in 4.75. In the particular case $c = h$, we get the estimate $D_a = 0.0549\,\mu h^4$ and this is an underestimate, 4.38. Combining this with 4.75 (4) we have

$$0.0549\,\mu h^4 < D < 0.0770\,\mu h^4 .$$

For the stress $\widehat{xz}$ we have

$$\widehat{xz} = \mu\tau \frac{\partial \Omega}{\partial y} = \mu\tau A\,[2y - \psi(a+b)].$$

Therefore from (2)

$$\widehat{xz} = A\,\mu\tau(a-b)\,\psi \text{ on } C_1 \text{ and } -A\,\mu\tau(a-b)\,\psi \text{ on } C_2 .$$

The maximum value of $|\widehat{xz}|$ therefore occurs when $t = \frac{1}{2}$ and is given by

$$|\widehat{xz}|_{\max} = \frac{1}{4}\mu\tau(a-b)\,|A| = \mu\tau(h_1 - h_2)\,|A| .$$

Now

$$|A| = \frac{1}{1 + \frac{16}{9}\frac{(h_1^2 - h_1 h_2 + h_2^2)}{c^2}} = \frac{105}{16}\frac{D_a}{\mu}\frac{1}{(h_1 - h_2)^3 c}.$$

Therefore

$$|\widehat{xz}|_{\max} = \frac{105}{16}\frac{M_a}{(h_1 - h_2)^2 c} = \frac{105}{16}\frac{M_a}{c h^2}$$

where $M_a = D_a \tau$ is the approximation to the twisting moment and $h = h_1 - h_2$.

4.87. Segmental cross-section

By a *segmental cross-section* we mean the area bounded by two intersecting arcs and a straight line. We can always take the point of intersection of the arcs as origin and the y-axis perpendicular to the line. We then get fig. 4.87 in which the straight line $B_1 B_2$ is $y = h$ and the arcs $O B_1$, $O B_2$ are

$$x = \theta_1(y) = \theta_1, \; x = \theta_2(y) = \theta_2 \tag{1}$$

respectively.

If we write

$$\Omega = (x - \theta_1)(x - \theta_2) f(y) \tag{2}$$

the boundary condition $\Omega = 0$ is automatically satisfied on the arcs OB_1, OB_2 and it will be satisfied on the chord $B_1 B_2$ if we arrange that

$$f(h) = 0. \tag{3}$$

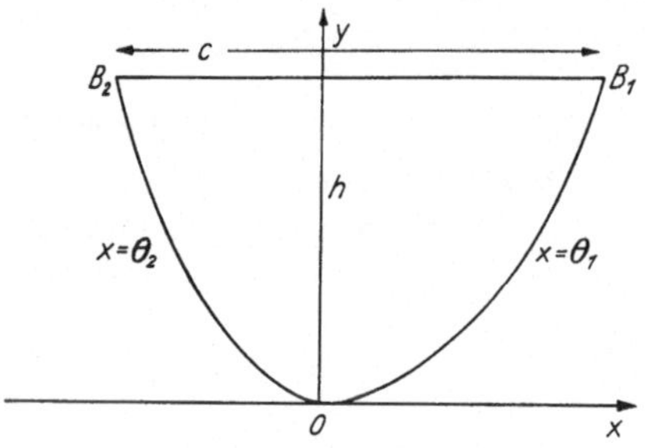

Fig. 4.87

Further we shall suppose that $f(0)$ is finite in order to ensure that $\Omega = 0$ at O.

If we succeed in choosing f so that Ω given by (2) satisfies 4.2 (4), the stress torsion function has been found and the problem is solved.

If we wish to obtain an approximation Ω_a to Ω, using the principle of virtual stresses, we replace (2) by

$$\Omega_a = (x - \theta_1)(x - \theta_2) f(y) \tag{4}$$

and then determine the value of f which renders the function K of 4.81 (7) stationary. From (4) $\Omega_a = [x^2 - x(\theta_1 + \theta_2) + \theta_1 \theta_2] f$ and therefore

$$\frac{\partial \Omega_a}{\partial x} = [2x - (\theta_1 + \theta_2)] f$$

$$\frac{\partial \Omega_a}{\partial y} = [-x(\theta_1' + \theta_2') + \theta_1' \theta_2 + \theta_1 \theta_2'] f + [x^2 - (\theta_1 + \theta_2) x + \theta_1 \theta_2] f'$$

and therefore

$$\left(\frac{\partial \Omega_a}{\partial x}\right)^2 + \left(\frac{\partial \Omega_a}{\partial y}\right)^2 = P f'^2 + 2 Q f f' + R f^2, \tag{5}$$

where P, Q, R are polynomials in x whose coefficients are functions of y.

Therefore

$$\begin{aligned} \int_S \left[\left(\frac{\partial \Omega_a}{\partial x}\right)^2 + \left(\frac{\partial \Omega_a}{\partial y}\right)^2\right] dS &= \int_0^h dy \int_{\theta_1}^{\theta_2} \left[\left(\frac{\partial \Omega_a}{\partial x}\right)^2 + \left(\frac{\partial \Omega_a}{\partial y}\right)^2\right] dx \\ &= \int_0^h (A f'^2 + 2 B f f' + C f^2)\, dy, \end{aligned} \tag{6}$$

where

$$A = \int_{\theta_1}^{\theta_2} P\, dx, \quad B = \int_{\theta_1}^{\theta_2} Q\, dx, \quad C = \int_{\theta_1}^{\theta_2} R\, dx$$

are known functions of y which can always be evaluated since P, Q, R are polynomials in x.

In the same way we get directly from (4)

$$\int_S 4\Omega_a dS = 4\int_0^h f\,dy \int_{\theta_1}^{\theta_2} [x^2-(\theta_1+\theta_2)x+\theta_1\theta_2]\,dx = 4\int_0^h Ef\,dy\,, \quad (7)$$

where E is a known function of y.

Therefore the variational equation for K, 4.81 (8), is

$$\delta\int_0^h [Af'^2+2Bff'+Cf^2-4Ef]\,dy=0 \quad (8)$$

and the variation is obtained solely by varying f. This is equivalent to

$$\int_0^h [(2Af'+2Bf)\,\delta f' + (2Bf'+2Cf-4E)\,\delta f]\,dy = 0\,.$$

Integrate by parts. Then

$$[(2Af'+2Bf)\,\delta f]_0^h + \int_0^h \left[-\frac{d}{dy}(2Af'+2Bf)+(2Bf'+2Cf-4E)\right]\delta f\,dy = 0.$$

Since δf is arbitrary let us postulate

$$[(2Af'+2Bf)\,\delta f]_0^h = 0 \quad (9)$$

so that

$$-\frac{d}{dy}(Af'+Bf)+Bf'+Cf-2E=0 \quad (10)$$

which is an ordinary differential equation of the second order to determine f.

4.88. Triangular cross-section

We regard the triangle OB_1B_2, of height h, as a segmental profile. Take the origin at the vertex O and the y-axis perpendicular to the side B_1B_2. Then in the notation of fig. 4.88, if

$$k=\tan\alpha,\ \lambda=-\tan\beta/\tan\alpha \quad (1)$$

the sides OB_1, OB_2 are

$$x=ky=\theta_1,\ x=\lambda ky=\theta_2\,. \quad (2)$$

We seek an approximate torsion stress function in the form 4.87 (4) or

$$\begin{aligned}\Omega_a &= [x^2-(1+\lambda)kxy+\lambda k^2y^2]f\\ &= [x^2-\mu xy+\nu y^2]f\end{aligned} \quad (3)$$

where

$$\mu=(1+\lambda)k,\ \nu=\lambda k^2\,. \quad (4)$$

Then

$$\frac{\partial \Omega_a}{\partial x} = (2x - \mu y) f, \frac{\partial \Omega_a}{\partial y} = (-\mu x + 2\nu y) f + (x^2 - \mu x y + \nu y^2) f'$$

whence we obtain 4.87 (6) where

$$A = \int_{ky}^{\lambda k y} (x^2 - \mu\, x y + \nu y^2)^2 d x = k^5 y^5 A_1$$

$$B = \int_{ky}^{\lambda k y} (x^2 - \mu\, x y + \nu y^2)\,(-\mu x + 2\nu y)\, d x = k^5 y^4 B_1$$

$$C = \int_{ky}^{\lambda k y} \{(2x - \mu y)^2 + (\mu x - 2\nu y)^2\} d x = k^3 y^3 C_1\,,$$

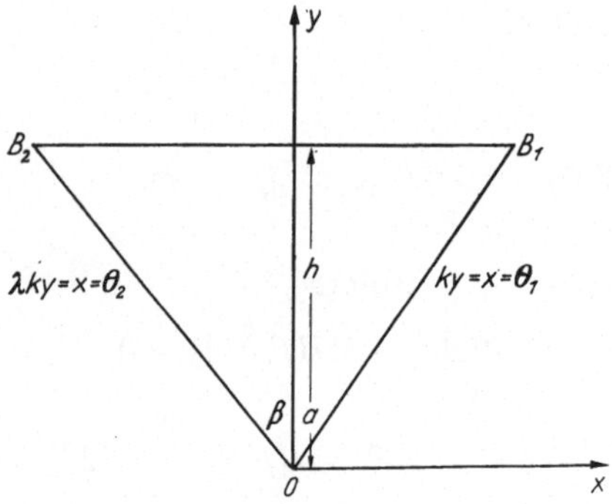

Fig. 4.88

where the constants A_1, B_1, C_1 will be found, after reduction, to be

$$A_1 = \frac{1}{30}(\lambda - 1)^5,\ B_1 = \frac{1}{12}(\lambda - 1)^5,\ C_1 = \frac{1}{3}(\lambda - 1)^3 (k^2\lambda^2 - k^2\lambda + k^2 + 1) \quad (5)$$

while 4.87 (7) yields

$$E = \int_{ky}^{\lambda k y} (x^2 - \mu\, x y + \nu y^2)\, d x = k^3 y^3 E_1$$

where

$$E_1 = -\frac{1}{6}(\lambda - 1)^3\,. \qquad (6)$$

The differential equation 4.87 (10) then reduces to

$$y^2 f'' + 5 y f' - F f - G = 0\,, \qquad (7)$$

where

$$F = \frac{10\,(k^2\lambda + 1)}{k^2(\lambda - 1)^2}\,,\ G = \frac{10}{k^2(\lambda - 1)^2}\,. \qquad (8)$$

The solution of this is of the form

$$P y^{n_1} + Q y^{n_2} + g(y)$$

where n_1, n_2 are the roots of the equation

$$n^2 + 4n - F = 0\,.$$

Let

$$n_1 = -2 + \sqrt{(4+F)}\,,\; n_2 = -2 - \sqrt{(4+F)}\,. \tag{9}$$

Since the solution must be finite at $y = 0$ we must have $Q = 0$. This also follows from 4.87 (9). Therefore

$$f = P y^{n_1} + g(y)\,.$$

Substitute this in (7) getting

$$y^2 g'' + 5 y g' - F g - G = 0$$

and we can satisfy this by the particular solution $g = -G/F$. Further $f(h) = 0$ by hypothesis. Therefore finally

$$f = -\frac{G}{F}\left\{1 - \left(\frac{y}{h}\right)^{n_1}\right\} = -\frac{1}{k^2\lambda + 1}\left\{1 - \left(\frac{y}{h}\right)^{n_1}\right\} \tag{10}$$

where

$$n_1 = -2 + \left\{4 + \frac{10(k^2\lambda + 1)}{k^2(\lambda - 1)^2}\right\}^{1/2}.$$

From 4.87 (7) we have for the approximation to the torsional rigidity

$$\frac{D_a}{\mu} = \int_S 2\Omega_a dS = 2\int_0^h E f\, dy = -\frac{k^3(\lambda - 1)^3}{3(k^2\lambda + 1)}\,\frac{n_1}{4(n_1 + 4)}\, h^4\,. \tag{11}$$

For an *isosceles triangle* $\lambda = -1$ and

$$\frac{D_a}{\mu} = \frac{2}{3}\,\frac{k^3}{(1 - k^2)}\,\frac{n_1}{n_1 + 4}\, h^4\,. \tag{12}$$

This fails for an isosceles right-angled triangle, for then $1 - k^2 = 0$. But in this case $n_1 = 0$ also, so that

$$\frac{D_a}{\mu} = \lim_{k\to 1} \frac{2}{3}\,\frac{k^3}{n_1 + 4}\left(\frac{n_1}{1 - k^2}\right) h^4 = \frac{h^4}{6} \lim_{k\to 1} \frac{d n_1/d k}{d(1 - k^2)/d k} = \frac{5h^4}{48}\,.$$

For an equilateral triangle $k = \frac{1}{\sqrt{3}}$, $\lambda = -1$. Therefore $n_1 = 1$ and

$$\Omega_a = \frac{1}{2h}(3x^2 - y^2)(y - h)$$

which coincides with the accurate value found in 4.53 having regard to the different axes of reference, and observing that $h = 3a$.

4.9. Torsion of a compound bar of isotropic materials

Fig. 4.9 (i) shows a curve K_1 which completely encloses a curve K_2. We suppose this diagram to represent the cross-section of a compound bar in which the material within the curve K_2 is different from that between the curve K_1 and K_2, i.e. a bar with a longitudinal hole K_2 which is completely filled with a bar of another material. We shall assume the materials to be welded together along the interface K_2.

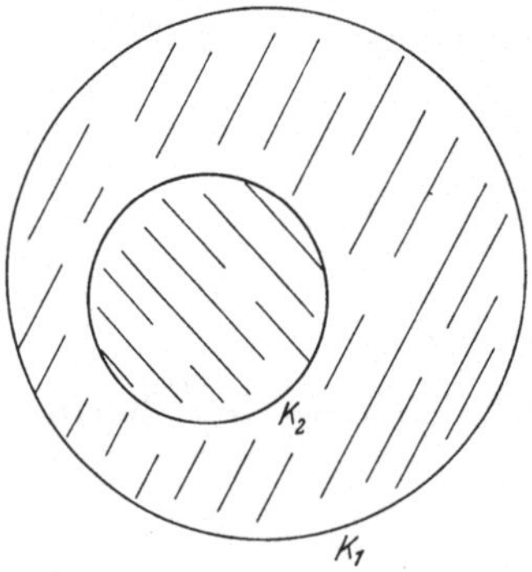

Fig. 4.9 (i)

Taking the case of isotropic materials, let the material within K_2 have rigidity μ_2 and that between K_1 and K_2 have rigidity μ_1. If the bar is subject to torsion we then have from 4.1

$$\frac{\widehat{xz} - i\,\widehat{yz}}{\mu_1 \tau} = \frac{d\,w(z)}{d\,z} - i\,\bar{z} \tag{1}$$

for the material of rigidity μ_1

$$\frac{\widehat{xz} - i\,\widehat{yz}}{\mu_2 \tau} = \frac{d\,w(z)}{d\,z} - i\,\bar{z} \tag{2}$$

for the material of rigidity μ_2.
Let

$$w(z) = \phi + i\,\psi \tag{3}$$

and denote by ϕ_1, ψ_1 or ϕ_2, ψ_2 the values of the functions ϕ, ψ for the materials μ_1, μ_2 respectively.

Assuming the lateral surface K_1 to be free of applied force we have

$$l\,\widehat{xz} + m\,\widehat{yz} = 0\,, \tag{4}$$

where

$$l = \frac{\partial y}{\partial s}\,,\; m = -\frac{\partial x}{\partial s} \tag{5}$$

are the direction cosines of the outwardly drawn normal to K_1. Therefore using 4.1 (5)

$$\mu_1 \tau \frac{\partial y}{\partial s}\left(\frac{\partial \psi_1}{\partial y} - y\right) - \mu_1 \tau \frac{\partial x}{\partial s}\left(-\frac{\partial \psi_1}{\partial x} + x\right) = 0 \text{ on } K_1\,.$$

Integrating along the boundary we have

$$\psi_1 - \frac{1}{2}(x^2 + y^2) = c_1 \text{ on } K_1 \tag{6}$$

where c_1 is an arbitrary constant.

At the interface K_2 between the materials, since they are welded together the stress $l\,\widehat{xz} + m\,\widehat{yz}$ must be continuous i.e. with an obvious notation

$$l\,\widehat{xz}_1 + m\,\widehat{yz}_1 = l\,\widehat{xz}_2 + m\,\widehat{yz}_2 \text{ on } K_2$$

or substituting for the stress from 4.1 the values

$$\frac{\widehat{xz}_1}{\mu_1\tau} = \frac{\partial\psi_1}{\partial y} - y, \quad \frac{\widehat{yz}_1}{\mu_1\tau} = -\frac{\partial\psi_1}{\partial x} + x, \quad \frac{\widehat{xz}_2}{\mu_2\tau} = \frac{\partial\psi_2}{\partial y} - y, \quad \frac{\widehat{yz}_2}{\mu_2\tau} = -\frac{\partial\psi_2}{\partial x} + x$$

and integrating along the boundary we get, in the manner of (6) above,

$$\mu_1\psi_1 - \mu_2\psi_2 - \frac{1}{2}(\mu_1 - \mu_2)(x^2 + y^2) = c_2 \text{ on } K_2\,, \tag{7}$$

where c_2 is a constant.

Further since the materials are welded together the displacement is continuous at the interface so that, from 4.22, $\tau\phi_1 = \tau\phi_2$ on K_2, and differentiating,

$$\frac{\partial\phi_1}{\partial s} = \frac{\partial\phi_2}{\partial s} \text{ on } K_2\,.$$

Therefore, using the Cauchy-Riemann equations we get

$$\frac{\partial\psi_1}{\partial n} = \frac{\partial\psi_2}{\partial n} \text{ on } K_2 \tag{8}$$

where $\partial/\partial n$ denotes differentiation along the normal to K_2.

Equations (6), (7), (8) constitute the boundary conditions and to solve the problem we must determine a harmonic function ψ such that $\psi = \psi_1$ or $\psi = \psi_2$ according to the region, and satisfies also (6), (7), (8) on the boundaries.

This problem can be solved when K_1 and K_2 can be mapped on concentric circles γ_1, γ_2 in such a way that the whole region within K_1 is mapped on the whole region within γ_1. Let the mapping function be

$$z = m(\zeta) \tag{9}$$

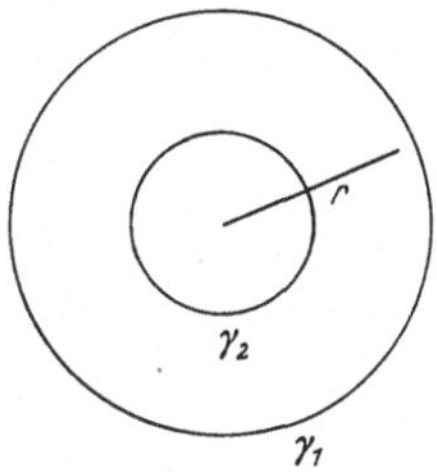

Fig. 4.9 (ii)

and let r_1, r_2 be the radii of the circles.

By the property of conformal mapping that angles are preserved, the element dn of normal to K_2 will map into an element dr_2 of normal to γ_2 and in fact

$$dn = |m'(\zeta)|\,dr_2$$

so that (8) can be replaced by

$$\frac{\partial\psi_1}{\partial r_2} = \frac{\partial\psi_2}{\partial r_2} \qquad \text{on } \gamma_2\,. \tag{10}$$

By means of (9), $w(z)$ can be expressed in terms of ζ. Say

$$w(z) = w[m(\zeta)] = W(\zeta) \tag{11}$$

and therefore

$$2i\psi = w(z) - \overline{w}(\overline{z}) = W(\zeta) - \overline{W}(\overline{\zeta})\,. \tag{12}$$

Now $W(\zeta)$ must be holomorphic in the entire region within γ_1 and so must have a Laurent expansion. Denote by $W_1(\zeta)$ the expansion for the part between γ_1 and γ_2 which therefore excludes $\zeta = 0$, and by $W_2(\zeta)$ the expansion in the region interior to γ_2. Therefore

$$W_1(\zeta) = \sum_{-\infty}^{\infty} a_k \zeta^k = a_0 + \sum_1^{\infty} (a_k \zeta^k + a_{-k}\zeta^{-k}), \; r_2 \leqq |\zeta| \leqq r_1 \tag{13}$$

$$W_2(\zeta) = \sum_0^{\infty} b_k \zeta^k = b_0 + \sum_1^{\infty} b_k \zeta^k, \quad 0 \leqq |\zeta| \leqq r_2 . \tag{14}$$

Using (12) we have

$$2i\,\psi_1 = a_0 - \bar{a}_0 + \sum_1^{\infty} (a_k \zeta^k + a_{-k}\zeta^{-k}) - \sum_1^{\infty} (\bar{a}_k \bar{\zeta}^k + \bar{a}_{-k}\bar{\zeta}^{-k}) \tag{15}$$

$$2i\,\psi_2 = b_0 - \bar{b}_0 + \sum_1^{\infty} (b_k \zeta^k - \bar{b}_k \bar{\zeta}^k) . \tag{16}$$

Now if $\zeta = re^{i\theta}$, $\dfrac{\partial \zeta}{\partial r} = \dfrac{\zeta}{r}$, $\dfrac{\partial \bar{\zeta}}{\partial r} = \dfrac{\bar{\zeta}}{r}$ and therefore (10) combined with (15) and (16) gives

$$\begin{aligned} &\sum_1^{\infty} k\,\{(a_k\zeta^k - a_{-k}\zeta^{-k}) - (\bar{a}_k\bar{\zeta}^k - \bar{a}_{-k}\bar{\zeta}^{-k})\} \\ &\qquad = \sum_1^{\infty} k\,(b_k\zeta^k - \bar{b}_k\bar{\zeta}^k) \text{ on } K_2 . \end{aligned} \tag{17}$$

But $\bar{\zeta} = r_2^2/\zeta$ on K_2. Substituting this in (17) and equating the coefficients of like powers of ζ on the two sides we get

$$a_k + \frac{\bar{a}_{-k}}{r_2^{2k}} = b_k . \tag{18}$$

In order to use the boundary conditions (6) and (7) we want an expansion for $z\bar{z}$ in powers of ζ when $r = r_1$ and $r = r_2$, where $\zeta = r e^{i\theta}$. Thus we want an expansion of the form

$$z\bar{z} = m(\zeta)\,\bar{m}(\bar{\zeta}) = M_0(r) + \sum_1^{\infty} [M_k(r)\,\zeta^k + \overline{M}_k(r)\,\bar{\zeta}^k] \tag{19}$$

where $M_k(r)$ is a coefficient which may depend on r.

Then from (6), (15) we get

$$\begin{aligned} &a_0 - \bar{a}_0 + \sum_1^{\infty} [(a_k - \bar{a}_{-k} r_1^{-2k})\zeta^k + (a_{-k} - \bar{a}_k r_1^{2k})\zeta^{-k}] \\ &\qquad = 2ic_1 + iM_0(r_1) + i\sum_1^{\infty} \left[M_k(r_1)\,\zeta^k + \overline{M}_k(r_1)\,\frac{r_1^{2k}}{\zeta^k}\right], \end{aligned} \tag{20}$$

since on γ_1, $\bar{\zeta} = r_1^2/\zeta$.

Equating the coefficients of like powers of ζ on the two sides of (20) we get

$$a_k - \frac{\bar{a}_{-k}}{r_1^{2k}} = i M_k(r_1) . \tag{21}$$

In the same way from (7), (15), (16) we get

$$\mu_1 (a_k - \bar{a}_{-k} r_2^{-2k}) - \mu_2 b_k = i(\mu_1 - \mu_2) M_k(r_2) \tag{22}$$

while $b_0 - \bar{b}_0$ remains arbitrary. Substituting for b_k from (18) we get

$$\frac{\mu_1 - \mu_2}{\mu_1 + \mu_2} a_k - \frac{\bar{a}_{-k}}{r_2^{2k}} = i \frac{\mu_1 - \mu_2}{\mu_1 + \mu_2} M_k(r_2) . \tag{23}$$

Thus the coefficients a_k, b_k are to be determined from (18), (21), (22). These are linear recurrence relations and the problem is solved in principle. The solution depends essentially on determining the coefficients M_k of (19).

4.92. Circular shaft with an inserted circular shaft

In fig. 4.9 (i) let K_1 be a circumference centre at $(h_1, 0)$, radius a and K_2 a circumference centre at $(h_2, 0)$, radius b. When these circumferences are concentric $w'(z)$ is constant and the problem is solved by

$$w(z) = i h_1 z . \tag{1}$$

Excluding this trivial case the mapping can be effected on a concentric annulus by

$$z = m(\zeta) = \frac{c\zeta}{c - \zeta} \tag{2}$$

where the radii r_1, r_2 of the concentric annulus and the constant c are related to h_1, h_2, a, b by

$$a = \frac{c^2 r_1}{c^2 - r_1^2}, \quad b = \frac{c^2 r_2}{c^2 - r_2^2}, \quad h_1 = \frac{c r_1^2}{c^2 - r_1^2}, \quad h_2 = \frac{c r_2^2}{c^2 - r_2^2} . \tag{3}$$

Now

$$\zeta \bar{\zeta} = r^2 . \tag{4}$$

Therefore

$$z\bar{z} = \frac{c^2 r^2}{(c - \zeta)(c - \bar{\zeta})} = \frac{c^2 r^2}{(c - \zeta)(c - r^2/\zeta)} = \frac{c^2 r^2}{c^2 - r^2} \left(\frac{c}{c - \zeta} + \frac{r^2}{\zeta c - r^2} \right)$$

$$= \frac{c^2 r^2}{c^2 - r^2} \left(-1 + \frac{c}{c - \zeta} + \frac{c}{c - \bar{\zeta}} \right) = \frac{c^2 r^2}{c^2 - r^2} \left[1 + \sum_1^\infty \left(\frac{\zeta^k}{c^k} + \frac{\bar{\zeta}^k}{c^k} \right) \right] .$$

Comparing this with 4.9 (19) we see that

$$M_k(r) = \frac{c^2 r^2}{c^2 - r^2} \cdot \frac{1}{c^k} . \tag{5}$$

By adding 4.9 (21), (23) to their conjugates we find that $a_k + \bar{a}_k = 0$ so that a_k is imaginary. Therefore from 4.9 (18) b_k is also imaginary. Write

$$a_k = i\,\alpha_k, \qquad b_k = i\,\beta_k\,. \tag{6}$$

Then the equations 4.9 (18), (21), (23) give

$$\alpha_k + \frac{\alpha_{-k}}{r_1^{2k}} = \frac{c^2 r_1^2}{c^2 - r_1^2}\,\frac{1}{c^k} = \frac{h_1}{c^{k-1}}$$

$$\lambda\,\alpha_k + \frac{\alpha_{-k}}{r_2^{2k}} = \frac{\lambda h_2}{c^{k-1}}\,,\ \lambda = \frac{\mu_1 - \mu_2}{\mu_1 + \mu_2}$$

$$\alpha_k - \frac{\alpha_{-k}}{r_2^{2k}} = \beta_k\,.$$

Solving these equations and writing

$$h = h_1 - h_2, \qquad \varrho = \left(\frac{r_2}{r_1}\right)^2 \tag{7}$$

we get

$$\alpha_k = \left(h_1 + \frac{\lambda h \varrho^k}{1 - \lambda\varrho^k}\right)\frac{1}{c^{k-1}}$$

$$\alpha_{-k} = \frac{-\lambda h r_2^{2k}}{1 - \lambda\varrho^k}\cdot\frac{1}{c^{k-1}}$$

$$\beta_k = \left[h_1 + \frac{\lambda h(1 + \varrho^k)}{1 - \lambda\varrho^k}\right]\frac{1}{c^{k-1}}\,.$$

Putting zero for the arbitrary constants a_0 and b_0 whose values do not affect the stresses, 4.9 (13), (14) give

$$\begin{aligned} W_1(\zeta) = \frac{i h_1 c\zeta}{c - \zeta} &+ i h\lambda\zeta \sum_{k=1}^{\infty} \frac{\varrho^k}{1 - \lambda\varrho^k}\left(\frac{\zeta}{c}\right)^{k-1} \\ &- i h\lambda c \sum_{k=1}^{\infty} \frac{r_2^{2k}}{1 - \lambda\varrho^k}\,\frac{1}{(c\zeta)^k}\,,\ r_2 \leqq |\zeta| \leqq r_1 \end{aligned} \tag{8}$$

$$W_2(\zeta) = \frac{i h_1 c\zeta}{c - \zeta} + i h\lambda\zeta \sum_{k=1}^{\infty} \frac{1 + \varrho^k}{1 - \lambda\varrho^k}\left(\frac{\zeta}{c}\right)^{k-1},\ 0 \leqq |\zeta| \leqq r_2\,. \tag{9}$$

We note that from (2)

$$\frac{i h_1 c\zeta}{c - \zeta} = i z h_1\,.$$

Therefore if $h = 0$, the concentric case, we have

$$w(z) = W_1(\zeta) = W_2(\zeta) = i h_1 z \tag{10}$$

in agreement with (1).

We get the same result when $\lambda = 0$ i.e. when $\mu_1 = \mu_2$ so that the bar is of the same material throughout.

EXAMPLES IV

1. In the torsion of an elliptic cylinder whose cross-section is $x^2/a^2 + y^2/b^2 = 1$, show that the stress at the point P of the boundary is

$$\frac{2\mu\tau ab}{a^2+b^2} r,$$

where r is the length of the semidiameter OQ which is parallel to the tangent at P.

2. Prove that the torsional rigidity of an elliptic beam is

$$\frac{\pi\mu a^3 b^3}{a^2+b^2},$$

where a, b, are the semi-axes of the cross-section.

Verify this result by calculating the torsional rigidity by use of the first hydrodynamic analogy.

3. In the torsion of an elliptic beam whose cross-section is the ellipse $x^2/a^2 + y^2/b^2 = 1$, prove that the warping w is given by

$$w = -\tau\frac{a^2-b^2}{a^2+b^2}xy.$$

By considering the sign of xy in each of the four quadrants, draw a diagram to show how in the warping some parts of a cross-section are elevated and others depressed, and show that the curves of constant warping are hyperbolas.

4. When the cross-section is one loop of Bernoulli's lemniscate with the notation of 4.46 show that on the boundary

$$z\bar{z} = c^2(1+\sigma)/\sigma^{1/2} = c^2\sigma^{1/2} + c^2\bar{\sigma}^{1/2}.$$

Explain why the $z\bar{z}$ method can not be used to infer that

$$w(z) = ic^2\zeta^{1/2}.$$

5. In the torsion of a beam whose cross-section is the equilateral triangle of 4.53 prove that the lines of stresses are given by the cubic curves

$$x^3 - 3xy^2 + 3a(x^2+y^2) = c,$$

where c is a constant.

Trace these curves for $c = \lambda a^3$, $\lambda = 1, 2, 3, 4$.

6. In the torsion of a beam whose cross-section is the equilateral triangle of 4.53 show that the warping displacement w is given by

$$w = \frac{\tau}{6a}(3x^2y - y^3).$$

Trace some of the curves $w =$ constant, and indicate on which parts of the cross-section there is elevation and on which there is depression of the material.

7. In the torsion of a beam whose cross-section is the equilateral triangle of 4.53 show that the torsional rigidity is

$$\frac{9\mu\sqrt{3}}{5}a^4 = \frac{\mu\Delta^2}{5\sqrt{3}},$$

where Δ is the area of the triangle.

8. In the torsion of an isotropic beam prove that the shear stress function furnishes shears on a cross-section whose resultant is zero, but that their moment reduces to a couple.

9. Prove that the semicircle in the upper half plane bounded by $x^2 + y^2 = a^2$, $y = 0$ is mapped on the unit circle $|\zeta| = 1$ by

$$\zeta = \frac{(z+a)^2 - i(z-a)^2}{(z+a)^2 + i(z-a)^2}.$$

Prove also that for a beam having this semicircular cross-section, the torsion problem is solved by $w(z)$ where

$$2\pi w(z) = \pi i z^2 + \frac{2a(z^2+a^2)}{z} + \frac{(z^2-a^2)^2}{z^2} \ln \frac{a-z}{a+z}.$$

Calculate the stress at the point $z = a$, and explain the result which you find.

10. Show that the lens-shaped region bounded by the circles

$$x^2 + (y-a)^2 = 2a^2,\ x^2 + (y+a)^2 = 2a^2$$

is mapped on the unit circle $|\zeta| = 1$ by

$$\zeta = \frac{2az}{z^2+a^2}.$$

Hence show that $w(z)$ for the torsion of a beam which has the above region for cross-section is

$$w(z) = i a^2 \frac{z^2-a^2}{z^2+a^2} - \frac{ia}{\pi z}\frac{(a^2-z^2)^2}{(a^2+z^2)} \ln \frac{a-z}{a+z}.$$

11. In the torsion of a beam whose cross-section is the lune of 4.54, calculate the resultant shear at the points A, B, C of Fig. 4.54, and prove that the maximum shearing stress occurs at the point B, and that it is nearly twice the stress on the boundary of a circular shaft of radius a without the notch.

12. In the torsion of a bar of rectangular cross-section prove that the maximum shearing stress occurs at the midpoint of the longer side of the rectangle, and that an approximate value is

$$2\mu\tau a\left(1 - \frac{8}{\pi^2}\operatorname{sech}\frac{\pi b}{2a}\right).$$

Compare this result numerically with that given by 4.74 (4).

13. In the torsion of a beam whose cross-section is that of Booth's lemniscate (4.56) prove that the torsional rigidity is

$$\frac{1}{2}\mu\pi k^4 \frac{a^8+1}{(a^4-1)^4}.$$

14. Prove that the torsional rigidity for a beam, whose cross-section is a cardioid, is

$$\frac{17\mu}{36\pi} S^2,$$

where S is the area of the cardioid.

15. Regarding the segment of a parabola cut off by a line perpendicular to the axis as a segmental cross-section, use the method of **4.87** to obtain an approximation to the torsion shear stress function and hence find approximations for the maximum shear and the torsional rigidity.

Determine also the approximate aspect of the lines of stress.

16. Apply the method of **4.87** to solve approximately the torsion problem for the half elliptic cross-section bounded by

$$\frac{x^2}{a^2}+\frac{y^2}{b^2}=1,\ y=0$$

in the upper half-plane, regarded as a segmental cross-section.

Deduce approximations for the maximum shear and the torsional rigidity. Discuss in particular the case of a semicircle.

17. In the torsion of a cylindrical or prismatic isotropic beam fixed at one end, show that the solution as an antiplane problem will hold right up to the fixed end provided that the condition of fixing takes proper account of the warping function $\tau\phi$, and prove further that the distribution of warping cannot be arbitrarily prescribed.

18. Use the hydrodynamic analogy to discuss the general effect of a groove of small semicircular section along the surface of a cylindrical bar subjected to torsion.

19. A cylindrical bar of material of rigidity μ_1 contains a longitudinal cylindrical hollow into which another cylindrical bar of material of rigidity μ_2 fits exactly. The two bars are welded along their common interface and subjected to torsional twist. Use the hydrodynamic analogy to prove that at the interface the warping function and also the normal component of shear must be continuous.

20. Use a physical argument to show that in 4.39 the potential energy is given by

$$V=\tau M\int_0^1 \lambda\, d\lambda.$$

21. In the torsion of a beam whose cross-section is the cardioid of 4.44 find the ratio of the shear stress at the ends of a chord which passes through the pole O at inclination θ to the axis OA.

22. Show that in the case of a hyperbolic segment whose base is of length c and whose height is h the torsional rigidity is

$$D = \frac{16\mu c h^3}{105\left(1+\frac{8}{3}\frac{h^2}{c^2}\right)}$$

and that the maximum shearing stress is given by

$$\frac{105 D \mu}{16 c h^2}\left(1+\frac{8}{3}\frac{h^2}{c^2}\right).$$

Prove that the torsional rigidity of a hyperbolic segment is less than that of a parabolic segment of the same base and height.

23. Comparing the homothetic elliptic ring of 4.52 with the confocal elliptic ring of 4.63 show that the ratio of the breadth of the ring at the end of the major to that at the end of the minor axes is greater than unity for the homothetic ring and less than unity for the confocal ring.

24. Find a general formula for the torsional rigidity of a beam whose cross-section is an epitrochoid (4.55) and apply it to the special cases $n = 1$ a circle and $n = 2$ an elliptic limaçon.

25. Establish the mapping 4.56 (1) for Booth's lemniscate, starting from its definition as the inverse of an ellipse with respect to its centre, and give interpretations of the constants k and a. Determine the breadth of the lemniscate at its "waist", fig. 4.56, and discuss the situation which arises when this breadth tends to zero.

26. For the torsion of a hollow beam whose cross-section is bounded by confocal ellipses calculate the expression for the torsional rigidity given by 4.63 (3).

27. Use 4.74 (4) to plot a graph to read values of $D_a/(4a^2\mu S)$, where S is the area of the cross-section, in terms of the thickness ration b/a.

28. For the aerofoil profile given by

$$\psi(t) = t^m(1-t^p)^q$$

prove that

$$\int_0^1 \psi^3\, dt = \frac{1}{p} B\left(\frac{3m+1}{p}, 3q+1\right)$$

and that

$$\int_0^1 \psi^3 \psi'^2\, dt = \frac{m^2}{p} B\left(\frac{5m-1}{p}, 5p-1\right)$$

$$-\frac{2m(m+pq)}{p} B\left(\frac{5m-1+p}{p}, 5q-1\right)$$

$$+\frac{(m+pq)^2}{p} B\left(\frac{5m-1+2p}{p}, 5q-1\right)$$

where $B(p, q)$ is the beta function,

$$B(p,q) = \int_0^1 x^{p-1}(1-x)^{q-1} dx.$$

29. In the notation of 4.3 (3) show that

$$T = \frac{1}{8} \oint_C [w(z) + \overline{w}(\overline{z})]\, d(z\overline{z}).$$

30. In the notation of 4.48 prove that

$$\int_{-1}^{1} m(\xi)\overline{m}(\xi) \left(\frac{1}{\xi - \zeta} + \frac{1}{\xi - \frac{1}{\zeta}} \right) d\xi + \int_{\delta} m(\sigma)\,\overline{m}(\sigma) \left(\frac{1}{\sigma - \zeta} + \frac{1}{\sigma - \frac{1}{\zeta}} \right) d\sigma$$

$$= 2\pi i\, m(\zeta)\, \overline{m}(\zeta)$$

and deduce that

$$W(\zeta) = i m(\zeta)\, \overline{m}(\zeta) + \frac{1}{2\pi} \int_{\delta} m(\sigma) \left[\overline{m}\left(\frac{1}{\sigma}\right) - \overline{m}(\sigma) \right] \left(\frac{1}{\sigma - \zeta} + \frac{1}{\sigma - \frac{1}{\zeta}} \right) d\sigma\,.$$

31. In 4.48 use the fact that $m(\zeta)\, \overline{m}(\zeta)\, m'(\zeta)$ and $W(\zeta) \frac{d}{d\zeta} [m(\zeta)\,\overline{m}(\zeta)]$ are holomorphic in the semicircle to prove that the path of integration along the diameter ε can be replaced by the path $-\delta$ and deduce that

$$I = -\frac{1}{4} i \int_{\delta} \left[\overline{m}^2\left(\frac{1}{\sigma}\right) - \overline{m}^2(\sigma) \right] m(\sigma)\, d[m(\sigma)]$$

$$T = \frac{1}{4} \int_{\delta} W(\sigma)\, d\left\{ m(\sigma) \left[\overline{m}\left(\frac{1}{\sigma}\right) - \overline{m}(\sigma) \right] \right\}.$$

32. For the torsion of a beam of semicircular cross-section, in the notation of 4.48 prove that

$$W(\zeta) = i a^2 \zeta^2 + \frac{a^2}{2\pi} \int_{\delta} (1 - \sigma^2) \left(\frac{1}{\sigma - \zeta} + \frac{1}{\sigma - \frac{1}{\zeta}} \right) d\sigma\,.$$

33. For a beam of semicircular cross-section of radius a prove that

$$T = a^4 \left(\frac{2}{\pi} - \frac{\pi}{8} \right)$$

and deduce the torsional rigidity.

Chapter V

The flexure of isotropic beams

In this chapter we consider the stress and displacement in an isotropic cantilever, that is to say, a prismatic beam fixed at one end, the root, and in equilibrium under the action of a force applied in the plane of the free end.

5.1. The flexure problem

We consider a cantilever in equilibrium under a force $\boldsymbol{P}$ applied in the plane of the free end. The determination of the stress and displacement at any point of the material constitutes the *flexure problem.*

As explained in detail in 3.1 we suppose this force to be applied, not as a concentrated force but as a distribution of shearing traction over the free end, the antiplane, so arranged as to be consistent with a state of antiplane stress.

Take the origin in the free end so that the antiplane is $R = 0$ and let the components of $\boldsymbol{P}$ be $(P_1, P_2, 0)$, while the moment of $\boldsymbol{P}$ about the origin is $(0, 0, -M)$.

Then from 3.2 (7)

$$\int\limits_S \widehat{xz}\, dS = -P_1, \quad \int\limits_S \widehat{yz}\, dS = -P_2, \qquad (1)$$

$$-M_3 = M = \int\limits_S (x \cdot \widehat{yz} - y \cdot \widehat{xz})\, dS$$

while

$$\widehat{zz} = 0 \text{ when } R = 0\,. \qquad (2)$$

As usual the integrals are all taken over the area of a cross-section.

The above system is equivalent to the single force $\boldsymbol{P}$ acting at the load point $(x_L, y_L, 0)$, where

$$M = x_L P_2 - y_L P_1\,. \qquad (3)$$

When M, P_1, P_2 are numerically assigned, (3) shows that the locus of possible positions of the load point is a straight line, namely, the line of action of the resultant force $\boldsymbol{P}$.

5.11. The stress component $\widehat{zz}$ in the isotropic case

We consider the pure antiplane problem, 3.2, with the loading $(P_1, P_2, 0)$ whose moment with respect to the origin is $(0, 0, -M)$. The surface traction is $(0, 0, Z_n)$ and the body-vector is $(0, 0, b_3)$.

From 5.1 (2) we have

$$\widehat{zz} = \widehat{zz}_a = 0 \text{ when } R = 0$$

and therefore from 3.2 (12), since $\widehat{zz}_p = 0$ by hypothesis,

$$A_2 x + B_2 y + C_2 m \equiv 0$$

so that

$$A_2 = B_2 = C_2 = 0 \qquad (1)$$

and

$$\widehat{zz} = E\,(A_1 x + B_1 y + C_1 m)\,R\,. \qquad (2)$$

In the notation of 3.24 let us write

$$Q_1 = E(k_3 - x_G k_1) = P_1 + \oint_C (x - x_G) Z_n ds - \int_S (x - x_G) b_3 dS$$
$$Q_2 = E(k_2 - y_G k_1) = P_2 + \oint_C (y - y_G) Z_n ds - \int_S (y - y_G) b_3 dS . \tag{3}$$

Then from 3.24 (9)

$$A_1 = \frac{-AQ_1 + HQ_2}{E(AB - H^2)}, \quad B_1 = \frac{-BQ_2 + HQ_1}{E(AB - H^2)} \tag{4}$$

where A, B, H are second moments of area with respect to parallel axes through the centroid.

Let us call

$$Q = Q_1 + i Q_2$$

the *reduced load*. Also let the affix of the centroid of the cross-section be denoted by

$$\xi = x_G + i y_G . \tag{5}$$

Then we have for the reduced load

$$Q = P + \oint_C (z - \xi) Z_n ds - \int_S (z - \xi) b_3 dS . \tag{6}$$

Note that when there is no surface or body-force $Q = P$. In terms of the reduced load we have

$$\beta = A_1 + i B_1 = \frac{-(A + B) Q + (B - A + 2iH) \bar{Q}}{2E(AB - H^2)} . \tag{7}$$

Since $2(A_1 x + B_1 y) = \beta \bar{z} + \bar{\beta} z$, we can express (2) in the form

$$\widehat{zz} = \frac{1}{2} ER(\beta \bar{z} + \bar{\beta} z + 2C_1 m) \tag{8}$$

where from 3.24 (6) and (7)

$$C_1 m = -\frac{1}{2}(\beta \bar{\xi} + \bar{\beta} \xi) - \frac{1}{ES} \oint_C Z_n ds + \frac{1}{ES} \int_S b_3 dS . \tag{9}$$

When the body-vector is derived from a potential, 2.24 (2) shows that $b_3 = E \varepsilon m$ and therefore

$$(C_1 - \varepsilon) m = -\frac{1}{2}(\beta \bar{\xi} + \bar{\beta} \xi) - \frac{1}{ES} \oint_C Z_n ds . \tag{10}$$

Thus if the origin is the centroid and if $Z_n = 0$, we have $(C_1 - \varepsilon) m = 0$. In the absence of body-force and lateral loading

$$C_1 m = -\frac{1}{2}(\beta \bar{\xi} + \bar{\beta} \xi) . \tag{11}$$

5.12. The geometrical parameters of the cross-section

We shall call the area S, the affix ξ of the centroid, and the second moments A, B, H with respect to axes through the centroid, the geometrical parameters of the cross-section. They are independent of the loading and the elastic parameters, but, with the exception of S, they depend on the position of the origin and the orientation of the axes of reference.

These parameters can be conveniently evaluated by means of the area theorem, 3.25, as follows

$$S = \int_S dS = -\frac{1}{2} i \oint_C \bar{z}\, dz \tag{1}$$

$$\xi S = \int_S z\, dS = -\frac{1}{2} i \oint_S \bar{z} z\, dz. \tag{2}$$

Let z_1 be the affix $x_1 + i y_1$ of the point z referred to parallel axes through the centroid G. Then using 5.11 (5) we have $z = z_1 + \xi$ and therefore

$$\int_S z\bar{z}\, dS = \int_S z_1 \bar{z}_1\, dS + \xi \int_S \bar{z}_1\, dS + \bar{\xi} \int_S z_1\, dS + \int_S \xi\bar{\xi}\, dS.$$

The third and fourth integrals vanish since the origin of z_1 is at the centroid. Also $z_1\bar{z}_1 = x_1^2 + y_1^2$ and therefore the value of the second integral is $A + B$. Therefore

$$A + B + \xi\bar{\xi} S = \int_S z\bar{z}\, dS = -\frac{1}{4} i \oint_C z\bar{z}^2\, dz. \tag{3}$$

Again

$$\int_S z_1^2\, dS = \int_S (x_1^2 - y_1^2 + 2 i x_1 y_1)\, dS = B - A + 2iH$$

while

$$\int_S z^2\, dS = \int_S z_1^2\, dS + 2\xi \int_S z_1\, dS + \int_S \xi^2\, dS$$

and the third integral vanishes. Therefore

$$B - A + 2iH + \xi^2 S = \int_S z^2\, dS = -\frac{1}{2} i \oint_C z^2 \bar{z}\, dz \tag{4}$$

the real part of which furnishes $B - A$ while the imaginary part furnishes H.

Moreover $\bar{z}$ is a function of z on C so that the integrands are all functions of z. The integrals can therefore be evaluated by Cauchy's residue theorem.

This last remark points out that the area theorem probably gives the simplest analytic process for evaluating plane surface integrals when the boundary is an analytic curve.

5.14. The shears $\widehat{xz}$, $\widehat{yz}$

From 3.2 (5) we have

$$\frac{\widehat{xz} - i\widehat{yz}}{\mu} = \Phi(z) - p\bar{z} - q\bar{z}^2 - r\, z\bar{z} \tag{1}$$

where

$$p = i\tau + (1+\eta)(C_1 - \varepsilon)m,\ q = \frac{1+2\eta}{4}\beta,\ r = \frac{1}{2}\bar{\beta} \tag{2}$$

and β is given in terms of the reduced loading by 5.11 (7).

The complete solution of the problem of finding $\Phi(z)$ in terms of conformal mapping, when the region defined by the cross-section is simply connected, has been given in 3.51, namely

$$\Phi(z)\, m'(\zeta) = \frac{1}{2\pi i}\oint_{\gamma} \frac{H(\sigma)}{\sigma - \zeta}\, d\sigma + \frac{1}{\mu\pi}\oint_{C} \frac{Z_n ds}{\sigma - \zeta} \tag{3}$$

where $H(\sigma)$ is defined by 3.51 (6) in terms of the mapping function $m(\zeta)$.

Knowing $\Phi(z)$ the stress, moment and displacement can be determined.

The presence of the real constant τ in (1) shows that flexure is, in general, accompanied by torsion. We can call the part of the solution which depends upon τ the *associated torsion problem* which can be solved independently by the methods of Chapter IV.

The *associated torsional twist* τ is not arbitrary when the load point is given, for equating the two values of the moment M given by 3.26 (4) and 5.1 (3) we have

$$\mu\tau I = x_L P_2 - y_L P_1 - \mu\left\{\frac{1}{4}(K+\bar{K}) + \frac{1}{48}(1-2\eta)(\beta\bar{J} + \bar{\beta}J)\right\} \tag{4}$$

which determines τ.

In particular if the load point is at the origin, we get τ by equating M to zero.

If the load is applied along an axis of symmetry of the cross-section, we must have $\tau = 0$.

5.16. The displacement

From 5.11 (1) in the flexure problem $A_2 = B_2 = C_2 = 0$ and therefore in 3.4 (8), $\delta = \theta_0 = 0$ while

$$\gamma = C_1 m = \varepsilon m - \frac{1}{2}(\beta\bar{\xi} + \bar{\beta}\xi) - \frac{1}{ES}\oint_{C} Z_n ds \tag{1}$$

from 5.11 (10).

Therefore 3.41 (6) and (7) give for the displacement

$$\begin{aligned} \upsilon = u + iv = -\frac{1}{6}\beta R^3 + \left\{-\frac{1}{2}\eta\,\bar{\beta} z^2 + (i\tau - \eta\gamma)z\right\} R \\ + (R\omega_2 - y\omega_3 + a_0) + i(x\omega_3 - R\omega_1 + b_0) \end{aligned} \tag{2}$$

$$\begin{aligned} w = \frac{1}{2}[\Phi(z) + \bar{\Phi}(\bar{z})] + \frac{1}{4}(\beta\bar{z} + \bar{\beta}z + 2\gamma)R^2 - \frac{1}{8}\beta\bar{z}^2 z - \frac{1}{8}\bar{\beta}z^2\bar{z} \\ + \frac{1}{2}[(1+\eta)\varepsilon m - \gamma]z\bar{z} + \omega_1 y - \omega_2 x + c_0 \,. \end{aligned} \tag{3}$$

The last terms in each expression correspond to a rigid body movement of small rotation (ω_1, ω_2, ω_3) and translation (a_0, b_0, c_0). These constants will be determined by the conditions of fixation at the root.

The axis of the beam, that is to say the line of centroids of the cross-sections in the undeformed state, deforms into a plane curve.

To see this note that the displacement of any point (x_G, y_G, R) on the axis of the beam is got by replacing, in (2) and (3), z by $\xi = x_G + iy_G$. We then have

$$\upsilon = -\frac{1}{6}\beta R^3 + kR + l,\; w = m\,, \tag{4}$$

where k, l, m are constants. Eliminate R^3 between (4) and its complex conjugate. Then

$$\bar{\beta}\upsilon - \beta\bar{\upsilon} = (\bar{\beta}k - \beta\bar{k})\,R + \bar{\beta}\,l - \beta\bar{l}\,. \tag{5}$$

Now the point (x_G, y_G, R) is displaced to (x', y', R') where $x' = x_G + u$, $y' = y_G + v$, $R' = R + m$, and therefore (5) is a linear relation between x', y', R' and so the deformed axis lies in a plane.

Observe that this plane does not, in general, contain the undeformed axis.

5.2. The centre of flexure

The component about the R-axis of the rotation $\nabla \wedge \boldsymbol{q}$ is

$$\Omega = \frac{\partial v}{\partial x} - \frac{\partial u}{\partial y} = -i\left(\frac{\partial \upsilon}{\partial z} - \frac{\partial \bar{\upsilon}}{\partial \bar{z}}\right) = [2\tau - i\eta(\beta\bar{z} - \bar{\beta}z)]R + 2\omega_3\,.$$

Following Stevenson we call the rate of change of Ω with respect to R the *local twist* ω so that

$$\omega = 2\tau - i\eta(\beta\bar{z} - \bar{\beta}z)\,. \tag{1}$$

The mean value ω_m of the local twist over the area of a cross-section is

$$\omega_m = 2\tau - i\eta(\beta\bar{\xi} - \bar{\beta}\xi),\; \xi = x_G + iy_G\,, \tag{2}$$

which is the same as the value of ω at the centroid of the section.

We now give Stevenson's definition of the centre of flexure for an isotropic beam.

Definition. The *centre of flexure* is the position of the load point when the local twist at the centroid is zero.

This occurs when $\tau = \tau_0$, where

$$\tau_0 = \frac{1}{2} i \eta (\beta \bar{\xi} - \bar{\beta} \xi) . \tag{3}$$

The corresponding value p_0 of p given by 3.2 (6), 5.11 (10) is

$$p_0 = -\frac{1}{2}(1 + 2\eta)\beta\bar{\xi} - \frac{1}{2}\bar{\beta}\xi - \frac{1}{2\mu S}\oint_C Z_n ds . \tag{4}$$

The corresponding moment $M_0 (= -M_3)$ as given by 3.26 is

$$M_0 = \mu\left\{\frac{1}{4}(K + \bar{K}) + \tau_0 I + \frac{1}{48}(1 - 2\eta)(\beta J + \overline{\beta J})\right\}, \tag{5}$$

where I, J, K are given by 3.26 (2) and I and J are independent of β and therefore of $P = P_1 + iP_2$.

Therefore from 3.27 the centre of flexure is given by $\bar{z}_f = x_f - iy_f = 2i\,\partial M_0/\partial P$. Thus

$$\begin{aligned} \bar{z}_f = 2i\mu\left\{\frac{1}{4}\frac{\partial(K+\bar{K})}{\partial\beta} + (1-2\eta)\frac{J}{48} + \frac{1}{2}i\eta I\bar{\xi}\right\}\frac{\partial\beta}{\partial P} \\ + 2i\mu\left\{\frac{1}{4}\frac{\partial(K+\bar{K})}{\partial\bar{\beta}} + (1-2\eta)\frac{\bar{J}}{48} - \frac{1}{2}i\eta I\xi\right\}\frac{\partial\bar{\beta}}{\partial P} . \end{aligned} \tag{6}$$

From the foregoing it appears that body force and loading on the lateral surface of the beam are accounted for by the parameters β and C_1, and therefore solutions of the flexure problem in terms of β and C_1 will have complete generality. Consequently investigation of problems on the assumption that body-force and lateral loading are absent, will economize in writing without introducing any essential restriction of principle.

5.22. Cross-section a circle

Let the cross-section be the circle $x^2 + y^2 = a^2$. This is mapped on the unit circle in the ζ-plane by

$$z = a\zeta = m(\zeta)$$

and therefore, with the notation of 3.51,

$$H(\sigma) = a^3(r + \bar{q}) + (p + \bar{p})\frac{a^2}{\sigma} + (q + \bar{r})\frac{a^3}{\sigma^2} .$$

Thus from 3.51 (6), (10), (11)

$$\Phi(z) \cdot a = a^3(r + \bar{q}) = a^3\bar{\beta}(3 + 2\eta)/4 . \tag{1}$$

Now for the circle

$$A = B = \frac{\pi a^4}{4}, \; H = 0$$

and therefore from 5.11 (7)

$$\beta = \frac{-2P}{\mu(1+\eta)\pi a^4}, \qquad P = P_1 + iP_2. \tag{2}$$

Also $\xi = 0$ so that

$$p = i\tau, \quad q = \frac{1}{4}(1+2\eta)\beta, \quad r = \frac{1}{2}\bar{\beta}$$

and therefore from 5.14 (1)

$$\widehat{xz} - i\widehat{yz} = -i\mu\tau\bar{z} - \frac{(3+2\eta)a^2 - 2z\bar{z}}{2\pi a^4(1+\eta)}\bar{P} + \frac{(1+2\eta)\bar{z}^2}{2\pi a^4(1+\eta)}P. \tag{3}$$

To find the moment we have, in the notation of 3.26,

$$I = \frac{1}{2}\pi a^4, \ J = \int_C a^2\bar{z}^2 dz, \ K = \int_C z\bar{z}\frac{a^2\bar{\beta}(3+2\eta)}{4}dz.$$

But on C, $z\bar{z} = a^2$. Therefore $J = K = 0$.

Thus we have

$$M = \mu\tau I = \frac{1}{2}\pi a^4\mu\tau, \tag{4}$$

so that if the load point is (x_L, y_L),

$$\frac{1}{2}\pi a^4\mu\tau = P_2 x_L - P_1 y_L. \tag{5}$$

If the load point is the centre, we have $\tau = 0$, and therefore the centre of the section is also the centre of flexure.

If the load is $(-P_1, 0)$ applied at the centre of the section we have from (3)

$$\widehat{xz} = \frac{P_1}{2\pi a^4(1+\eta)}\{(3+2\eta)(a^2 - x^2) - (1-2\eta)y^2\} \tag{6}$$

$$\widehat{yz} = \frac{-P_1(1+2\eta)}{\pi a^4(1+\eta)}xy. \tag{7}$$

These results (6) and (7) can be regarded as the general case for the circular section when the load point is at the centre, for we can always choose the direction of the x-axis to coincide with the direction of the given resultant load.

As in the case of torsion a line of shearing stress is a line such that its tangent at each point is in the direction of the resultant shearing stress at that point. Thus for a line of shearing stress

$$\frac{dx}{\widehat{xz}} = \frac{dy}{\widehat{yz}}.$$

The boundary of the cross-section must itself be a line of shearing stress when the boundary is unloaded, for the normal component of shearing stress is then zero.

In the case of the circular boundary loaded as just described to give (6) and (7), the differential equation of the lines of shearing stress becomes.

$$(1+2\eta)\,2xy\,dx+[(3+2\eta)\,(a^2-x^2)-(1-2\eta)\,y^2]\,dy=0. \quad (9)$$

Now (9) is of the form $M\,dx+N\,dy$ where

$$\frac{1}{M}\left(\frac{\partial M}{\partial y}-\frac{\partial N}{\partial x}\right)=\frac{4+4\eta}{1+2\eta}\cdot\frac{1}{y}$$

and therefore has the integrating factor

$$\exp\left(\int-\frac{4+4\eta}{1+2\eta}\,\frac{dy}{y}\right)=y^{-\left(\frac{4+4\eta}{1+2\eta}\right)}.$$

Applying this factor we get the equation of the lines of shearing stress in the form

$$a^2-x^2-y^2=C\,y^{\frac{3+2\eta}{1+2\eta}}, \quad (10)$$

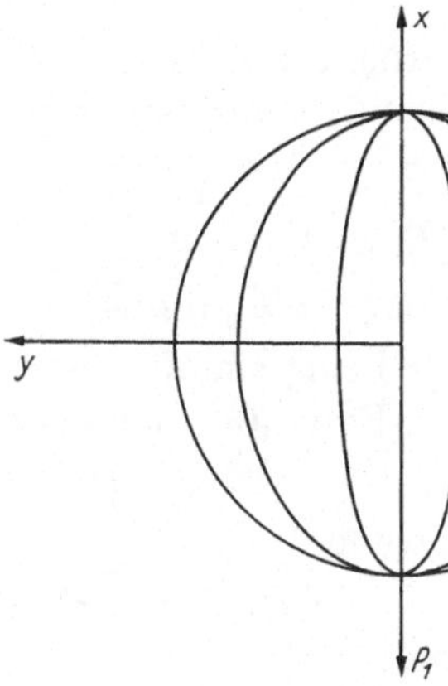

Fig. 5.22

where C is a constant.
Fig. 5.22 shows some of the lines of shearing stress. The oval curves would be the ellipses

$$\frac{x^2}{a^2}+\frac{y^2}{a^2(1+C)^{-1}}=1$$

if the value of η were $\frac{1}{2}$.

5.23. Cross-section a circular annulus

Let the cross-section be the annulus comprised between the concentric circumferences γ_1 of radius a and γ_2 of radius b. On γ_1, $z=a\,e^{i\theta}$, $s=a\,\theta$ and so

$$\frac{dz}{ds}=i\,e^{i\theta}=\frac{iz}{a}.$$

Therefore the boundary condition 3.5 (3) gives

$$\Phi(z)+\frac{a^2}{z^2}\,\bar{\Phi}\left(\frac{a^2}{z}\right)=a^2A+\frac{a^4\bar{A}}{z^2}\text{ on }\gamma_1, \quad (1)$$

$$\Phi(z)+\frac{b^2}{z^2}\,\bar{\Phi}\left(\frac{b^2}{z}\right)=b^2A+\frac{b^4\bar{A}}{z^2}\text{ on }\gamma_2, \quad (2)$$

$$\text{where } A=\bar{q}+r, \quad (3)$$

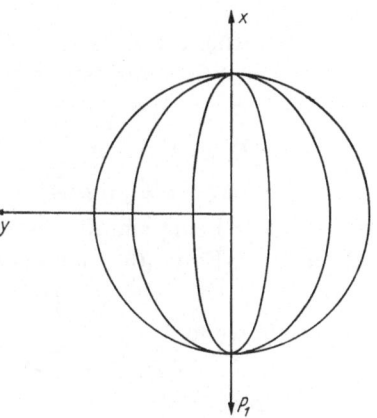

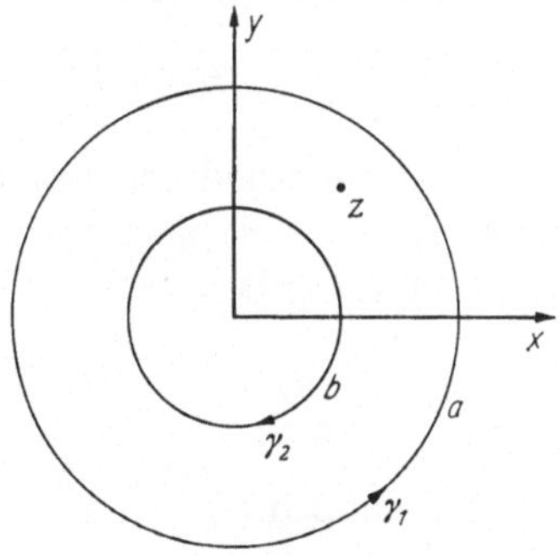

Fig. 5.23

and where we have used the condition $p+\bar{p}=0$, for $p=i\tau$ since the origin is at the centroid.

Since $\Phi(z)$ is holomorphic in the region covered by the annulus, we have a Laurent expansion, valid in this region

$$\Phi(z) = \sum_{n=-\infty}^{\infty} C_n z^n . \tag{4}$$

Substituting this in (1) and (2) and equating coefficients we find that all coefficients C_n vanish except C_{-2}, C_{-1}, C_0 which satisfy the equations

$$C_0 + a^{-2}\bar{C}_{-2} = a^2 A, \qquad C_0 + b^{-2}\bar{C}_{-2} = b^2 A \tag{5}$$

$$C_{-1} + \bar{C}_{-1} = 0 \tag{6}$$

$$C_{-2} + a^2\bar{C}_0 = a^4\bar{A}, \qquad C_{-2} + b^2\bar{C}_0 = b^4\bar{A},$$

the last pair being simply the complex conjugates of (5) and so yield nothing new. From (6) it follows that C_{-1} is imaginary say

$$C_{-1} = i\lambda, \ \lambda \text{ real} . \tag{7}$$

Solving (5) we find

$$C_0 = (a^2 + b^2) A, \qquad C_{-2} = -a^2 b^2 \bar{A}$$

and therefore

$$\Phi(z) = (a^2 + b^2) A + \frac{i\lambda}{z} - \frac{a^2 b^2 \bar{A}}{z^2} . \tag{8}$$

To find the moment we have, 3.26 (2),

$$I = \pi(a^2 - b^2)\frac{1}{2}(a^2 + b^2) = \frac{1}{2}\pi(a^4 - b^4)$$

$$J = \oint_{\gamma_1 + \gamma_2} z\bar{z}^3\, dz = \oint_{\gamma_1 + \gamma_2} r^5 i e^{-i\theta}\, d\theta = 0$$

$$K = 2i\int\left((a^2 + b^2) A z + i\lambda - \frac{a^2 b^2 \bar{A}}{z}\right) dS = -2\lambda\pi(a^2 - b^2) .$$

Therefore

$$2M = \mu\{\tau(a^2 + b^2) - 2\lambda\}\pi(a^2 - b^2) \tag{9}$$

and this relation determines λ in terms of the moment and the torsion. In particular if the load is applied at the centre, $M = \tau = \lambda = 0$ and

$$\Phi(z) = \left(\frac{3}{4} + \frac{1}{2}\eta\right)\left\{\beta(a^2 + b^2) - \frac{a^2 b^2 \beta}{z^2}\right\} . \tag{10}$$

The same method (cf. 4.9) can be applied to any ring-shaped region which can be mapped on a concentric circular annulus. For example, the ring comprised between confocal ellipses

$$\frac{x^2}{a_1^2} + \frac{y^2}{b_1^2} = 1, \qquad \frac{x^2}{a_2^2} + \frac{y^2}{b_2^2} = 1$$

$$a_1^2 - b_1^2 = 4c^2 = a_2^2 - b_2^2$$

can be mapped by

$$z = \zeta + \frac{c^2}{\zeta}$$

on the region between concentric circumferences of radii $a = \frac{1}{2}(a_1 + b_1)$, $b = \frac{1}{2}(a_2 + b_2)$.

5.24. Cross-section a limaçon

In terms of non-negative parameters b and c Pascal's limaçons are the curves of the family

$$r = b + 2c\cos\theta \tag{1}$$

If $b < 2c$, the curve has a double point at the origin. We exclude this case by the condition $b \geqq 2c$.

If $b = 2c$ we have the cardioid $r = 2c(1 + \cos\theta)$.

If $c = 0$ we have the circle $r = b$.

Fig. 5.24 shows a circle, a cardioid and a limaçon of the type for which $b > 2c$.

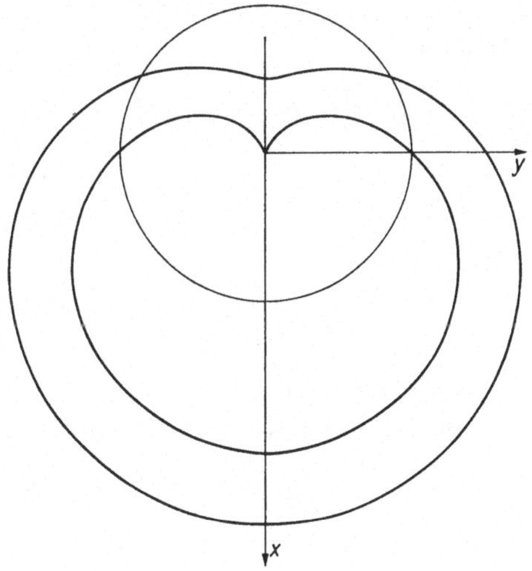

Fig. 5.24

Multiplying (1) by $\sigma = e^{i\theta}$ we get

$$z = c\sigma^2 + b\sigma + c = m(\sigma) \tag{2}$$

and therefore (2) maps the limaçon on the circumference $\sigma = e^{i\theta}$ of the unit circle $|\zeta| \leqq 1$. The general point of the circular disc corresponds to z by

$$z = m(\zeta)\,. \tag{3}$$

To solve the flexure problem for a beam whose cross-section is the limaçon we have from (2)

$$m(\sigma) = c\sigma^2 + b\sigma + c, \qquad \overline{m}\left(\frac{1}{\sigma}\right) = c + \frac{b}{\sigma} + \frac{c}{\sigma^2} = \frac{m(\sigma)}{\sigma^2}$$

$$m'(\sigma) = 2c\sigma + b, \qquad \overline{m}'\left(\frac{1}{\sigma}\right) = b + \frac{2c}{\sigma}\,.$$

Substitute these in 3.51 (6) and use 3.51 (10). Then we get

$$\begin{aligned} H_1(\sigma) = 2rc^3\sigma^3 + (5r + \overline{q})bc^2\sigma^2 + [2pc^2 + (2q + 2\overline{q} + 4r)c^3 \\ + (4r + 2\overline{q})b^2c]\sigma \\ + (3p + \overline{p})bc + (5q + 6\overline{q} + 6r + \overline{r})bc^2 + (\overline{q} + r)b^3\,. \end{aligned}$$

Therefore from 3.51 (11)

$$\begin{aligned} \Phi(z)\,(b + 2c\zeta) = 2rc^3\zeta^3 + (5r + \overline{q})bc^2\zeta^2 + [2pc^2 + (2q + 2\overline{q} + 4r)c^3 \\ + (4r + 2\overline{q})b^2c]\zeta \\ + (3p + \overline{p})bc + (5q + 6\overline{q} + 6r + \overline{r})bc^2 + (\overline{q} + r)b^3\,. \end{aligned} \tag{4}$$

Equations (3) and (4) constitute the solution of the problem.

If we put $c = 0$, we get the result already obtained in 5.22 for the circle, namely $\Phi(z) = b^2(\overline{q} + r)$.

We note that $\Phi(z)$ is infinite if $b + 2c\zeta = 0$, i.e., if $\zeta = -\frac{b}{2c}$. This point is outside the unit circle unless $b = 2c$. Thus the stress is nowhere infinite for the limaçon, unless $b = 2c$. When $b = 2c$ the curve becomes a cardioid and $\zeta = -1$ corresponds to the cusp. At this point the stress becomes infinite. This is to be expected at a re-entrant point and the situation is solved physically by plastic yielding.

For the area S and the position of the centroid we have by the area theorem

$$\begin{aligned} S &= -\frac{1}{2} i \oint_C \bar{z}\, dz = -\frac{1}{2} i \int \bar{m}\left(\frac{1}{\sigma}\right) m'(\sigma)\, d\sigma = \pi(2c^2 + b^2) \\ \xi S &= -\frac{1}{2} i \oint_C z\bar{z}\, dz = -\frac{1}{2} i \int m(\sigma)\, \bar{m}\left(\frac{1}{\sigma}\right) m'(\sigma)\, d\sigma = 2\pi c(b^2 + c^2), \end{aligned} \tag{5}$$

so that

$$\xi = \frac{2c(b^2 + c^2)}{2c^2 + b^2}. \tag{6}$$

Substituting $z = m(\sigma)$, $\bar{z} = \bar{m}(1/\sigma)$ in **5.12** (3), (4) and evaluating the contour integrals, we find without difficulty

$$A + B = \left[(b^2 + 2c^2)^2 + \frac{2c^4(b^2 - 2c^2)}{b^2 + 2c^2}\right]\frac{\pi}{2} \tag{7}$$

$$B - A + 2iH = -\frac{b^4 c^2}{b^2 + 2c^2}\pi. \tag{8}$$

Therefore $H = 0$ which was obvious from the symmetry, so that

$$A = \{(b^2 + 2c^2)^2 + 2c^2(b^2 - c^2)\}\frac{\pi}{4}. \tag{9}$$

To find the moment we evaluate the integrals I, J, K of **3.26** (2)

$$I = A + B + \xi\bar{\xi} S = \frac{\pi}{2}(b^4 + 12b^2c^2 + 6c^4)$$

$$\frac{J}{2\pi i} = \frac{1}{2\pi i}\int_C z\bar{z}^3\, dz = \frac{1}{2\pi i}\int_\gamma (c\sigma^2 + b\sigma + c)\left(c + \frac{b}{\sigma} + \frac{c}{\sigma^2}\right)^3 (2c\sigma + b)\, d\sigma$$

$$= \frac{1}{2\pi i}\int_\gamma \frac{(c\sigma^2 + b\sigma + c)^4}{\sigma^6}(2c\sigma + b)\, d\sigma.$$

Picking out the residue, the coefficient of $1/\sigma$, we get

$$J = 2\pi i(6b^4 + 36b^2c^2 + 12c^4)c.$$

Using (3) and (4) we have

$$K = \int_C z\bar{z}\,\Phi(z)\, dz = \int_\gamma \frac{m(\sigma)\, m(\sigma)}{\sigma^2(2\sigma c + b)} H_1(\sigma)\,(2\sigma c + b)\, d\sigma$$

whence, picking out the coefficient of $1/\sigma$, we have

$$K = 4\pi i c\{p(c^3 + 3b^2c) + \bar{p}b^2c + q(c^4 + 5b^2c^2) + \bar{q}(c^4 + 7b^2c^2 + b^4) + r(2c^4 + 8b^2c^2 + b^4) + \bar{r}b^2c^2\}.$$

Therefore from 3.26 (4) the moment $M = -M_3$ is given by

$$\frac{2M}{\pi\mu} = \tau(b^4 + 4b^2c^2 + 2c^4) - i(\beta - \bar{\beta})c\{(b^4 + 5b^2c^2 + c^4) + \eta(2b^4 + 8b^2c^2 + 2c^4)\}.$$

To find the centre of flexure we have from (6) and 5.2 (3)

$$\tau_0 = i\eta(\beta - \bar{\beta})c\frac{b^2 + c^2}{b^2 + 2c^2}$$

so that from 5.2 (5)

$$\frac{2M_0}{\pi\mu} = -Nic(\beta - \bar{\beta}),$$

where

$$N = b^4 + 5b^2c^2 + c^4 + \eta\frac{b^6 + 7b^4c^2 + 12b^2c^4 + 2c^6}{b^2 + 2c^2}. \tag{11}$$

From 5.2 the centre of flexure is given by

$$\bar{z}_f = 2i\frac{\partial M_0}{\partial P} = \pi\mu Nc\frac{\partial}{\partial P}(\beta - \bar{\beta}).$$

Now from 5.11 (7) putting $-Q = P = P_1 + iP_2$ and remembering that $H = 0$ we have

$$\beta - \bar{\beta} = \frac{2BP - 2B\bar{P}}{2EAB}$$

and therefore

$$\frac{\partial(\beta - \bar{\beta})}{\partial P} = \frac{1}{EA}.$$

Therefore

$$\bar{z}_f = \frac{\pi\mu Nc}{EA} = \frac{\pi Nc}{2A(1+\eta)} \tag{12}$$

where N is given by (11) and A by (9). Thus the centre of flexure lies on the axis of symmetry at the point $(x_f, 0)$.

In the case of the *cardioid* we have $b = 2c$ and therefore

$$x_f = \frac{111 + 113\eta}{63(1+\eta)}c$$

which agrees with the result obtained by Sokolnikoff by a different method.

In the case of the circle we have $c = 0$ and therefore $x_f = 0$, so that the centre of flexure is at the centre of the circle.

5.26. Cross-section one loop of Bernoulli's lemniscate

One loop of Bernoulli's lemniscate (see fig. 4.46 (i))

$$r^2 = 2c^2 \cos 2\theta \tag{1}$$

is mapped on the unit circumference γ by

$$z = c(1+\sigma)^{1/2}, \sigma = e^{i\eta} .$$

Thus

$$m(\sigma) = c(1+\sigma)^{1/2} \tag{2}$$

is not a rational function so that the simple method of 3.51, note (iv), is not available. We have in fact

$$\begin{aligned} H(\sigma) = {} & \frac{1}{2} p c^2 \sigma^{-1/2} + \frac{1}{2} q c^3 (1+\sigma)^{1/2} \sigma^{-1} + \frac{1}{2} r c^3 (1+\sigma)^{1/2} \sigma^{-1/2} \\ & + \frac{1}{2} \bar{p} c^2 \sigma^{-3/2} + \frac{1}{2} \bar{q} c^3 (1+\sigma)^{1/2} \sigma^{-3/2} + \frac{1}{2} \bar{r} c^3 (1+\sigma)^{1/2} \sigma^{-2} . \end{aligned} \tag{3}$$

Using the identities

$$\frac{1}{\sigma(\sigma-\zeta)} = \frac{1}{\zeta}\left(\frac{1}{\sigma-\zeta} - \frac{1}{\sigma}\right)$$

$$\frac{1}{\sigma^2(\sigma-\zeta)} = \frac{1}{\zeta^2}\left(\frac{1}{\sigma-\zeta} - \frac{1}{\sigma}\right) - \frac{1}{\zeta}\cdot\frac{1}{\sigma^2}$$

the evaluation of

$$\Phi(z)\, m'(\zeta) = \frac{1}{2\pi i}\oint_\gamma \frac{H(\sigma)\, d\sigma}{\sigma-\zeta} \tag{4}$$

reduces effectively to calculating

$$f_1(\zeta) = \frac{1}{2\pi i}\oint_\gamma \frac{(1+\sigma)^{1/2} d\sigma}{\sigma-\zeta}, \quad f_2(\zeta) = \frac{1}{2\pi i}\oint_\gamma \frac{\sigma^{1/2} d\sigma}{\sigma-\zeta},$$

$$f_3(\zeta) = \frac{1}{2\pi i}\oint_\gamma \frac{(1+\sigma)^{1/2}\sigma^{1/2} d\sigma}{\sigma-\zeta},$$

for

$$\frac{1}{2\pi i}\oint_\gamma \frac{(1+\sigma)^{1/2} d\sigma}{\sigma} = f_1(0), \quad \frac{1}{2\pi i}\oint_\gamma \frac{(1+\sigma)^{1/2} d\sigma}{\sigma^2} = f_1'(0)$$

$$\frac{1}{2\pi i}\oint_\gamma \frac{d\sigma}{\sigma^{1/2}} = f_2(0), \quad \frac{1}{2\pi i}\oint_\gamma \frac{d\sigma}{\sigma^{3/2}} = f_2'(0)$$

$$\frac{1}{2\pi i}\oint_\gamma \frac{(1+\sigma)^{1/2} d\sigma}{\sigma^{1/2}} = f_3(0), \quad \frac{1}{2\pi i}\oint_\gamma \frac{(1+\sigma)^{1/2} d\sigma}{\sigma^{3/2}} = f_3'(0) .$$

Since $(1+\sigma)^{1/2}$ has no singularity inside the circle Cauchy's formula gives

$$f_1(\zeta) = (1+\zeta)^{1/2}, \tag{5}$$

where we choose the determination of the square root to be that which is equal to unity when $\zeta = 0$.

The integrands of $f_2(\zeta)$, $f_3(\zeta)$ have a branch point at $\sigma = 0$. To evalute them we use the contour $ABCDEF$ shown in fig. 4.46 (ii) (for more details of the computation which follows see 4.46). We denote by L the limit of the path $BCDEF$ when the radius r of the circle CDE tends to zero. Within this contour the integrands of $f_2(\zeta)$ and $f_3(\zeta)$ are holomorphic except at ζ and therefore by Cauchy's theorem

$$\frac{1}{2\pi i}\oint_{\gamma}\frac{(1+\sigma)^{1/2}\sigma^{1/2}\,d\sigma}{\sigma-\zeta}+\frac{1}{2\pi i}\int_{L}\frac{(1+\sigma)^{1/2}\sigma^{1/2}\,d\sigma}{\sigma-\zeta}=\zeta^{1/2}(1+\zeta)^{1/2}.$$

Now

$$-\int_{L}\frac{(1+\sigma)^{1/2}\sigma^{1/2}\,d\sigma}{\sigma-\zeta}=2i\int_{0}^{1}\frac{t^{1/2}(1-t)^{1/2}\,dt}{t+\zeta},$$

where t is a real variable. This integral is easily evaluated by the substitution $t = 1/(1+u^2)$ and we get finally

$$f_3(\zeta)=\frac{1}{2}(1+2\zeta). \tag{6}$$

The same contour and the substitution $t = u^2$ for the integral on L gives, as in 4.46,

$$f_2(\zeta)=\frac{2}{\pi}\{1+\zeta^{1/2}\tan^{-1}\zeta^{1/2}\}. \tag{7}$$

Thus finally we get from (4)

$$\frac{\frac{1}{2}c\Phi(z)}{(1+\zeta)^{1/2}}=\left(\frac{pc^2}{\pi}\right)\frac{1}{\zeta^{1/2}}\tan^{-1}\zeta^{1/2}$$

$$+\left(\frac{\bar{p}c^2}{\pi}\right)\left\{-\frac{1}{\zeta}+\frac{1}{\zeta^{3/2}}\tan^{-1}\zeta^{1/2}\right\}+\frac{1}{2}\frac{qc^3}{\zeta}\{-1+(1+\zeta)^{1/2}\}$$

$$+\frac{1}{2}rc^3+\frac{1}{2}\frac{\bar{r}c^3}{\zeta^2}\left\{-1-\frac{1}{2}\zeta+(1+\zeta)^{1/2}\right\}.$$

Near $\zeta = 0$ we have the expansions

$$\tan^{-1}\zeta^{1/2}=\zeta^{1/2}-\frac{1}{3}\zeta^{3/2}+\cdots,\quad (1+\zeta)^{1/2}=1+\frac{1}{2}\zeta+\cdots$$

from which it follows that $\Phi(z)$ has no singularity at $\zeta = 0$. Also at $\zeta = -1$, the vertex of the loop, $\Phi(z) = 0$ and the stress vanishes as it should at this salient point.

5.3. Half-sections

Suppose we have a cross-section for which the x-axis is an axis of symmetry. Let this section be mapped on the unit circle $|t| \leq 1$ in a t-plane by

$$z = m(t) \tag{1}$$

in such a way that the axis of symmetry OA of the section maps on to the diameter OA on the real axis in the t-plane. Then the same mapping function (1) maps the *half-section* OAB in the z-plane on the semicircle OAB in the t-plane.

This semicircle can in turn be mapped on the unit circle in the ζ-plane by

$$\frac{\zeta - 1}{\zeta + 1} = -i\left(\frac{t-1}{t+1}\right)^2 \tag{2}$$

and these two mappings combine to map the half-section in the z-plane on the unit circle on the ζ-plane, so that we can also solve the flexure

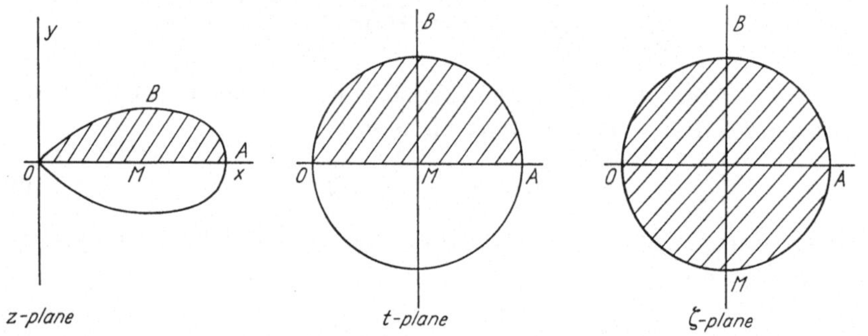

Fig. 5.3

problem for the half-section when we know the mapping function (1) for the whole section. See also 4.48.

5.4. Shear stress functions

Remembering that in the isotropic case $k_{33} = 1/E$ the shear components $\widehat{xz}$, $\widehat{yz}$ are given by 2.24 (3) in the form

$$\widehat{xz} = \frac{\partial \psi}{\partial y} - \mu(1+\eta)\{A_1 x^2 + (C_1 - \varepsilon) m x\} \tag{1}$$

$$\widehat{yz} = -\frac{\partial \psi}{\partial x} - \mu(1+\eta)\{B_1 y^2 + (C_1 - \varepsilon) m y\} \tag{2}$$

in terms of the *shear stress function* ψ.

From 2.9 (4), ψ satisfies the equation

$$\frac{1}{\mu}\nabla^2 \psi = 2\eta(A_1 y - B_1 x) - 2\tau\,. \tag{3}$$

Let us introduce the real-valued stress function $F = F(x, y)$ defined by

$$\mu F = \psi + \mu B_1(1+\eta)\,{}^{\backprime}s(x) - \mu A_1(1+\eta)\,{}^{\backprime}t(y)\,, \tag{4}$$

where $s(x)$, $t(y)$ are arbitrary functions of x and y respectively. Corresponding to any choice of $s(x)$ and $t(y)$, (4) will define a shear stress function F in terms of which

$$\widehat{xz} = \mu\left\{\frac{\partial F}{\partial y} + (1+\eta)\,[A_1\{t(y) - x^2\} - (C_1 - \varepsilon)\,m\,x]\right\} \tag{5}$$

$$\widehat{yz} = \mu\left\{-\frac{\partial F}{\partial x} + (1+\eta)\,[B_1\{s(x) - y^2\} - (C_1 - \varepsilon)\,m\,y]\right\} \tag{6}$$

also from (3)

$$\nabla^2 F = A_1[2\eta y - (1+\eta)\,t'(y)] - B_1[2\eta x - (1+\eta)\,s'(x)] - 2\tau. \tag{7}$$

The boundary condition in the case of pure antiplane stress is

$$l\,\widehat{xz} + m\,\widehat{yz} = 0, \qquad l = \frac{dy}{ds}, \qquad m = -\frac{dx}{ds} \text{ on } C \tag{8}$$

l, m being the direction cosines of the outward normal to C.

Therefore (5), (6), (8) give the boundary condition

$$\begin{aligned}\frac{1}{1+\eta}\frac{dF}{ds} = &-[A_1\{t(y) - x^2\} - (C_1 - \varepsilon)\,m\,x]\,\frac{dy}{ds}\\ &+ [B_1\{s(x) - y^2\} - (C_1 - \varepsilon)\,m\,y]\,\frac{dx}{ds} \text{ on } C\,.\end{aligned} \tag{9}$$

Therefore in terms of the shear stress function F the problem is to find the solution of (7) which satisfies the boundary condition (9).

Note that the functions $s(x)$, $t(y)$ are entirely at our disposal. We can, for example, so choose them that (9) simplifies to $dF/ds = 0$, or alternatively we can choose them so that $\nabla^2 F = 0$.

We can relate ψ and therefore F to χ_2 the imaginary part of $'\Phi(z)$ as follows:

From (1) and (2), using 5.11 (10) we can write

$$\begin{aligned}\frac{\widehat{xz} - i\,\widehat{yz}}{\mu} = \frac{\partial}{\partial z}\Big\{&\frac{2i\psi}{\mu} - \frac{1+\eta}{12}(\beta z^3 - \bar\beta \bar z^3) - \frac{1+\eta}{4}(\beta\bar z + \bar\beta z)\,z\bar z\\ &+ \frac{1}{2}(1+\eta)\,(\beta\bar\xi + \bar\beta\xi)\,z\bar z\Big\}.\end{aligned}$$

Now let

$$'\Phi(z) = \chi_1 + i\,\chi_2 \tag{10}$$

so that χ_1 and χ_2 are conjugate functions.

Then from 3.2 (5)

$$\begin{aligned}\frac{\widehat{xz} - i\,\widehat{yz}}{\mu} = \frac{\partial}{\partial z}\Big\{&2i\,\chi_2 - \left[i\tau - \frac{1}{2}(1+\eta)\,(\beta\bar\xi + \bar\beta\xi)\right] z\bar z\\ &- \frac{1}{4}(1+2\eta)\,\beta\bar z^2 z - \frac{1}{4}\bar\beta z^2\bar z\Big\}.\end{aligned}$$

Equating the expressions of $\widehat{xz} - i\,\widehat{yz}$ we have, save for a real constant, for χ_2 and ψ are real-valued,

$$\chi_2 = \frac{\psi}{\mu} + \frac{1}{2}\tau z\bar z + \frac{1}{8}\,i\,\eta\,(\bar\beta z - \beta\bar z)\,z\bar z + \frac{i(1+\eta)}{24}(\beta z^3 - \bar\beta\bar z^3). \tag{11}$$

5.41. Timoshenko's stress function

In 5.4 (9) let us choose the arbitrary functions $s(x)$, $t(y)$ such that the right-hand side is zero. Denoting the stress function by $T = T(x, y)$ we will then have

$$\frac{dT}{ds} = 0 \qquad \text{on } C \tag{1}$$

and therefore T is constant on C. Without loss of generality we can take this constant to be zero for a simply connected region. Thus the boundary condition is

$$T = 0 \text{ on } C\,. \tag{2}$$

The function T is *Timoshenko's stress function.*

Assuming that dx/ds and dy/ds are both different from zero we can satisfy (1) and therefore (2) by choosing $s(x)$ and $t(y)$ so that on C

$$A_1\{t(y) - x^2\} - (C_1 - \varepsilon)mx = 0 \tag{3}$$

$$B_1\{s(x) - y^2\} - (C_1 - \varepsilon)my = 0\,. \tag{4}$$

Observe, however, that for any part of the boundary along which $dy/ds = 0$ no condition need be imposed on $t(y)$. Similarly, for any part of the boundary along which $dx/ds = 0$, no condition need be imposed on $s(x)$.

If the equation of the boundary C can be written in the form $x = f_1(y)$, we can take

$$A_1 t(y) \equiv A_1[f_1(y)]^2 + (C_1 - \varepsilon)mf_1(y)\,. \tag{5}$$

Similarly if the equation of the boundary can be written in the form $y = f_2(x)$, (4) will be satisfied by

$$B_1 s(x) \equiv B_1[f_2(x)]^2 + (C_1 - \varepsilon)mf_2(x)\,. \tag{6}$$

Observe also that the functions $t(y)$, $s(x)$ defined by (5) and (6) apply throughout the material, not only on the boundary.

The shears $\widehat{xz}, \widehat{yz}$ are given by 5.4 (5), (6) in which T is written for F, while T satisfies 5.4 (7). In the case where $B_1 = 0$ the differential equation satisfied by T may be written

$$\nabla^2 T = A_1[2\eta y - (1+\eta)t'(y)] - 2\tau = -Q(y) \tag{7}$$

the boundary condition being

$$T = 0\,. \tag{8}$$

The equations which give the sag w of a membrane fixed at the edge and under pressure q, are, from 4.77 (4),

$$\nabla^2 w + q = 0, \qquad w = 0 \qquad \text{on } C\,. \tag{9}$$

Comparing these with (8) and (9), we have the *membrane analogy for flexure*, namely that Timoshenko's stress function is proportional to the sag of a membrane stretched by uniform tension, fixed at its edge, which coincides with the boundary of the cross-section and which is subjected to a variable pressure proportional to $Q(y)$.

5.42. Cross-section an ellipse

Let the boundary be the ellipse

$$\frac{x^2}{a^2} + \frac{y^2}{b^2} = 1. \tag{1}$$

Let the loading be $(-W_x, 0, 0)$ applied at the centre of the free end. Then from 5.11 (7), since $B = \pi a^3 b/4$,

$$A_1 = \frac{2W_x}{\pi\mu(1+\eta)a^3 b}, \quad B_1 = 0 \tag{2}$$

while if we omit body force

$$C_1 - \varepsilon = 0. \tag{3}$$

To use Timoshenko's stress function we must satisfy 5.41 (3) and (4), the second of which is satisfied identically on account of (2) and (3). We satisfy 5.41 (3) by observing that from (1)

$$x^2 = a^2\left(1 - \frac{y^2}{b^2}\right) \text{ on } C \tag{4}$$

and therefore we take

$$t(y) = a^2\left(1 - \frac{y^2}{b^2}\right) \text{ everywhere} \tag{5}$$

and therefore from 5.4 (7), since $\tau = 0$, for the load is applied along an axis of symmetry,

$$\nabla^2 T = 2A_1 y\left\{\eta + (1+\eta)\frac{a^2}{b^2}\right\}. \tag{6}$$

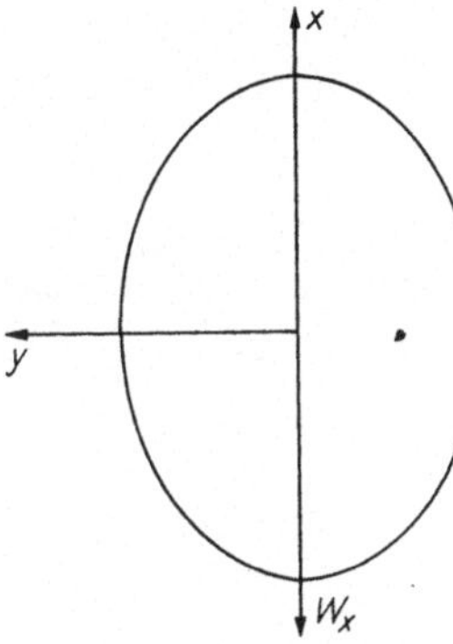

Fig. 5.42

Since $\nabla^2 T$ is proportional to y and $T = 0$ on the boundary, we can satisfy (6) by taking

$$T = ky\left(\frac{x^2}{a^2} + \frac{y^2}{b^2} - 1\right), \tag{7}$$

where k is a constant given by

$$k\left(\frac{2}{a^2} + \frac{6}{b^2}\right) = 2A_1\left\{\eta + (1+\eta)\frac{a^2}{b^2}\right\}$$

or

$$k = \frac{2W_x}{\pi\mu(1+\eta)ab} \cdot \frac{[b^2\eta + a^2(1+\eta)]}{b^2 + 3a^2}. \tag{8}$$

Therefore from 5.4 (5), (6)

$$\widehat{xz} = \frac{2W_x}{\pi(1+\eta)ab}\left\{\frac{b^2\eta + a^2(1+\eta)}{b^2+3a^2}\left(\frac{x^2}{a^2}+\frac{3y^2}{b^2}-1\right)\right.$$

$$\left. -(1+\eta)\left(\frac{x^2}{a^2}+\frac{y^2}{b^2}-1\right)\right\}, \tag{9}$$

$$\widehat{yz} = -\frac{W_x}{\pi(1+\eta)a^3b}\cdot\frac{b^2\eta + a^2(1+\eta)}{b^2+3a^2}4xy,$$

which agree with 5.22 (6), (7) when $b = a$, and $P_1 = W_x$.

5.44. Cross-section an equilateral triangle

Let the cross-section be the equilateral triangle of fig. 4.53 whose boundary has the equation

$$f(x,y) = (x-a)(x-y\sqrt{3}+2a)(x+y\sqrt{3}+2a) = 0 \tag{1}$$

and let the load W_y be applied along the negative y-axis.

On the side $x = a$, $dx/ds = 0$, and therefore in 5.4 (9) no condition is imposed along this side. In the present case $A_1 = 0$, $C_1 - \varepsilon = 0$ in the absence of body force so that to enforce the boundary condition $T = 0$ on Timoshenko's stress function, we have only the requirement

$$B_1\{s(x) - y^2\} = 0 \text{ on the oblique sides} \tag{2}$$

namely $y = \pm(x+2a)/\sqrt{3}$. Therefore we must take

$$s(x) = \frac{1}{3}(x+2a)^2. \tag{3}$$

Also T must satisfy 5.4 (7), with $\tau = 0$, so that

$$\nabla^2 T = -\frac{4}{3}(1+\eta)B_1\left\{\frac{\left(\eta - \frac{1}{2}\right)x}{1+\eta} - a\right\}. \tag{4}$$

In the special case when Poisson's ratio η is $\frac{1}{2}$ this equation will be satisfied by

$$T = kB_1 f(x,y),$$

where substitution in (4) gives $k = \frac{1}{6}$. Thus when $\eta = \frac{1}{2}$

$$T = \frac{1}{6}B_1(x-a)\left[(x+2a)^2 - 3y^2\right] \tag{5}$$

where B_1 is given by 5.11 (7).

5.46. Gauss's theorem and integration by parts

Let τ be the region interior to a closed surface S and let $\boldsymbol{n}$ be the outwards drawn unit normal vector to S at the element of area dS.

Let X be a differentiable vector or scalar function of position. Then Gauss's theorem [MILNE-THOMSON (5)] states that

$$\int_\tau \nabla \circ X \, d\tau = \int_S \boldsymbol{n} \circ X \, dS \qquad (1)$$

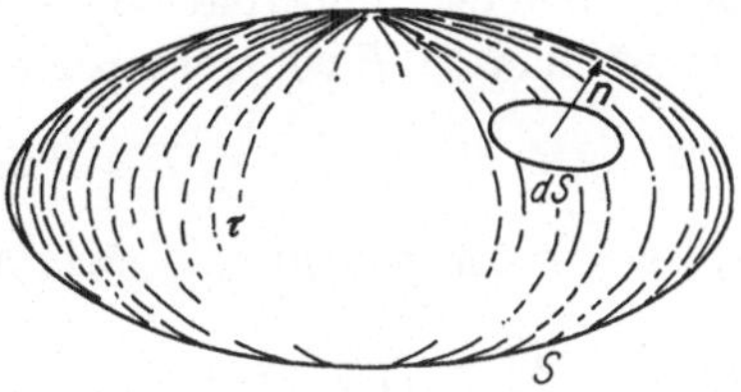

Fig. 5.46

where the small circle denotes scalar, vector, or dyadic multiplication, whichever is applicable and appropriate.

If we put

$$X = \boldsymbol{q}\phi, \qquad (2)$$

where $\boldsymbol{q}$ is a vector and ϕ a scalar field, we have

$$\nabla \cdot (\boldsymbol{q}\phi) = \boldsymbol{q} \cdot \nabla\phi + \phi \nabla \cdot \boldsymbol{q} \qquad (3)$$

and therefore from Gauss's theorem

$$\int_S \boldsymbol{n} \cdot \boldsymbol{q}\phi \, dS = \int_\tau \boldsymbol{q} \cdot \nabla\phi \, d\tau + \int_\tau \phi \nabla \cdot \boldsymbol{q} \, d\tau$$

or rearranging

$$\int_\tau \boldsymbol{q} \cdot \nabla\phi \, d\tau = \int_S (\boldsymbol{n} \cdot \boldsymbol{q})\phi \, dS - \int_\tau \phi \, \nabla \cdot \boldsymbol{q} \, d\tau . \qquad (4)$$

This is a formula for *integration by parts* by means of which the roles of $\boldsymbol{q}$ and ϕ in the integrand can be interchanged.

To interpret (4) in components let

$$\boldsymbol{n} = \boldsymbol{i}\,l + \boldsymbol{j}m + \boldsymbol{k}n, \; \boldsymbol{q} = \boldsymbol{i}u + \boldsymbol{j}v + \boldsymbol{k}w, \; \nabla = \boldsymbol{i}\frac{\partial}{\partial x} + \boldsymbol{j}\frac{\partial}{\partial y} + \boldsymbol{k}\frac{\partial}{\partial R} .$$

Then (4) becomes

$$\int_\tau \left(u\frac{\partial\phi}{\partial x} + v\frac{\partial\phi}{\partial y} + w\frac{\partial\phi}{\partial R}\right) d\tau = \int_S \phi\,(lu + mv + nw)\, dS - \int_\tau \phi\left(\frac{\partial u}{\partial x} + \frac{\partial v}{\partial y} + \frac{\partial w}{\partial R}\right) d\tau . \qquad (5)$$

Here l, m, n are the direction cosines of the outward normal.

In two dimensions, with a similar notation,

$$\int_S \left(u\frac{\partial\phi}{\partial x} + v\frac{\partial\phi}{\partial y}\right) dS = \oint_C \phi\,(lu + mv)\, ds - \int_S \phi\left(\frac{\partial u}{\partial x} + \frac{\partial v}{\partial y}\right) dS . \qquad (6)$$

Here the plane area S replaces the volume τ, and the contour C which encloses S replaces the three-dimensional surface S in (5).

5.47. The principle of virtual stresses applied to Timoshenko's stress function

We deduce the equation 5.41 (7) satisfied by Timoshenko's stress function from the principle of virtual stresses, 4.8. For simplicity of writing we shall suppose that $B_1 = 0$, that the origin is at the centroid of the free end, and that body-force is absent. Then using Timoshenko's stress function T which satisfies the condition

$$T = 0 \text{ on the boundary} \tag{1}$$

the stress components 5.4, 5.11 are

$$\widehat{xz} = \mu\left\{\frac{\partial T}{\partial y} + (1+\eta)\,A_1\,[t(y) - x^2]\right\},\ \widehat{yz} = -\,\mu\frac{\partial T}{\partial x},\ \widehat{zz} = E A_1 x R\,. \tag{2}$$

We shall take the load to be $(-W_x, 0, 0)$ along a principal axis of inertia, the x-axis. Then from 5.11 (7)

$$A_1 = \frac{W_x}{EB},\qquad B = \int_S x^2\, dS,\qquad B_1 = 0\,. \tag{3}$$

Therefore from 1.82 and 1.98 (5) the strain energy function is

$$U = \frac{1}{2E}\,\widehat{zz}^2 + \frac{1}{2\mu}\,(\widehat{yz}^2 + \widehat{xz}^2)\,. \tag{4}$$

Therefore the potential energy is

$$V = \int_0^L\int_S U\, dS\, dR = \frac{W_x^2 L^3}{6EB} + \frac{1}{2}\,\mu L J_1 \tag{5}$$

where

$$J_1 = \int_S \left[\left(\frac{\partial T}{\partial x}\right)^2 + \left\{\frac{\partial T}{\partial y} + (1+\eta)\,A_1\,[t(y) - x^2]\right\}^2\right] dS\,.$$

Therefore

$$V = \frac{1}{2}\,\mu L J_2 + V_0$$

where

$$J_2 = \int_S \left\{\left(\frac{\partial T}{\partial x}\right)^2 + \left(\frac{\partial T}{\partial y}\right)^2 + 2A_1(1+\eta)\,[t(y) - x^2]\,\frac{\partial T}{\partial y}\right\} dS \tag{6}$$

and V_0 is independent of the stress function T.

Since the stress variation arises only from varying the stress function T we have

$$\delta\,\widehat{xx} = \delta\,\widehat{xy} = \delta\,\widehat{yy} = \delta\,\widehat{zz} = 0$$

$$\delta(\widehat{xz}) = \delta\left(\mu\frac{\partial T}{\partial y}\right) = \mu\frac{\partial(\delta T)}{\partial y},\ \delta\,\widehat{yz} = -\,\delta\left(\mu\frac{\partial T}{\partial x}\right) = -\,\mu\frac{\partial(\delta T)}{\partial x}\,. \tag{7}$$

The direction cosines of the outward normal to the element $d\Sigma$ of the lateral surface are $(l, m, 0)$ where

$$l = \frac{dy}{ds}, \quad m = -\frac{dx}{ds},$$

We take the variation δT such that the lateral surface of the beam, which is free from external applied force remains free of applied force when the stress is varied. Therefore on the lateral surface

$$0 = \delta(l\,\widehat{xz} + m\,\widehat{yz}) = l\frac{\partial(\delta T)}{\partial y} - m\frac{\partial(\delta T)}{\partial x} = \frac{\partial(\delta T)}{\partial s}$$

and therefore we must take

$$\delta T = 0 \text{ on the lateral surface}. \tag{8}$$

In the elastic case the complementary potential energy coincides with the potential energy and therefore from 4.8 (9)

$$\delta V = \int_{\Sigma} \boldsymbol{q} \cdot \delta \boldsymbol{F}_n d\Sigma \tag{9}$$

the integral being taken over the entire surface of the beam.

Since the lateral surface remains free of applied force, $\delta \boldsymbol{F}_n = 0$ on the lateral surface and therefore the only contributions to (9) come from the free end and the root.

Now we have

$$\int_{\Sigma} \boldsymbol{q} \cdot \delta \boldsymbol{F}_n d\Sigma = \int_{\Sigma} (u\delta\,\widehat{xz} + v\delta\,\widehat{yz} + w\delta\,\widehat{zz})\,d\Sigma = \int_{\Sigma} (u\delta\,\widehat{xz} + v\delta\,\widehat{yz})\,d\Sigma$$

since from (7), $\delta\,\widehat{zz} = 0$.

Also on the free end ($R = 0$), from 5.16 we have $u = v = 0$ apart from a rigid-body movement, and so the free end contributes nothing. Therefore the principle of virtual stresses gives

$$\delta V = \int_{\Sigma} \boldsymbol{q} \cdot \delta \boldsymbol{F}_n\, d\Sigma = \int_{S} (u\delta\,\widehat{xz} + v\delta\,\widehat{yz})\,dS \tag{10}$$

the integral being taken over the root. Now

$$\int_{S} (u\delta\,\widehat{xz} + v\delta\,\widehat{yz})\,dS = \mu\int_{S} \left[u\frac{\partial(\delta T)}{\partial y} - v\frac{\partial(\delta T)}{\partial x}\right] dS$$

$$= \mu\oint_{C} \delta T[-lv + mu]\,ds - \mu\int_{S} \delta T\left[-\frac{\partial v}{\partial x} + \frac{\partial u}{\partial y}\right] dS$$

on integrating by parts, 5.46 (6). The first integral on the right vanishes since $\delta T = 0$ on C, while from 5.16 on the root $R = L$ we have

$$\frac{\partial u}{\partial y} - \frac{\partial v}{\partial x} = 2\eta A_1 L y - 2L\tau.$$

Therefore finally

$$\int_{\Sigma} q \cdot \delta \mathbf{F}_n \, d\Sigma = 2\mu L \tau \int_S \delta T \, dS - 2\mu \eta A_1 L \int_S y \delta T \, dS. \tag{11}$$

Also from (5) and (6) we have

$$\begin{aligned} \delta V &= \frac{1}{2} \mu L \delta J_2 \\ &= \mu L \int_S \left[\frac{\partial T}{\partial x} \frac{\partial (\delta T)}{\partial x} + \left\{ \frac{\partial T}{\partial y} + (1+\eta) A_1 [t(y) - x^2] \right\} \frac{\partial (\delta T)}{\partial y} \right] dS \\ &= \mu L \oint_C \delta T \left[\frac{\partial T}{\partial x} + (1+\eta) A_1 [t(y) - x^2] \, m \right] ds \\ &\quad - \mu L \int_S \delta T \left[\nabla^2 T + (1+\eta) A_1 t'(y) \right] dS \end{aligned} \tag{12}$$

on integrating by parts, 5.46 (6). The first integral vanishes, for $\delta T = 0$ on C.

Putting (11) and (12) in (9) we have

$$\int_S \delta T \left[\nabla^2 T - 2\eta A_1 y + (1+\eta) A_1 t'(y) + 2\tau \right] dS = 0. \tag{13}$$

Since δT is zero on the surface and is arbitrary inside the material we have from (13)

$$\nabla^2 T = A_1 [2\eta y - (1+\eta) t'(y)] - 2\tau \tag{14}$$

which is precisely 5.41 (7).

Thus Castigliano's principle applied to Timoshenko's stress function T in the presence of the boundary condition $\delta T = 0$ leads to the equation (14) satisfied by T.

Combining (9), (11), (6) with 4.8 (14) we see that Timoshenko's stress function T for the present problem minimizes the integral

$$\begin{aligned} J_3 = \int_S \Bigg\{ & \left(\frac{\partial T}{\partial x} \right)^2 + \left(\frac{\partial T}{\partial y} \right)^2 \\ & + 2A_1 (1+\eta) [t(y) - x^2] \frac{\partial T}{\partial y} + 4T(\eta A_1 y - \tau) \Bigg\} dS. \end{aligned} \tag{15}$$

We can use this fact to find an approximate stress function T_a which must be chosen beforehand to satisfy the condition

$$T_a = 0 \qquad \text{on } C. \tag{16}$$

If we choose T_a to obey this condition and to contain one or more arbitrary constants $a, b, c, \ldots$, we obtain the values of these constants from the conditions that J_3 is a minimum when T_a is substituted for T, namely

$$\frac{\partial J_3}{\partial a} = 0, \frac{\partial J_3}{\partial b} = 0, \frac{\partial J_3}{\partial c} = 0, \ldots \tag{17}$$

Observe also that it is not necessary that T_a should be chosen beforehand to satisfy (14), for the condition of minimizing (15) will lead to the approximate satisfaction of (14), the more accurately, the better the approximation T_a.

5.48. Cross-section an ellipse (virtual stresses)

Let the cross-section be bounded by the ellipse

$$\frac{x^2}{a^2} + \frac{y^2}{b^2} = 1 \tag{1}$$

and let the load be $(-W_x, 0, 0)$ applied at the centroid of the free end, fig. 5.42. In this case since the x-axis is an axis of symmetry there is no torsion, $\tau = 0$.

Take as an approximation to Timoshenko's stress function

$$T_a = ky\left(\frac{x^2}{a^2} + \frac{y^2}{b^2} - 1\right), \; k \text{ constant.} \tag{2}$$

On account of (1) this vanishes on the boundary C.

Substitution of (2) in 5.47 (15) with $t(y) = a^2\left(1 - \frac{y^2}{b^2}\right)$ leads to

$$J_3 = \int\limits_S \left\{k^2\left[\frac{x^4}{a^4} + \frac{9y^4}{b^4} + x^2y^2\left(\frac{4}{a^4} + \frac{6}{a^2b^2}\right) - \frac{2x^2}{a^2} - \frac{6y^2}{b^2} + 1\right]\right.$$
$$+ 2kA_1\left[a^2(1+\eta)\left(-\frac{x^4}{a^4} - \frac{3y^4}{b^4} - \frac{4x^2y^2}{a^2b^2} + \frac{2x^2}{a^2} + \frac{4y^2}{b^2} - 1\right)\right.$$
$$\left.\left. + 2\eta\left(\frac{y^4}{b^2} + \frac{x^2y^2}{a^2} - y^2\right)\right]\right\} dS.$$

Now if $S = \pi ab$ is the area of the ellipse we have from Ex. V. 26

$$\int\limits_S x^4\, dS = \frac{Sa^4}{8}, \int\limits_S y^4\, dS = \frac{Sb^4}{8}, \int\limits_S x^2y^2\, dS = \frac{Sa^2b^2}{24},$$

while

$$\int\limits_S x^2\, dS = \frac{Sa^2}{4}, \int\limits_S y^2\, dS = \frac{Sb^2}{4}$$

from the well-known rule for moments of inertia.

Therefore we have

$$J_3 = Sk^2\frac{(b^2 + 3a^2)}{6a^2} - 2kSA_1\frac{a^2(1+\eta) + b^2\eta}{6}$$

and J_3 is a minimum when

$$\frac{\partial J_3}{\partial k} = 0$$

whence

$$k = \frac{2 W_x}{\mu(1+\eta) S} \cdot \frac{b^2 \eta + a^2 (1+\eta)}{b^2 + 3a^2}.$$

in agreement with 5.42.

With this value of k the approximation coincides with the exact result.

5.5. de St. Venant's flexure function

We now introduce de St. Venant's flexure function as defined below. In order not to complicate the algebra too much we shall make the following assumptions concerning the flexure problem for an isotropic cylindrical beam fixed at one end and bent by a transverse force applied at the free end.

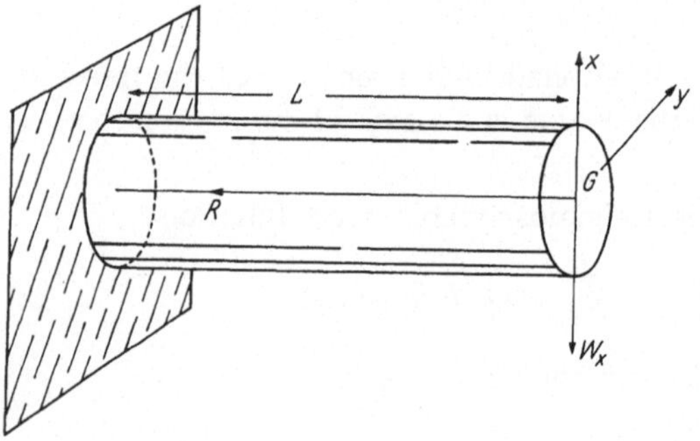

Fig. 5.5

(i) The beam is regarded as weightless.

(ii) The lateral surface is free of applied force.

(iii) The resultant of the shear stress distribution applied to the free end acts along a *principal axis* of inertia of that end, and at the centroid.

We take the origin at the centroid G of the free end and let the x- and y-axes be principal axes of inertia of that end.

Let the resultant applied load W_x act along the negative x-axis, G being the load point.

We shall write

$$A = \int y^2 dS, \; B = \int x^2 dS \tag{1}$$

for the principal second moments of area about the axes. In the notation of 5.11 we have

$$\xi = 0, \; \beta = \frac{W_x}{EB} = \frac{W_x}{2\mu(1+\eta) B} = \bar{\beta}, \tag{2}$$

$$\widehat{zz} = \frac{Rx}{B} W_x. \tag{3}$$

Then from 5.14

$$\frac{\widehat{xz} - i\widehat{yz}}{\mu} = \Phi(z) - i\tau\bar{z} - q\bar{z}^2 - r z\bar{z}, \tag{4}$$

where

$$q = \frac{1}{4}(1+2\eta)\beta = \frac{(1+2\eta) W_x}{4EB}, \; r = \frac{1}{2}\beta = \frac{W_x}{2EB}. \tag{5}$$

Introducing the torsion function $w(z)$ of 4.1, we define a holomorphic function $f(z)$ by

$$\Phi(z) = \tau\, w(z) - \beta\left[f(z) - \frac{1}{4} z^3\right]. \qquad (6)$$

If we now write

$$f(z) = \chi + i\chi', \qquad (7)$$

where χ and χ' are real conjugate harmonic functions, the *harmonic function* $\chi = \chi(x, y)$ is, by definition, *de St. Venant's flexure function.*

In terms of the flexure and torsion functions we have

$$\widehat{xz} = \mu\tau\left(\frac{\partial\phi}{\partial x} - y\right) - \mu\beta\left(\frac{\partial\chi}{\partial x} + \alpha_1\right), \ \alpha_1 = \frac{1}{2}\eta\, x^2 + \left(1 - \frac{1}{2}\eta\right) y^2 \qquad (8)$$

$$\widehat{yz} = \mu\tau\left(\frac{\partial\phi}{\partial y} + x\right) - \mu\beta\left(\frac{\partial\chi}{\partial y} + \alpha_2\right), \ \alpha_2 = (2 + \eta)\, x y. \qquad (9)$$

Here, from their definitions, ϕ and χ are harmonic functions

$$\nabla^2\phi = 0, \ \nabla^2\chi = 0. \qquad (10)$$

The twist per unit length τ is determined by the fact that the origin is the load point and therefore the moment about the origin vanishes, i.e.,

$$\int_S (x \cdot \widehat{yz} - y \cdot \widehat{xz})\, dS = 0. \qquad (11)$$

The torsion function ϕ is determined by the methods of Chapter IV. In the case where the line of action of the load is an axis of symmetry of the section, we shall have $\tau = 0$ and so the torsion problem will not arise.

Since the lateral surface of the beam is unloaded, the boundary condition is

$$l\,\widehat{xz} + m\,\widehat{yz} = 0 \text{ on the boundary } C, \qquad (12)$$

where $l = dy/ds$ and $m = -dx/ds$.

But from 4.6 (3) the torsion function ϕ satisfies the condition

$$l\frac{\partial\phi}{\partial x} + m\frac{\partial\phi}{\partial y} = \frac{\partial\phi}{\partial n} = l y - m x \text{ on the boundary } C \qquad (13)$$

and therefore

$$l\frac{\partial\chi}{\partial x} + m\frac{\partial\chi}{\partial y} = \frac{\partial\chi}{\partial n} = -l\alpha_1 - m\alpha_2 \text{ on the boundary } C. \qquad (14)$$

The conditions (13) and (14) are the statical boundary conditions defined in 4.7.

From 1.82 and 1.98 the strain-energy function is

$$U = \frac{1}{2E}\frac{x^2 R^2}{B^2} W_x^2 + \frac{1}{2\mu}(\widehat{xz}^2 + \widehat{yz}^2). \qquad (15)$$

Integrating this through the volume of the beam, we get the potential energy

$$V = \int_0^L dR \int_S U\,dS = \frac{W_x^2 L^3}{6EB} + \frac{L}{2\mu}\int_S (\widehat{xz}^2 + \widehat{yz}^2)\,dS\,, \tag{16}$$

where L is the length of the beam, and $\widehat{xz}$, $\widehat{yz}$ are given in terms of ϕ and χ by (8), (9). From 5.16 the components of displacement are, omitting a rigid body displacement,

$$\begin{aligned} u &= -\tau y R - \beta\left[\frac{1}{6}R^3 + \frac{1}{2}\eta R(x^2 - y^2)\right] \\ v &= \tau x R - \beta\eta\, xyR \\ w &= \tau\phi - \beta\chi - \beta\left(xy^2 + \frac{1}{2}xR^2\right). \end{aligned} \tag{17}$$

If instead of W_x along the negative x-axis the load is W_y along the negative y-axis, we should have $\beta = iW_y/(EA)$ and de St. Venant's flexure function χ_1 for this situation would be defined by

$$`\Phi(z) = \tau\, w(z) + i\beta\left[g(z) - \frac{1}{4}iz^3\right], \tag{18}$$

where

$$g(z) = \chi_1 + i\chi_1'\,.$$

The stresses are now given by

$$\begin{aligned} \widehat{xz} &= \mu\tau\left(\frac{\partial\phi}{\partial x} - y\right) + \mu i\beta\left(\frac{\partial\chi_1}{\partial x} + \alpha_3\right), \quad \alpha_3 = (2+\eta)\,xy \\ \widehat{yz} &= \mu\tau\left(\frac{\partial\phi}{\partial x} + x\right) + \mu i\beta\left(\frac{\partial\chi_1}{\partial y} + \alpha_4\right), \quad \alpha_4 = \frac{1}{2}\eta y^2 + \left(1 - \frac{1}{2}\eta\right)x^2. \end{aligned} \tag{19}$$

5.51. Cross-section a rectangle

Let the cross-section of the beam be bounded by the lines

$$x = \pm a, \qquad y = \pm b$$

and let the load be W_x applied in the negative direction of the x-axis at the centre of the free end. The boundary conditions are

$$\widehat{xz} = 0,\ x = \pm a;\ \widehat{yz} = 0,\ y = \pm b$$

and from the symmetry $\tau = 0$. Therefore using de St. Venant's flexure function 5.5 (8), (9) give

$$\frac{\partial\chi}{\partial x} = -\frac{1}{2}\eta a^2 - \left(1 - \frac{1}{2}\eta\right)y^2 \text{ on } x = \pm a \tag{1}$$

$$\frac{\partial\chi}{\partial y} = -(2+\eta)(\pm b)\,x \qquad \text{on } y = \pm b\,. \tag{2}$$

Now χ is a harmonic function and so therefore is

$$\omega = \chi - \frac{1}{6}(2+\eta)(x^3 - 3xy^2) \tag{3}$$

and moreover, on account of (2),

$$\frac{\partial \omega}{\partial y} = 0 \text{ on } y = \pm b\,. \tag{4}$$

Therefore we seek a harmonic function ω such that

$$\frac{\partial \omega}{\partial x} = -(1+\eta)a^2 + \eta y^2 \text{ on } x = \pm a\,. \tag{5}$$

Now in the interval $-b < y < b$ we have, as is easily verified, the Fourier expansion

$$y^2 = \frac{1}{3}b^2 + \frac{4b^2}{\pi^2}\sum_{n=1}^{\infty}\frac{(-1)^n}{n^2}\cos\frac{n\pi y}{b}\,. \tag{6}$$

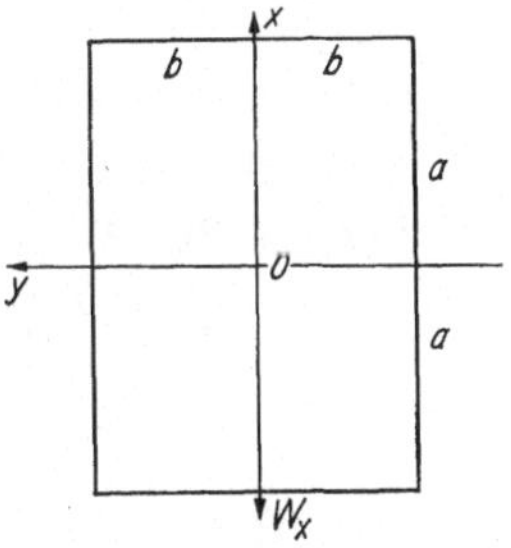

Fig. 5.51

Also the function

$$\sinh\frac{n\pi x}{b}\cos\frac{n\pi y}{b}$$

is harmonic. Therefore we attempt to satisfy (5) by taking

$$\omega = A_0 x + \sum_{n=1}^{\infty} A_n \sinh\frac{n\pi x}{b}\cos\frac{n\pi y}{b}\,.$$

Substitute this and (6) in (5) and equate the coefficients of $\cos n\pi y/b$. Then we get

$$A_0 = -(1+\eta)a^2 + \frac{1}{3}\eta b^2$$

$$\frac{n\pi}{b}A_n\cosh\frac{n\pi a}{b} = \frac{4b^2}{\pi^2}\frac{(-1)^n}{n^2}\eta$$

so that finally

$$\chi = \frac{1}{6}(2+\eta)(x^3 - 3xy^2) + \left\{-(1+\eta)a^2 + \frac{1}{3}\eta b^2\right\}x$$

$$+\,\eta\frac{4b^3}{\pi^3}\sum_{n=1}^{\infty}\frac{(-1)^n}{n^3}\frac{\sinh\dfrac{n\pi x}{b}}{\cosh\dfrac{n\pi a}{b}}\cos\frac{n\pi y}{b}\,.$$

Stevenson has given the solution of the flexure problem for the circular sector, the isosceles right-angled triangle, and half a loop of Bernoulli's lemniscate.

5.53. Application of the principle of virtual work

From 5.5 (17) we see that the virtual displacement $(\delta u, \delta v, \delta w)$ of the point (x, y, R) can arise only by varying ϕ and χ. Assuming principal axes we have therefore

$$\delta u = 0,\ \delta v = 0,\ \delta w = \tau\delta\phi - \beta\delta\chi\,, \tag{1}$$

where the variations $\delta\phi$, $\delta\chi$ are perfectly arbitrary both inside the material and on the contour.

Assume that body-force is absent, that the lateral surface is free of applied force and that the load on the free end is $(-W_x, 0, 0)$. Then the only contribution to the virtual work arises from the stress distribution $(\widehat{xz}, \widehat{yz}, 0)$ over the free end and the stress distribution $(\widehat{xz}, \widehat{yz}, \widehat{zz})$ over the root. Thus from (1) the only contribution arises from the root, namely

$$\int_S \widehat{zz}(\tau\delta\phi - \beta\delta\chi)\,dS, \qquad \widehat{zz} = \frac{xLW_x}{B}. \tag{2}$$

Therefore the equation of virtual work as expressed by 4.71 (10) is

$$\delta V - \frac{LW_x}{B}\int_S x(\tau\delta\phi - \beta\delta\chi)\,dS = 0. \tag{3}$$

Using 5.5 (8), (9), (16) we have

$$V = \frac{1}{2}L\mu\tau^2 P + \frac{1}{2}L\mu\beta^2 Q - L\mu\beta\tau T + \frac{W_x^2 L^3}{6EB},$$

where

$$\left.\begin{aligned} P &= \int_S \left[\left(\frac{\partial\phi}{\partial x} - y\right)^2 + \left(\frac{\partial\phi}{\partial y} + x\right)^2\right] dS,\\ Q &= \int_S \left[\left(\frac{\partial\chi}{\partial x} + \alpha_1\right)^2 + \left(\frac{\partial\chi}{\partial y} + \alpha_2\right)^2\right] dS,\\ T &= \int_S \left[\left(\frac{\partial\phi}{\partial x} - y\right)\left(\frac{\partial\chi}{\partial x} + \alpha_1\right) + \left(\frac{\partial\phi}{\partial y} + x\right)\left(\frac{\partial\chi}{\partial y} + \alpha_2\right)\right] dS \end{aligned}\right\}. \tag{4}$$

Therefore the principle of virtual work, (3), states that

$$\frac{1}{2}\tau^2\delta P + \frac{1}{2}\beta^2\delta Q - \beta\tau\delta T - 2\beta(1+\eta)\int_S (\tau\delta\phi - \beta\delta\chi)\,x\,dS = 0. \tag{5}$$

This variational equation must hold for all loadings and in particular that for which $\beta = 0$, $\tau \neq 0$ which yield $\delta P = 0$ and therefore coincides, as it should, with 4.72 (4) for the solution of the torsion problem. On the other hand the variational equation for determining the flexure function χ, the pure flexure problem, is obtained by putting $\tau = 0$, and we then get

$$\delta Q + 4(1+\eta)\int_S x\,\delta\chi\,dS = 0 \tag{6}$$

so that the flexure function χ minimizes the integral

$$\int_S \left\{\left(\frac{\partial\chi}{\partial x} + \alpha_1\right)^2 + \left(\frac{\partial\chi}{\partial y} + \alpha_2\right)^2 + 4(1+\eta)\,x\chi\right\} dS. \tag{7}$$

It is possible therefore to obtain approximations to the solution of this flexure problem by assuming a form for $\chi(x, y)$ and then minimizing (7). In 4.71 it was proved that the statical boundary condition, 5.5 (14), is a consequence of the variational equation. There is therefore no limitation imposed on the form which may be assumed for χ. The boundary condition will be automatically satisfied with greater accuracy, the more accurately we take our approximation for χ.

When the load is W_y along the negative y-axis (which is a principal axis) corresponding to 5.5 (18), (19) we can determine χ_1 by minimizing the integral

$$\int_S \left\{\left(\frac{\partial \chi_1}{\partial x} + \alpha_3\right)^2 + \left(\frac{\partial \chi_1}{\partial y} + \alpha_4\right)^4 + 4(1+\eta)\, y\, \chi_1\right\} dS. \tag{8}$$

5.54. Cross-section an ellipse

Let the cross-section be the ellipse

$$\frac{x^2}{a^2} + \frac{y^2}{b^2} = 1$$

the bending force W_x being applied at the centre along the negative x-axis, fig. 5.42.

Let us try de St. Venant's flexure function in the form

$$\chi = Px + Q(x^3 - 3xy^2) \tag{1}$$

which satisfies Laplace's equation if P and Q are constants whose values are to be found.

Substitute (1) in 5.53 (7). Then we find after a slight reduction that we have to minimize the expression

$$J = 8P^2 + 12\,PQ(a^2 - b^2) + 3Q^2(3a^4 + 2a^2b^2 + 3b^4)$$
$$+ 2P\{(4+5\eta)a^2 + (2-\eta)b^2\} + Q\{(4+7\eta)a^4 - 10(1+\eta)a^2b^2$$
$$-(6-3\eta)b^4\}.$$

The equations

$$\frac{\partial J}{\partial P} = 0, \quad \frac{\partial J}{\partial Q} = 0$$

lead to

$$P = \frac{-a^2\{b^2 + 2a^2(1+\eta)\}}{3a^2 + b^2}, \quad Q = \frac{2a^2 + b^2 + \frac{1}{2}\eta(a^2 - b^2)}{3(3a^2 + b^2)} \tag{2}$$

and the value of χ found from (1) is in agreement with the exact solution obtained in 5.42.

Since the force acts along one of the two orthogonal axes of symmetry, there is no torsion, $\tau = 0$.

5.56. Cross-section a rectangle (virtual work)

Let the cross-section be the rectangle bounded by $x^2 = a^2$, $y^2 = b^2$, fig. 5.51.

We seek an approximate solution for which de St. Venant's flexure functions is assumed to be of the form

$$\chi = Px + Qx^3 + Txy^2 , \tag{1}$$

where P, Q, T are constants to be determined. This function is not a solution of Laplace's equation.

Substitute (1) in 5.53 (7) and minimize the resulting expression J by the equations

$$\frac{\partial J}{\partial P} = 0, \frac{\partial J}{\partial Q} = 0, \frac{\partial J}{\partial T} = 0 .$$

We then find that

$$\begin{aligned} P &= \frac{-a^2\{15(1+\eta)a^2 + (3-2\eta)b^2\}}{3(5a^2+b^2)}, \quad Q = \frac{2+\eta}{6} \\ T &= \frac{-5(2+\eta)a^2 - (2-\eta)b^2}{2(5a^2+b^2)} . \end{aligned} \tag{2}$$

The area of the rectangle is $S = 4ab$, while $B = \frac{1}{3} a^2 S$. We then have

$$\widehat{xz} = \frac{-3W_x}{2S} \cdot \frac{P + \left(\frac{1}{2}\eta + 3Q\right) x^2 + \left(1 - \frac{1}{2}\eta + T\right) y^2}{a^2(1+\eta)} .$$

At the centre (0, 0) we have

$$\widehat{xz} = \widehat{xz}_0 = \frac{3W_x}{2S} \cdot \frac{15(1+\eta)a^2 + (3-2\eta)b^2}{3(1+\eta)(5a^2+b^2)} = k \cdot \frac{3W_x}{2S} .$$

The table, due to Leibenzon, gives values k_a of k from the approximate and k_e of k from the exact solution, 5.51, for $\eta = 0.25$ and different ratios a/b.

$\frac{a}{b}$	2	1	$\frac{1}{2}$	$\frac{1}{4}$
k_a	0.984	0.944	0.852	0.720
k_e	0.983	0.940	0.856	0.805

Thus the approximation to $\widehat{xz}_0$ is satisfactory when $\frac{1}{2} < \frac{a}{b} < 2$.

EXAMPLES V

1. In the flexure of an isotropic beam show that the deformed line of centroids of cross-sections lies in a plane, *the plane of bending*.

2. In the flexure of an isotropic beam show that the greatest deflection of the deformed central line occurs at the loaded end.

3. Show that the *plane of the load*, that is the plane which contains the undeformed line of centroids and which is parallel to the line of action of the load, does not in general coincide with the plane of bending.

4. Prove that filaments whose lengths are not altered during flexure lie in a plane, *the neutral plane*.

5. Prove that the neutral plane is at right angles to the plane of bending.

6. Let $\varkappa$ be the curvature of the deformed central line and let M_x, M_y be the bending moments. Prove that

$$M_x = \frac{-W_y}{\sqrt{\beta\bar{\beta}}}\varkappa, \; M_y = \frac{W_x}{\sqrt{\beta\bar{\beta}}}\varkappa .$$

7. An isotropic beam whose cross-section is bounded by the circle $x^2 + y^2 = a^2$ is bent by a terminal force W_x parallel to the negative x-axis; prove that

$$\widehat{xz} = \frac{(3+2\eta)\,W_x}{2\pi a^4(1+\eta)}\left(a^2 - x^2 - \frac{1-2\eta}{3+2\eta}\,y^2\right)$$

$$\widehat{yz} = -\frac{(1+2\eta)\,W_x}{\pi\,a^4(1+\eta)}\,x y$$

$$\widehat{zz} = \frac{4\,W_x}{\pi a^4}\,Rx .$$

Find corresponding formulae for a force W_y parallel to the negative y-axis.

8. An isotropic beam whose cross-section is bounded by the circumference $x^2 + y^2 = a^2$ is bent by a terminal load W_x along the x-axis. Show that on the diameter $x = 0$ of a cross-section, the shear $\widehat{yz}$ vanishes and that the shear $\widehat{xz}$ is greatest at the centre of the section.

If S is the total shear at the end of the diameter $x = 0$, prove that

$$\frac{\widehat{xz}_{\max}}{S} = \frac{3+2\eta}{2+4\eta} .$$

9. An isotropic beam whose cross-section is bounded by the circumference $x^2 + y^2 = a^2$ is bent by a terminal load W_x along the x-axis. Prove that the equation of the lines of shearing stress is

$$x^2 + y^2 = a^2 + c\,y^{\frac{3+2\eta}{1+2\eta}}$$

where c is an arbitrary constant. Sketch these lines when $\eta = 0.25$.

10. Find the position of the centre of flexure when the cross-section is one loop of Bernoulli's lemniscate.

11. Verify that ψ as defined by 5.4 (11) satisfies 5.4 (3).

12. For the elliptic cross-section bounded by

$$\frac{x^2}{a^2} + \frac{y^2}{b^2} = 1$$

with the load $(-W_x, 0, 0)$ applied at the centre of the free end of the beam, prove that de St. Venant's flexure function is

$$\chi = \frac{-a^2\{2(1+\eta)a^2+b^2\}}{3a^2+b^2}x + \frac{1}{3}\cdot\frac{2a^2+b^2+\frac{1}{2}\eta(a^2-b^2)}{3a^2+b^2}(x^3-3xy^2).$$

13. An isotropic beam whose cross-section is bounded by the ellipse $x^2/a^2+y^2/b^2=1$ is bent by a load W_x along the negative x-axis. Prove that the shear at any point $(0, y)$ of the diameter $x=0$ is

$$\frac{2W_x}{\pi a^3 b}\frac{2(1+\eta)a^2+b^2}{(1+\eta)(3a^2+b^2)}\left[a^2-\frac{(1-2\eta)a^2}{2(1+\eta)a^2+b^2}y^2\right]$$

and find the ratio of the maximum value of the shear to its value at the ends of the diameter.

14. Calculate the value of B_1 in the formula 5.44 (5).

15. Prove that de St. Venant's flexure function χ for the circle and the ellipse can be obtained by suitably choosing the constant P and Q in

$$\chi = Px + Q(x^3-3xy^2),$$

the loading being assumed to be along the x-axis.

16. Show that the flexure function of Ex. 15 can be adapted to the cross-section

$$\frac{y^2}{b^2} = \left(1-\frac{x^2}{a^2}\right)^{2\eta}, \quad (-a < x < a)$$

by choosing $P = -a^2$, $Q = \frac{1}{3}\left(1-\frac{1}{2}\eta\right)$.

17. Show that the flexure function of Ex. 15 can be adapted to the cross-section bounded by two arcs of the hyperbola $x^2(1+\eta)-\eta y^2=a^2$ and the two lines $y=\pm a$ by taking

$$P = -a^2,\ Q = \frac{1}{6}(2+\eta).$$

18. Show that de St. Venant's flexure function when the load is $(-W_x, -W_y)$ along principal axes at the centroid is given by

$$\chi = \mathrm{Re}[\beta\{f(z)-ig(z)\}],$$

where

$$\beta = \frac{W_x}{EB} + i\frac{W_y}{EA},$$

and $f(z)$, $g(z)$ define the flexure function for the loads $(-W_x, 0)$, $(0, -W_y)$ respectively.

19. For a cross-section bounded by the confocal ellipses $\xi = \xi_0$, $\xi = \xi_1$, $(\xi_0 > \xi_1)$ in the net $z = c \cosh \zeta$, $\zeta = \xi + i\eta$, prove that de St. Venant's flexure function is

$$\chi = c \cos \eta \left[\left(\frac{1}{4} - \frac{1}{8} \eta_0 \right) \cosh \xi \right.$$
$$- \left(\frac{3}{4} + \frac{1}{2} \eta_0 \right) \{ \cosh \xi_0 \cosh \xi_1 \cosh (\xi_0 + \xi_1) \cosh \xi$$
$$\left. - \sinh \xi_0 \sinh \xi_1 \sinh (\xi_0 + \xi_1) \sinh \xi \} \right]$$
$$+ c^3 \cos 3\eta \left[\frac{1}{16} \cosh 3\xi \right.$$
$$\left. - \left(\frac{5}{16} + \frac{1}{8} \eta_0 \right) \frac{\sinh \xi_0 \cosh 3(\xi - \xi_1) - \sinh \xi_1 \cosh 3(\xi_0 - \xi)}{3 \sinh 3(\xi_0 - \xi_1)} \right]$$

the load being along the x-axis, and η_0 denoting Poisson's ratio.

20. Use the formula for integration by parts

$$\int_S \left[u \frac{\partial (\delta \omega)}{\partial x} + v \frac{\partial (\delta \omega)}{\partial y} \right] dS$$
$$= \oint_C (l u + m v)(\delta \omega)\, ds - \int_S \left(\frac{\partial u}{\partial x} + \frac{\partial v}{\partial y} \right) (\delta \omega)\, dS$$

to show that the equation 5.53 (5) reduces to

$$\oint_C \left[\tau^2 \left(\frac{\partial \phi}{\partial n} - l y + m x \right) - \beta \tau \left(\frac{\partial \chi}{\partial n} + l \alpha_1 + m \alpha_2 \right) \right] \delta \phi\, ds$$
$$+ \oint_C \left[\beta^2 \left(\frac{\partial \chi}{\partial n} + l \alpha_1 + m \alpha_2 \right) - \beta \tau \left(\frac{\partial \phi}{\partial n} - l y + m x \right) \right] \delta \chi\, ds$$
$$- \int_S [\tau^2 \nabla^2 \phi - \beta \tau \nabla^2 \chi] \delta \phi\, dS - \int_S [\beta^2 \nabla^2 \chi - \beta \tau \nabla^2 \chi] \delta \chi\, dS = 0 .$$

21. Use the result of the preceding example to show that the variation equation 5.53 (5) leads to

$$\nabla^2 \phi = 0, \; \nabla^2 \chi = 0$$

and to the boundary conditions

$$\frac{\partial \phi}{\partial n} - l y + m x = 0, \; \frac{\partial \chi}{\partial n} + l \alpha_1 + m \alpha_2 = 0 .$$

22. Obtain the variational integral of 5.53 (8) by an argument similar to that which lead to 5.53 (7).

23. Relate the displacements (u, v, w) of 5.16 to the corresponding components in 2.3 (23).

24. An isotropic cantilever is bent by a force $(0, -W_y, 0)$ applied at the centroid of the free end along the y-axis, which is a principal axis of inertia of the end section. Prove that Timoshenko's stress function T minimizes the integral

$$\int_S \left\{\left(\frac{\partial T}{\partial x}\right)^2 + \left(\frac{\partial T}{\partial y}\right)^2 + 2B_1(1+\eta)\,[s(x) - y^2]\,\frac{\partial T}{\partial x} + 4T(\eta B_1 x - \tau)\right\} dS$$

subject to the conditions.

(i) $T = 0$ on the boundary.

(ii) $s(x) - y^2 = 0$ on the periphery of the cross-section.

25. Apply the area theorem to show that

$$16\int_S x^4 dS = \int_S (z+\bar{z})^4 dS = -\frac{1}{10}\, i \oint_C (z+\bar{z})^5 dz\,,$$

$$16\int_S x^2 y^2 dS = -\int_S (z^2-\bar{z}^2)^2 dS = \frac{1}{2}\, i \oint_C \left(z^4\bar{z} - \frac{2}{3} z^2\bar{z}^3 + \frac{1}{5}\bar{z}^5\right) dz\,.$$

26. In the case of the ellipse, by mapping the periphery

$$\frac{x^2}{a^2} + \frac{y^2}{b^2} = 1$$

on the unit circumference $\sigma = e^{i\theta}$ by the mapping function

$$z = c\left(\sigma + \frac{m}{\sigma}\right),\ a = c(1+m),\ b = c(1-m)\,,$$

use Ex. 25 to show that

$$\int_S x^4 dS = \frac{\pi a^5 b}{8}\,,\ \int_S y^4 dS = \frac{\pi a b^5}{8}\,,\ \int_S x^2 y^2 dS = \frac{\pi a^3 b^3}{24}\,.$$

Chapter VI

Antiplane of elastic symmetry

In this chapter we show how the presence of an antiplane of elastic symmetry (6.5) permits the reduction of antiplane problems concerning anisotropic material to corresponding problems concerning isotropic material.

Before doing this, however, it is advantageous to study certain situations which will throw light on the subsequent development.

6.1. Bending by couples

Let a cantilever be bent by a couple M applied at the free end in a principal plane, that is to say a plane which contains the axis, OR, of the beam and a principal axis of inertia Oy, of the free end, where the origin O is taken to be the centroid of the free end.

If $\widehat{zz}$ is the normal stress over the free end which produces the couple (in the sense R to y) about the x-axis we must have

$$M = \int_S y\,\widehat{zz}\,dS, \qquad \int_S \widehat{zz}\,dS = 0\,. \tag{1}$$

Since the origin is the centroid the second of these conditions can be fulfilled by taking

$$\widehat{zz} = \alpha y, \qquad \alpha \text{ constant}$$

and then the first gives (cf. 3.3 (ii))

$$M = \alpha \int_S y^2 dS = A\,\alpha$$

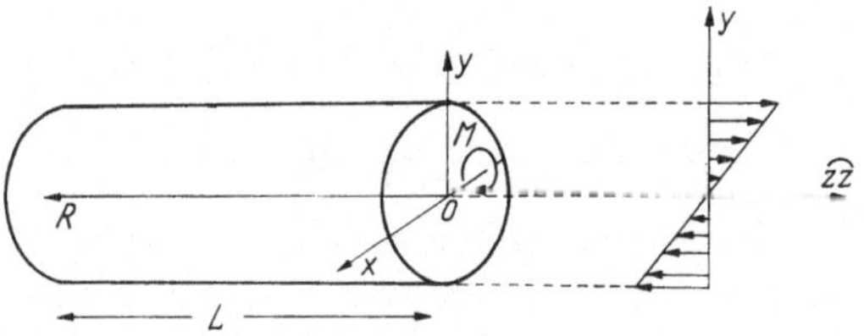

Fig. 6-1

where A is the second moment of area of the cross-section about $O\,x$. Thus

$$\alpha = \frac{M}{A}, \qquad \widehat{zz} = \frac{M\,y}{A} \tag{2}$$

while the remaining stress components vanish (cf. 3.3 (ii)).

Using the matrix form of Hooke's law, 1.72, we get

$$\begin{aligned}
\frac{\partial u}{\partial x} &= \alpha k_{13}\,y, & \frac{\partial w}{\partial y} + \frac{\partial v}{\partial R} &= \alpha k_{34}\,y\,,\\
\frac{\partial v}{\partial y} &= \alpha k_{23}\,y, & \frac{\partial u}{\partial R} + \frac{\partial w}{\partial x} &= \alpha k_{35}\,y\,,\\
\frac{\partial w}{\partial R} &= \alpha k_{33}\,y, & \frac{\partial u}{\partial y} + \frac{\partial v}{\partial x} &= \alpha k_{36}\,y\,.
\end{aligned} \tag{3}$$

Integrate these equations under the conditions that an element at $(0, 0, L)$ of the axis of the beam is fixed by

$$\begin{gathered}
u = 0, \quad v = 0, \quad w = 0\\
\frac{\partial u}{\partial R} = 0, \quad \frac{\partial v}{\partial R} = 0, \quad \frac{\partial v}{\partial x} - \frac{\partial u}{\partial y} = 0,
\end{gathered} \tag{4}$$

the last condition being that of no rotation of the element about the axis of the beam.

Write $R_1 = R - L$; then the above conditions have to be satisfied at $x = 0$, $y = 0$, $R_1 = 0$, while $\partial/\partial R = \partial/\partial R_1$. Evidently (4) will be satisfied if u, v, w are homogeneous quadratic functions of x, y, R_1. Having regard to the left-hand column of equation (3) we can write

$$u = \alpha k_{13} x y + C_1 y^2 + E y R_1 + F_1 R_1^2$$

$$v = \frac{1}{2} \alpha k_{23} y^2 + A_1 x^2 + D x R_1 + F_2 R_1^2$$

$$w = \alpha k_{33} y R_1 + A_2 x^2 + B x y + C_2 y^2 .$$

Substitute those in the right-hand column of equations (3). Then equating coefficients in the resulting identities we find the constants A_1, A_2, B, C_1, C_2, D, E, F_1, F_2 giving

$$u = \frac{M}{2A} \{2 k_{13} x y + k_{36} y^2 + k_{35} y (R - L)\} ,$$

$$v = \frac{M}{2A} \{- k_{13} x^2 + k_{23} y^2 - k_{33} (R - L)^2 - k_{35} x (R - L)\} , \tag{5}$$

$$w = \frac{M}{2A} \{k_{35} x y + k_{34} y^2 + 2 k_{33} y (R - L)\} .$$

The last formula shows that the cross-section is *warped* into a surface of the second degree. If, however, $k_{34} = k_{35} = 0$, which will happen if the material has a plane of elastic symmetry perpendicular to the axis of the beam, the cross-sections will remain plane.

The displacement of any point $(0, 0, R)$ on the axis of the beam is

$$u = w = 0, \qquad v = -\frac{M}{2A} k_{33} (R - L)^2$$

and therefore the bent axis of the beam is the parabola

$$x = 0, \quad y = -\frac{M}{2A} k_{33} (R - L)^2 . \tag{6}$$

For small bending this is a circular arc. For this arc, $d^2 y / d R^2 = - M k_{33} / A$ so that if ϱ is the radius of curvature,

$$\frac{1}{\varrho} = \frac{M k_{33}}{A} .$$

Thus Euler's law of bending holds also for the case of general anisotropy, the role of Young's modulus being played by $1/k_{33}$.

The maximum deflection of the central line occurs at the free end $R = 0$, its magnitude being

$$k_{33} \frac{M L^2}{2A} . \tag{8}$$

From (5) we have

$$\frac{\partial v}{\partial x} - \frac{\partial u}{\partial y} = \frac{M}{A} (-2 k_{13} x - k_{36} y - k_{35} (R - L))$$

and therefore the local twist, 5.2, is

$$\frac{\partial}{\partial R} \left(\frac{\partial v}{\partial x} - \frac{\partial u}{\partial y} \right) = -\frac{M}{A} k_{35} . \tag{9}$$

Thus bending is accompanied by twisting unless $k_{35}=0$, which would be the case if the planes perpendicular to the axis of the beam were planes of elastic symmetry.

6.2. Boundary conditions

With the notation of fig. 3.1 (ii) we have from 3.1, (1)–(4) on the antiplane $R=0$

$$\int_S \widehat{xz}\, dS + P_1 = 0\,, \tag{1}$$

$$\int_S \widehat{yz}\, dS + P_2 = 0\,, \tag{2}$$

$$\int_S (x\cdot\widehat{yz} - y\cdot\widehat{xz})\, dS + M_3 = 0\,, \tag{3}$$

$$\int_S \widehat{zz}_0\, dS + P_3 = 0\,, \tag{4}$$

$$\int_S y\cdot\widehat{zz}_0 + M_1 = 0\,, \tag{5}$$

$$\int_S x\cdot\widehat{zz}_0 - M_2 = 0\,, \tag{6}$$

where $\widehat{zz}_0$ is the value of $\widehat{zz}$ on the antiplane $R=0$.

On the lateral surface we have

$$l\,\widehat{xx} + m\,\widehat{xy} = X_n \tag{7}$$

$$l\,\widehat{xy} + m\,\widehat{yy} = Y_n \tag{8}$$

$$l\,\widehat{xz} + m\,\widehat{yz} = Z_n\,. \tag{9}$$

Here $(l, m, 0)$ are the direction cosines of the outward normal to the lateral surface, so that if ds is an element of arc of the boundary C of a cross-section

$$l = \frac{dy}{ds}\,,\quad m = -\frac{dx}{ds}\,. \tag{10}$$

The expressions for the stress components given in 2.52 combine with (7)–(10) to give, on the boundary

$$\begin{aligned}
&\left(\frac{\partial^2\chi}{\partial y^2} + V\right)\frac{dy}{ds} - \frac{\partial^2\chi}{\partial x\,\partial y}\left(-\frac{dx}{ds}\right) = \frac{d}{ds}\left(\frac{\partial\chi}{\partial y}\right) + V\frac{dy}{ds} = X_n\,,\\
&-\frac{\partial^2\chi}{\partial x\,\partial y}\frac{dy}{ds} + \left(\frac{\partial^2\chi}{\partial x^2} + V\right)\left(-\frac{dx}{ds}\right) = -\frac{d}{ds}\left(\frac{\partial\chi}{\partial x}\right) - V\frac{dx}{ds} = Y_n\,,\\
&\frac{d\psi}{ds} - \frac{1}{2k_{33}}\left\{[A_1x^2 + (C_1-\varepsilon)\,m\,x]\frac{dy}{ds} - [B_1y^2 + (C_1-\varepsilon)\,m\,y]\frac{dx}{ds}\right\} = Z_n\,.
\end{aligned} \tag{11}$$

Integration along C from 0 to s gives

$$\frac{\partial \chi}{\partial x} = -\int_0^s \left(Y_n + V\frac{dx}{ds}\right) ds + c_1$$
$$\frac{\partial \chi}{\partial y} = \int_0^s \left(X_n - V\frac{dy}{ds}\right) ds + c_2 \tag{12}$$

$$\psi = \int_0^s \Big[Z_n + \frac{1}{2k_{33}}\Big\{[A_1 x^2 + (C_1 - \varepsilon) m x] \frac{dy}{ds} - [B_1 y^2 + (C_1 - \varepsilon) m y] \frac{dx}{ds}\Big\}\Big] ds + c_3 . \tag{13}$$

In particular we note that when body-force is absent, and there is no axial loading on the lateral surface, $Z_n = 0$, we have

$$\psi = \frac{1}{2k_{33}} \int_0^s \left\{(A_1 x^2 + C_1 m x) \frac{dy}{ds} - (B_1 y^2 + C_1 m y) \frac{dx}{ds}\right\} ds + c_3$$

on the boundary

If, in addition, $A_1 = B_1 = C_1 = 0$,

$$\psi = c_3 \text{ on the boundary} .$$

When the boundary encloses a simply connected region we can take $c_3 = 0$.

Here the arbitrary constants c_1, c_2, c_3 merely point to the arbitrary position of the origin for the arcs of the contour C.

Where the displacement (u^*, v^*, w^*) is given on the boundary, the boundary conditions are simply

$$u = u^*, v = v^*, w = w^* \text{ on the boundary} . \tag{14}$$

6.3. A device for transforming integrals

Let $\frac{\partial p}{\partial x} + \frac{\partial q}{\partial y} = \varrho$,

then

$$\int_S p\, dS = \int_S \left[p + x\left(\frac{\partial p}{\partial x} + \frac{\partial q}{\partial y}\right) - x\varrho\right] dS$$
$$= \int_S \left[\frac{\partial (p x)}{\partial x} + \frac{\partial (q x)}{\partial y}\right] dS - \int_S x\varrho\, dS \tag{1}$$
$$= \oint_C x(lp + mq)\, ds - \int_S x\varrho\, dS$$

by Gauss's theorem, 5.46.

In the case of antiplane stress we can apply this to the stress components bearing in mind that

$$\frac{\partial \widehat{xx}}{\partial x} + \frac{\partial \widehat{xy}}{\partial y} = \frac{\partial V}{\partial x} = b_1 \tag{2}$$

$$\frac{\partial \widehat{xy}}{\partial x} + \frac{\partial \widehat{yy}}{\partial y} = \frac{\partial V}{\partial y} = b_2 \tag{3}$$

$$\frac{\partial \widehat{xz}}{\partial x} + \frac{\partial \widehat{yz}}{\partial y} = -\frac{1}{k_{33}} [A_1 x + B_1 y + (C_1 - \varepsilon) m] = B_3 \tag{4}$$

as in 2.1 (7).

6.32. Relations between the applied forces

Using the transformation of 6.3 we obtain the following expressions

$$\left.\begin{aligned} \int_S \widehat{xx}\, dS &= \oint_C x(l\widehat{xx} + m\widehat{yx})\, ds - \int_S x b_1 dS = \oint_C x X_n ds - \int_S x b_1 dS, \\ \int_S \widehat{yy}\, dS &= \oint_C y Y_n ds - \int_S y b_2 dS, \\ \int_S \widehat{xy}\, dS &= \oint_C x Y_n ds - \int_S x b_2 dS = \oint_C y X_n ds - \int_S y b_1 dS, \end{aligned}\right\} \tag{1}$$

$$\left.\begin{aligned} \int_S \widehat{yz}\, dS &= \oint_C y Z_n ds - \int_S y B_3 dS, \\ \int_S \widehat{xz}\, dS &= \oint_C x Z_n ds - \int_S x B_3 dS, \end{aligned}\right\} \tag{2}$$

wherein, when the body-field derives from a potential,

$$b_1 = \frac{\partial V}{\partial x}, \quad b_2 = \frac{\partial V}{\partial y}, \quad B_3 = -\frac{A_1 x + B_1 y + (C_1 - \varepsilon) m}{k_{33}}. \tag{3}$$

Now consider $\widehat{zz}$. From 2.32 (2), 2.2 (8), and 2.3 (14) we have

$$\widehat{zz} = \frac{e_{zz}}{k_{33}} - \nu_1 \widehat{xx} - \nu_2 \widehat{yy} - \nu_4 \widehat{yz} - \nu_5 \widehat{xz} - \nu_6 \widehat{xy}, \; \nu_\alpha = \frac{k_{\alpha 3}}{k_{33}} \tag{4}$$

$$e_{zz} = e_{zz}^0 + (A_1 x + B_1 y + C_1 m) R \tag{5}$$

$$\begin{aligned} e_{zz}^0 = \frac{1}{2} [\nu_5 A_1 x^2 + (\nu_4 A_1 + \nu_5 B_1)\, xy + \nu_4 B_1 y^2] \\ + A_2 x + B_2 y + C_2 m. \end{aligned} \tag{6}$$

Therefore

$$
\begin{aligned}
k_{33}\int_S \widehat{zz}\,dS - R\int_S (A_1x+B_1y+C_1m)\,dS \\
= \int_C e^0_{zz}\,dS - k_{13}\oint_C xX_n\,ds + k_{13}\int_S xb_1\,dS \\
- k_{23}\oint_C yY_n\,ds + k_{23}\int_S yb_2\,dS - k_{63}\oint_C yX_n\,ds + k_{63}\int_S yb_1\,dS \qquad (7)\\
- k_{43}\oint_C yZ_n\,ds + k_{43}\int_S yB_3\,dS - k_{53}\oint_C xZ_n\,ds + k_{53}\int_S xB_3\,dS \\
= -k_{33}P_3
\end{aligned}
$$

from 6.2 (4), since the left-hand side becomes $k_{33}\int_S \widehat{zz}_0\,dS$ when $R=0$, and the right-hand side is independent of R.

When body-force and surface loading are absent, (3), (6), (7) give

$$
\begin{aligned}
(A_2x_G + B_2y_G + C_2m)\,S = &- k_{33}P_3 \\
&+ \frac{1}{2}\left[\nu_5A_1B + (\nu_4A_1 + \nu_5B_1)\,H + \nu_4B_1A\right] \\
&+ C_1m(\nu_5x_G + \nu_4y_G)\,S\,.
\end{aligned} \qquad (8)
$$

From (2), (3) and 6.3 (1), (2) we get

$$
\left.
\begin{aligned}
\frac{1}{k_{33}}\left[BA_1 + HB_1 + (C_1-\varepsilon)\,m\,x_G S\right] &= -P_1 - \oint_C xZ_n\,ds \\
\frac{1}{k_{33}}\left[HA_1 + AB_1 + (C_1-\varepsilon)\,m\,y_G S\right] &= -P_2 - \oint_C yZ_n\,ds\,.
\end{aligned}
\right\} \qquad (9)
$$

By considering the equilibrium of the portion of the beam between $R=0$, $R=R_1$ and resolving in the direction of the axis we get, fig. 3.22,

$$
\int_S (\widehat{zz}_1 - \widehat{zz}_0)\,dS - \int_S b_3R_1\,dS + \oint_C Z_nR_1\,ds = 0\,.
$$

But $\widehat{zz}_1 - \widehat{zz}_0 = R_1(A_1x + B_1y + C_1m)/k_{33}$. Therefore

$$
S\left[A_1x_G + B_1y_G + (C_1-\varepsilon)\,m\right] = -k_{33}\oint_C Z_n\,ds \qquad (10)
$$

since $b_3 = \varepsilon m/k_{33}$ from 2.24 (2).

Thus (9), (10) furnish sufficient equations to determine A_1, B_1, C_1.

6.33. Properties of the couples M_1, M_2, M_3

From 6.2 (5) and (6) combined with 6.32 (4) we get

$$\begin{aligned} -k_{33}M_1 = \frac{1}{2}\int_S [\nu_5 A_1 x^2 y + (\nu_4 A_1 + \nu_5 B_1) x y^2 + \nu_4 B_1 y^3]\, dS \\ + A_2 H + B_2 A + C_2 m S y_G \\ - \int_S y (k_{13}\widehat{xx} + k_{23}\widehat{yy} + k_{43}\widehat{yz} + k_{53}\widehat{zx} + k_{63}\widehat{xy})\, dS \end{aligned} \tag{1}$$

$$\begin{aligned} k_{33}M_2 = \frac{1}{2}\int_S [\nu_5 A_1 x^3 + (\nu_4 A_1 + \nu_3 B_1) x^2 y + \nu_4 B_1 x y^2]\, dS \\ + A_2 B + B_2 H + C_2 m S x_G \\ - \int_S x (k_{13}\widehat{xx} + k_{23}\widehat{yy} + k_{43}\widehat{yz} + k_{53}\widehat{zx} + k_{63}\widehat{xy})\, dS. \end{aligned} \tag{2}$$

We have also the identities

$$\begin{aligned}
x \cdot \widehat{xx} &= \frac{\partial}{\partial x}\left(\frac{1}{2}x^2 \cdot \widehat{xx}\right) + \frac{\partial}{\partial y}\left(\frac{1}{2}x^2 \cdot \widehat{xy}\right) - \frac{1}{2}x^2\left(\frac{\partial \widehat{xx}}{\partial x} + \frac{\partial \widehat{xy}}{\partial y}\right) \\
x \cdot \widehat{yy} &= \frac{\partial}{\partial x}\left(xy \cdot \widehat{xy} - \frac{1}{2}y^2 \cdot \widehat{xx}\right) + \frac{\partial}{\partial y}\left(xy \cdot \widehat{yy} - \frac{1}{2}y^2 \cdot \widehat{xy}\right) \\
&\quad - xy\left(\frac{\partial \widehat{xy}}{\partial x} + \frac{\partial \widehat{yy}}{\partial y}\right) + \frac{1}{2}y^2\left(\frac{\partial \widehat{xx}}{\partial x} + \frac{\partial \widehat{xy}}{\partial y}\right) \\
x \cdot \widehat{xy} &= \frac{\partial}{\partial x}\left(\frac{1}{2}x^2 \cdot \widehat{xy}\right) + \frac{\partial}{\partial y}\left(\frac{1}{2}x^2 \cdot \widehat{yy}\right) - \frac{1}{2}x^2\left(\frac{\partial \widehat{xy}}{\partial x} + \frac{\partial \widehat{yy}}{\partial y}\right) \\
x \cdot \widehat{xz} &= \frac{\partial}{\partial x}\left(\frac{1}{2}x^2 \cdot \widehat{xz}\right) + \frac{\partial}{\partial y}\left(\frac{1}{2}x^2 \cdot \widehat{yz}\right) - \frac{1}{2}x^2\left(\frac{\partial \widehat{xz}}{\partial x} + \frac{\partial \widehat{yz}}{\partial y}\right) \\
2x \cdot \widehat{yz} &= \frac{\partial}{\partial x}(xy \cdot \widehat{xz}) + \frac{\partial}{\partial y}(xy \cdot \widehat{yz}) - xy\left(\frac{\partial \widehat{xz}}{\partial x} + \frac{\partial \widehat{yz}}{\partial y}\right) \\
&\quad + (x \cdot \widehat{yz} - y \cdot \widehat{xz}).
\end{aligned}$$

Substitute in (2) and use 6.2 (7)—(9). Then

$$\begin{aligned}
&A_2 B + B_2 H + C_2 m S x_G \\
&= k_{33}M_2 - \frac{1}{2}k_{43}M_3 - \frac{1}{2}\int_S [\nu_5 A_1 x^3 + (\nu_4 A_1 + \nu_5 B_1) x^2 y + \nu_4 B_1 x y^2]\, dS \\
&+ k_{13}\left\{\oint_C \frac{1}{2}x^2 X_n\, ds - \int_S \frac{1}{2}x^2 b_1\, dS\right\} \\
&+ k_{23}\left\{\oint_C \left(xy Y_n - \frac{1}{2}y^2 X_n\right) ds - \int_S \left(xy b_2 - \frac{1}{2}y^2 b_1\right) dS\right\} \\
&+ k_{63}\left\{\oint_C \frac{1}{2}x^2 Y_n\, ds - \int_S \frac{1}{2}x^2 b_2\, dS\right\} \\
&+ k_{43}\left\{\oint_C \frac{1}{2}xy Z_n\, ds - \int_S \frac{1}{2}xy B_3\, dS\right\} \\
&+ k_{53}\left\{\oint_C \frac{1}{2}x^2 Z_n\, ds - \int_S \frac{1}{2}x^2 B_3\, dS\right\}.
\end{aligned} \tag{3}$$

When body-field and surface loading are absent we get

$$A_2 B + B_2 H + C_2 m S x_G$$
$$= k_{33} M_2 - \frac{1}{2} k_{43} M_3 + \frac{1}{2} C_1 m (\nu_4 H + \nu_5 B) \tag{4}$$

and similarly

$$A_2 H + B_2 A + C_2 m S y_G$$
$$= - k_{33} M_1 + \frac{1}{2} k_{53} M_3 + \frac{1}{2} C_1 m (\nu_4 A + \nu_5 H) \,. \tag{5}$$

Observe that (4) and (5) together with 6.32 (8) namely

$$(A_2 x_G + B_2 y_G + C_2 m) S = - k_{33} P_3$$
$$+ \frac{1}{2} [\nu_5 A_1 B + (\nu_4 A_1 + \nu_5 B_1) H + \nu_4 B_1 A] + C_1 m (\nu_5 x_G + \nu_4 y_G) \tag{6}$$

serve to determine A_2, B_2, C_2 when body-force and surface loading are absent.

Again, using the expressions of 2.52 for $\widehat{xz}$, $\widehat{yz}$,

$$- M_3 = \int_S (x \cdot \widehat{yz} - y \cdot \widehat{xz}) \, dS$$
$$= \int_S \left[2\psi - \frac{\partial}{\partial x} (\psi x) - \frac{\partial}{\partial y} (\psi y) - \frac{1}{2k_{33}} (B_1 x y^2 - A_1 x^2 y) \right] dS$$
$$= \int_S 2\psi \, dS - \frac{1}{2k_{33}} \int_S (B_1 x y^2 - A_1 x^2 y) \, dS - \oint_C \psi (l x + m y) \, ds \,,$$

where ψ is given by 6.2 (13).

In particular when body-force and surface loading are absent and $A_1 = B_1 = C_1 = 0$, we can take $\psi = 0$ on the boundary and then

$$M = - M_3 = \int_S 2 \psi \, dS \,. \tag{7}$$

6.4. Simplifying assumptions

In the general treatment of 6.2—6.33 there is no restriction on the position of the origin and the orientation of the axes in the antiplane. In what follows we shall obtain notable simplification without restricting the principles of solution by making the following assumptions.

(i) The origin is at the centroid G of the antiplane.
(ii) The x- and y-axes are principal axes of inertia of the antiplane.
(iii) The lateral surface of the beam is free of applied force.
(iv) Body-force is absent.

Denote the principal second moments of area with respect to the x- and y-axis by I_x, I_y respectively. Then

$$I_x = \int_S y^2 \, dS, \quad I_y = \int_S x^2 \, dS, \quad \int_S x y \, dS = 0$$

and since the origin is at the centroid

$$\int_S x \, dS = 0, \quad \int_S y \, dS = 0 .$$

6.41. Twisting by an axial couple

As in the isotropic case we suppose the beam to be fixed at one end and twisted by a couple M applied (here in the sense y to x) by a suitable

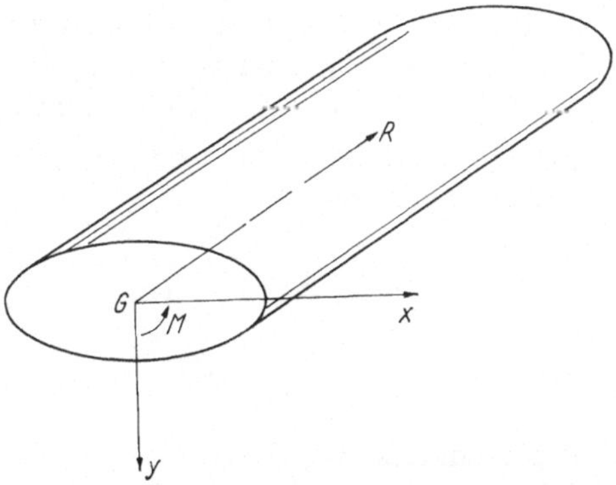

Fig. 6.41

traction distribution on the plane of the other end. This free end is taken as the antiplane $R = 0$ and the origin and axes are as described in 6.4.

We omit body-force and surface loading, so that, with our previous notations

$$\left.\begin{aligned} &V = 0, \quad X_n = Y_n = Z_n = 0, \quad P_1 = P_2 = P_3 = 0 \\ &M_1 = M_2 = 0, \quad M_3 = -M \neq 0 \end{aligned}\right\} \tag{1}$$

From 6.32 (9), (10) it follows that

$$A_1 = B_1 = C_1 = 0 \tag{2}$$

and therefore from 6.32 (8)

$$C_2 = 0 .$$

It now follows that

$$e_{zz} = e^0_{zz} = A_2 x + B_2 y ,$$

where from 6.33

$$A_2 = \frac{k_{43} M}{2 I_y}, \quad B_2 = -\frac{k_{53} M}{2 I_x} . \tag{3}$$

From 6.33 (7) the twisting moment M is given by

$$M = \int_S 2\,\psi\, d\,S. \tag{4}$$

From 2.3 (23) the displacements are

$$u = u_1 - \frac{1}{4}\frac{k_{43}M}{I_y}R^2 - \tau y R + \alpha + \omega_2 R - \omega_3 y \tag{5}$$

$$v = v_1 + \frac{1}{4}\frac{k_{53}M}{I_x}R^2 + \tau x R + \beta + \omega_3 x - \omega_1 R \tag{6}$$

$$w = w_1 + \frac{1}{2}M\left\{\frac{k_{43}}{I_y}x - \frac{k_{53}}{I_x}y\right\}R + \gamma + \omega_1 y - \omega_2 x\,. \tag{7}$$

The last 3 terms on the right correspond with an arbitrary rigid body movement, and u_1, v_1, w_1 satisfy equations 2.32 (6), (7).

It follows from (5) and (6) that if the end $(0, 0, L)$ of the axis of the rod is rigidly fixed, the projections of the bent axis on the planes $y = 0$, $x = 0$ have the respective equations

$$\begin{aligned} x' &= -\frac{M}{4I_y}k_{43}(L-R)^2 \\ y' &= \frac{M}{4I_x}k_{53}(L-R)^2 \end{aligned} \tag{8}$$

and therefore the displacements of the origin G, the free end of the axis, in the directions of the x- and y-axes are respectively

$$-\frac{M}{4I_y}k_{43}L^2, \qquad \frac{M}{4I_x}k_{53}L^2\,.$$

This means that twisting is, in general, accompanied by bending of the axis of the beam.

Also observe that if $k_{43} = k_{53} = 0$, which is always the case when the antiplane $R = 0$ is a plane of elastic symmetry, 1.92, the axis of the beam remains straight and no bending of the axis accompanies the twisting.

To discuss the stress we observe that the stress functions χ and ψ of 2.52 satisfy the equations

$$\begin{aligned} &D^{(4)}\chi - D^{(3)}\psi = 0 \\ &D^{(3)}\chi - D^{(2)}\psi = 2\tau - \frac{M}{2k_{33}}\left(\frac{k_{43}^2}{I_y} + \frac{k_{53}^2}{I_x}\right) = \gamma_1\,, \end{aligned} \tag{9}$$

where

$$D^{(2)} \equiv l_{44}\frac{\partial^2}{\partial x^2} - 2l_{45}\frac{\partial^2}{\partial x\,\partial y} + l_{55}\frac{\partial^2}{\partial y^2}\,.$$

Since the operator $D^{(3)}$ will annihilate a quadratic polynomial, a particular set of solutions of these equations is

$$\chi_0 = 0,\ \psi_0 = a x^2 + 2 b x y + c y^2$$

where

$$-\gamma_1 = D^{(2)}\psi_0 = 2(l_{44}a - 2l_{45}b + l_{55}c) \tag{10}$$

which we can satisfy by taking

$$a = k l_{55},\quad b = k l_{45},\quad c = k l_{44}\,,$$

where

$$4k(l_{44}l_{55} - l_{45}^2) = -\gamma_1\,.$$

Thus

$$\psi_0 = \left[-2\tau + \frac{M}{2k_{33}}\left(\frac{k_{43}^2}{I_y} + \frac{k_{35}^2}{I_x}\right)\right]\frac{l_{55}x^2 + 2l_{45}xy + l_{44}y^2}{4(l_{44}l_{55} - l_{45}^2)}\,. \tag{11}$$

The stress components are

$$\widehat{xx} = \frac{\partial^2\chi}{\partial y^2},\quad \widehat{xy} = -\frac{\partial^2\chi}{\partial x\,\partial y},\quad \widehat{yy} = \frac{\partial^2\chi}{\partial x^2}$$

$$\widehat{xz} = \frac{\partial\psi}{\partial y},\quad \widehat{yz} = -\frac{\partial\psi}{\partial x} \tag{12}$$

$$\widehat{zz} = \frac{1}{2}M\left(\frac{\nu_4 x}{I_y} - \frac{\nu_5 y}{I_x}\right) - (\nu_1\widehat{xx} + \nu_2\widehat{yy} + \nu_4\widehat{yz} + \nu_5\widehat{xz} + \nu_6\widehat{xy})\,.$$

We observe that $\widehat{zz}$ is in general different from zero but that, since $P_3 = 0$,

$$\int_S \widehat{zz}_0\,dS = 0\,.$$

From 6.2 (11)—(13), the boundary conditions are

$$\frac{\partial\chi}{\partial x} = c_1,\quad \frac{\partial\chi}{\partial y} = c_2,\quad \psi = c_3 \text{ on the boundary} \tag{13}$$

and therefore from 2.6 (19), (20), in terms of the complex stresses,

$$\left.\begin{aligned}
&\frac{1}{4}\sum_{\nu=1}^{3}\{{}'W_\nu(z_\nu) + {}'\overline{W}_\nu(\bar{z}_\nu)\} = c_1\\
&\frac{1}{4}\sum_{\nu=1}^{3}\{\lambda_\nu{}'W_\nu(z_\nu) + \bar{\lambda}_\nu{}'\overline{W}_\nu(\bar{z}_\nu)\} = c_2\\
&\frac{1}{4}\sum_{\nu=1}^{3}\{\mu_\nu{}'W_\nu(z_\nu) + \bar{\mu}_\nu{}'\overline{W}_\nu(\bar{z}_\nu)\} = c_3 + \psi_0
\end{aligned}\right\} \tag{14}$$

on the boundary.

Since from 2.6 (20)

$$\psi = \psi_0 + \frac{1}{4}\sum_{\nu=1}^{3}\{\mu_\nu{}'W_\nu(z_\nu) + \bar{\mu}_\nu\overline{W}_\nu(\bar{z}_\nu)\} \tag{15}$$

we write, in view of (9)

$$\psi = \frac{1}{2}\gamma_1 \psi^* = \left[\tau - \frac{M}{4k_{33}}\left(\frac{k_{43}^2}{I_y} + \frac{k_{53}^2}{I_x}\right)\right]\psi^* . \tag{16}$$

Thus, when $D^{(3)}\chi = 0$, ψ^* satisfies the equation

$$l_{44}\frac{\partial^2 \psi^*}{\partial x^2} - 2l_{45}\frac{\partial^2 \psi^*}{\partial x\,\partial y} + l_{55}\frac{\partial^2 \psi^*}{\partial y^2} = -2 \tag{17}$$

and depends only on the modified inverse moduli and the geometry of the cross-section.

The twist per unit length τ is determined by

$$M = 2\int \psi\, dS = \gamma_1 \int \psi^* dS$$

which gives

$$M = \left[\tau - \frac{M}{4k_{33}}\left(\frac{k_{43}^2}{I_y} + \frac{k_{53}^2}{I_x}\right)\right] D^*, \quad D^* = 2\int_S \psi^* dS . \tag{18}$$

Therefore for the torsional rigidity D (4.34) we have

$$\tau = M\left[\frac{1}{D^*} + \frac{1}{4k_{33}}\left(\frac{k_{43}^2}{I_y} + \frac{k_{53}^2}{I_x}\right)\right] = \frac{M}{D} \tag{19}$$

or

$$D = \frac{D^*}{1 + \dfrac{D^*}{4k_{33}}\left(\dfrac{k_{43}^2}{I_y} + \dfrac{k_{53}^2}{I_x}\right)} \tag{20}$$

and

$$D \leqq D^* . \tag{21}$$

Thus D^* is an upper bound for the torsional rigidity.

6.42. Twisting with simultaneous bending couples

If in addition to the simple twisting couple M about the axis of the beam, couples M_1 about Gx and M_2 about Gy are applied, it appears from 6.33 that

$$A_2 = \frac{1}{2}k_{43}\frac{M}{I_y} + k_{33}\frac{M_2}{I_y}, \; B_2 = -\frac{1}{2}k_{53}\frac{M}{I_x} - k_{33}\frac{M_1}{I_x}, \; C_2 = 0 \tag{1}$$

while from 2.32 (2)

$$\widehat{zz} = \frac{\frac{1}{2}M\nu_4 + M_2}{I_y}x - \frac{\frac{1}{2}M\nu_5 + M_1}{I_x}y - (\nu_1\widehat{xx} + \nu_2\widehat{yy} + \nu_4\widehat{yz} + \nu_5\widehat{xz} + \nu_6\widehat{xy}) . \tag{2}$$

The twist τ per unit length is determined from

$$M = 2\int_S \psi\, dS . \tag{3}$$

If ψ^* is the function defined as in 6.4 (16), we have now

$$\psi = \left[\tau - \frac{M}{4k_{33}}\left(\frac{k_{43}^2}{I_y} + \frac{k_{53}^2}{I_x}\right) - \frac{M_1}{2I_x}k_{53} - \frac{M_2}{2I_y}k_{43}\right]\psi^* \tag{4}$$

and therefore

$$\tau = M\left[\frac{1}{D^*} + \frac{1}{4k_{33}}\left(\frac{k_{43}^2}{I_y} + \frac{k_{53}^2}{I_x}\right)\right] + \frac{M_1}{2I_x}k_{53} + \frac{M_2}{2I_y}k_{43}\,. \tag{5}$$

The components of displacement are given by 2.3 (23). If we fix the end (0, 0, L) of the axis of the beam, we have for the displacements on the axis

$$u = -\frac{1}{2}A_2(L-R)^2,\; v = -\frac{1}{2}B_2(L-R)^2$$

and therefore the equations of the projection of the bent axis on the planes $y = 0$, $x = 0$ are respectively

$$\left.\begin{aligned} x' &= \frac{1}{2I_y}\left(-\frac{1}{2}Mk_{43} - M_2k_{33}\right)(L-R)^2 \\ y' &= \frac{1}{2I_x}\left(\frac{1}{2}Mk_{53} + M_1k_{33}\right)(L-R)^2 \end{aligned}\right\}. \tag{6}$$

Thus the bent axis lies in the plane $A_2y' - B_2x' = 0$; cf. 3.7 (9).

Torsion without bending or *pure torsion* can be achieved by keeping the axis straight that is by choosing the bending couples M_1 and M_2 so that $x' = 0 = y'$. This condition gives

$$M_1 = -\frac{1}{2}\frac{k_{53}}{k_{33}}M, \quad M_2 = -\frac{1}{2}\frac{k_{43}}{k_{33}}M\,. \tag{7}$$

Substitution in (5) gives

$$\tau = \frac{M}{D^*} \tag{8}$$

so that the torsional rigidity is now D^* which is an increase over the value when the axis is distorted.

This result indicates that the introduction of bending couples, which tend to inhibit deflection of the axis of the beam, also tends to increase the torsional rigidity.

We can also achieve *bending without twisting* by so choosing the moments M_1 and M_2 that the beam will bend in a principal plane. If we choose the plane of the bending to be $x = 0$, we put $x' = 0$ in (6) and $\tau = 0$ in (5).

6.44. Twisting of an elliptic cylinder by an axial couple

We consider the cylinder of 6.42 having now an elliptic cross-section bounded by

$$\frac{x^2}{a^2} + \frac{y^2}{b^2} = 1 \tag{1}$$

so that

$$I_x = \pi ab\cdot\frac{b^2}{4}, \quad I_y = \pi ab\cdot\frac{a^2}{4}\,. \tag{2}$$

Since the cross-section is simply connected, we can take the constants c_1, c_2, c_3 to be zero, and then the boundary conditions 6.4 (13) will be satisfied by

$$\chi = 0,\ \psi = \alpha\left(1 - \frac{x^2}{a^2} - \frac{y^2}{b^2}\right). \tag{3}$$

Since

$$M = 2\int_S \psi\, dS.$$

we have

$$M = 2\alpha\left(\pi ab - \frac{\pi ab}{4} - \frac{\pi ab}{4}\right) = \pi ab\,\alpha$$

and therefore

$$\psi = \frac{M}{\pi ab}\left(1 - \frac{x^2}{a^2} - \frac{y^2}{b^2}\right). \tag{4}$$

Substitution in 6.41 (12) shows that $\widehat{zz} = 0$. Again from 6.41 (9) ψ satisfies the equation

$$l_{44}\frac{\partial^2\psi}{\partial x^2} - 2l_{45}\frac{\partial^2\psi}{\partial x\,\partial y} + l_{55}\frac{\partial^2\psi}{\partial y^2} = -\gamma_1.$$

Substitute (4) for ψ. Then

$$\frac{M}{\pi ab}\left(2\frac{l_{44}}{a^2} + 2\frac{l_{55}}{b^2}\right) = \gamma_1.$$

Therefore from 6.41 (16)

$$\frac{M}{\pi ab}\,\psi^*\left(\frac{l_{44}}{a^2} + \frac{l_{55}}{b^2}\right) = \psi.$$

Therefore

$$D^*\cdot\frac{M}{\pi ab}\cdot\left(\frac{l_{44}}{a^2} + \frac{l_{55}}{b^2}\right) = M. \tag{5}$$

Therefore from 6.41 (20) the torsional rigidity is

$$D = \frac{\dfrac{\pi a^3 b^3}{b^2 l_{44} + a^2 l_{55}}}{1 + \dfrac{a^2b^2}{k_{33}(b^2 l_{44} + a^2 l_{55})}\left(\dfrac{k_{43}^2}{a^2} + \dfrac{k_{53}^2}{b^2}\right)}. \tag{6}$$

6.46. Pure torsion of an elliptic cylinder

From 6.42 (7) the axis will remain straight if we apply couples

$$M_1 = -\frac{1}{2}\frac{k_{53}}{k_{33}}M,\ M_2 = -\frac{1}{2}\frac{k_{43}}{k_{33}}M$$

in addition to the twisting couple M, and therefore, from 6.42 (5),

$$\tau = \frac{M}{D^*}$$

so that the torsional rigidity is, from 6.44 (5).

$$D^* = \frac{\pi a^3 b^3}{b^2 l_{44} + a^2 l_{55}}$$

6.5. Antiplane of elastic symmetry

For the rest of this chapter we shall confine attention to beams with an antiplane of elastic symmetry that is to say homogeneous cylindrical beams (fig. 3.22) for which the antiplane, $R = 0$, is a plane of elastic symmetry. Then all cross-sections will likewise be planes of elastic symmetry and therefore, 1.92,

$$k_{14} = k_{15} = k_{24} = k_{25} = k_{34} = k_{35} = k_{46} = k_{56} = 0\,, \qquad \nu_4 = \nu_5 = 0\,, \tag{1}$$

where $\nu_\alpha = k_{\alpha 3}/k_{33}$.

It follows from (1) that of the modified inverse moduli of 2.32

$$l_{14} = l_{15} = l_{24} = l_{25} = l_{34} = l_{35} = l_{46} = l_{56} = 0\,, \qquad l_{44} = k_{44}, \quad l_{55} = k_{55}, \quad l_{45} = k_{45}\,. \tag{2}$$

Now from 2.52 (8), (10), in the absence of a body-field, the stress functions χ and ψ satisfy the equations

$$D^{(4)}\chi = 0 \tag{3}$$

$$D^{(2)}\psi = (-\nu_6 A_1 + 2\nu_1 B_1)\,x + (\nu_6 B_1 - 2\nu_2 A_1)\,y - 2\tau + \frac{l_{45}}{k_{33}}(-A_1 x + B_1 y)\,. \tag{4}$$

These equations are independent, the first leading to the plane anisotropic system discussed in [MILNE-THOMSON (6)]. For the present purpose we can take, without loss of generality,

$$\chi = 0 \tag{5}$$

so that

$$\widehat{xx} = \widehat{xy} = \widehat{yy} = 0 \tag{6}$$

thus obtaining the Filon antiplane problems of 2.0 (6).

Therefore the Filon antiplane problem, or *pure antiplane problem*, for which every plane parallel to the antiplane is a plane of elastic symmetry reduces to solving equation (4) namely

$$k_{44}\frac{\partial^2\psi}{\partial x^2} - 2k_{45}\frac{\partial^2\psi}{\partial x\,\partial y} + k_{55}\frac{\partial^2\psi}{\partial y^2} = (-\nu_6 A_1 + 2\nu_1 B_1)\,x + (\nu_6 B_1 - 2\nu_2 A_1)\,y - 2\tau + \frac{k_{45}}{k_{33}}(-A_1 x + B_1 y) \tag{7}$$

subject to the appropriate boundary conditions.

6.52. An affine change of variables

In the equation 6.5 (7) satisfied by ψ namely

$$k_{44}\frac{\partial^2\psi}{\partial x^2} - 2k_{45}\frac{\partial^2\psi}{\partial x\,\partial y} + k_{55}\frac{\partial^2\psi}{\partial y^2} = Px + Qy - 2\tau\,, \tag{1}$$

where

$$P = -\nu_6 A_1 + 2\nu_1 B_1 - \frac{A_1 k_{45}}{k_{33}},\; Q = \nu_6 B_1 - 2\nu_2 A_1 + \frac{B_1 k_{45}}{k_{33}}, \tag{2}$$

make the affine change of variables

$$\xi = \alpha x + \beta y,\; \eta = \gamma x + \delta y\,. \tag{3}$$

We shall suppose that α, β, γ, δ are dimensionless constants so that (ξ, η) have the same dimensions as (x, y). Then

$$\frac{\partial}{\partial x} = \alpha\frac{\partial}{\partial\xi} + \gamma\frac{\partial}{\partial\eta},\; \frac{\partial}{\partial y} = \beta\frac{\partial}{\partial\xi} + \delta\frac{\partial}{\partial\eta} \tag{4}$$

so that (1) becomes

$$\begin{aligned}&(k_{44}\alpha^2 - 2k_{45}\alpha\beta + k_{55}\beta^2)\frac{\partial^2\psi}{\partial\xi^2}\\ &+ 2\left[k_{44}\alpha\gamma - k_{45}(\alpha\delta + \beta\gamma) + k_{55}\beta\delta\right]\frac{\partial^2\psi}{\partial\xi\,\partial\eta}\\ &+ (k_{44}\gamma^2 - 2k_{45}\gamma\delta + k_{55}\delta^2)\frac{\partial^2\psi}{\partial\eta^2} = P'\xi + Q'\eta - 2\tau\end{aligned} \tag{5}$$

where

$$P' = \frac{P\delta - \gamma Q}{\alpha\delta - \beta\gamma},\quad Q' = \frac{-P\beta + \alpha Q}{\alpha\delta - \beta\gamma}\,. \tag{6}$$

Equation (5) will reduce to the form

$$\frac{1}{\mu}\left(\frac{\partial^2\psi}{\partial\xi^2} + \frac{\partial^2\psi}{\partial\eta^2}\right) = P'\xi + Q'\eta - 2\tau \tag{7}$$

where μ has the dimensions of a stress, provided that

$$k_{44}\alpha^2 - 2k_{45}\alpha\beta + k_{55}\beta^2 = k_{44}\gamma^2 - 2k_{45}\gamma\delta + k_{55}\delta^2 = \frac{1}{\mu} \tag{8}$$

$$k_{44}\alpha\gamma - k_{45}(\alpha\delta + \beta\gamma) + k_{55}\beta\delta = 0 \tag{9}$$

Write

$$i\alpha + \gamma = L,\quad i\beta + \delta = M\,. \tag{10}$$

Then from (8) and (9) we get

$$k_{44}L^2 - 2k_{45}LM + k_{55}M^2 = 0\,, \tag{11}$$

whence

$$\frac{L}{M} = \frac{k_{45} + ip}{k_{44}},\quad p^2 = k_{44}k_{55} - k_{45}^2\,. \tag{12}$$

We arbitrarily choose the positive value of p. Then we can write

$$L = \frac{1}{\varepsilon}(k_{45} + ip),\quad M = \frac{1}{\varepsilon}k_{44} \tag{13}$$

where ε is a constant.

Therefore from (10)

$$\alpha = \frac{p}{\varepsilon}, \gamma = \frac{k_{45}}{\varepsilon}, \beta = 0, \delta = \frac{k_{44}}{\varepsilon}$$

so that from (3)

$$x = \frac{\varepsilon \xi}{p}, \quad y = -\frac{\varepsilon k_{45} \xi}{k_{44} p} + \frac{\varepsilon \eta}{k_{44}}. \tag{14}$$

Also from (8)

$$\mu = \frac{\varepsilon^2}{p^2 k_{44}}. \tag{15}$$

Now by hypothesis α, β, γ, δ are dimensionless and so therefore are ε/p, ε/k_{44}. There are infinitely many ways of achieving this by replacing ε by a homogeneous function of degree one in the variables k_{44}, k_{45}, k_{55}. A simple method is to write

$$\varepsilon = p \sqrt{\frac{2k_{44}}{k_{44} + k_{55}}} \tag{16}$$

and then we get

$$x = \xi \sqrt{\frac{2k_{44}}{k_{44} + k_{55}}}, \quad y = \left(-\xi \frac{k_{45}}{k_{44}} + \frac{p}{k_{44}} \eta\right) \sqrt{\frac{2k_{44}}{k_{44} + k_{55}}}, \tag{17}$$

$$\mu = \frac{2}{k_{44} + k_{55}}, \qquad p^2 = k_{44} k_{55} - k_{45}^2 \tag{18}$$

$$\xi = x \sqrt{\frac{k_{44} + k_{55}}{2k_{44}}}, \qquad \eta = \frac{k_{45} x + k_{44} y}{p} \sqrt{\frac{k_{44} + k_{55}}{2k_{44}}}. \tag{19}$$

When the material is orthotropic $k_{45} = 0$ and (17) becomes

$$x = \xi \sqrt{\frac{2k_{44}}{k_{44} + k_{55}}}, y = \eta \sqrt{\frac{2k_{55}}{k_{44} + k_{55}}}. \tag{20}$$

Note that in the case of isotropy $k_{44} = k_{55} = 1/\mu$ which is consistent with (18).

6.53. Reduction to the isotropic case

Consider equation 2.9 (4), wherein we temporarily replace Poisson's ratio by η_0, and A_1, B_1, τ by A_1', B_1', τ'. Also write (ξ, η) for (x, y). Then the stress function satisfies, in the isotropic case,

$$\frac{1}{\mu}\left(\frac{\partial^2 \psi}{\partial \xi^2} + \frac{\partial^2 \psi}{\partial \eta^2}\right) = 2\eta_0 (-B_1' \xi + A_1' \eta) - 2\tau'. \tag{1}$$

Now in 6.52 we have shown that in the anisotropic case the equation 6.52 (1) can be reduced to

$$\frac{1}{\mu}\left(\frac{\partial^2 \psi}{\partial \xi^2} + \frac{\partial^2 \psi}{\partial \eta^2}\right) = \frac{\varepsilon}{p}\left(P - Q\frac{k_{45}}{k_{44}}\right)\xi + \frac{\varepsilon}{k_{44}} Q \eta - 2\tau \tag{2}$$

by the affine substitutions 6.52 (17).

Comparing (1) and (2) and using 6.52 (2) we see that the equation satisfied by ψ in the anisotropic case with coordinates (x, y) is the same as the equation satisfied by ψ in an isotropic problem in which (2.9 (4))

$$2\eta_0 B_1' = \frac{\varepsilon}{p}\left\{\left(\nu_6 + \frac{k_{45}}{k_{33}} - 2\nu_2 \frac{k_{45}}{k_{44}}\right) A_1 + \left(-2\nu_1 + \nu_6 \frac{k_{45}}{k_{44}} + \frac{k_{45}^2}{k_{33}k_{44}}\right) B_1\right\} \tag{3}$$

$$2\eta_0 A_1' = \frac{\varepsilon}{k_{44}}\left\{-2\nu_2 A_1 + \left(\nu_6 + \frac{k_{45}}{k_{33}}\right) B_1\right\} \tag{4}$$

$$\tau' = \tau\,. \tag{5}$$

Let the boundary in the anisotropic problem be

$$f(x, y) = 0\,. \tag{6}$$

Then the boundary in the corresponding isotropic problem is

$$F(\xi, \eta) = 0\,, \tag{7}$$

where (7) is deduced from (6) by the substitutions 6.52 (17).

The boundary condition appropriate to the isotropic problem must be investigated in each case but the condition

$$\psi = 0 \text{ on the boundary} \tag{8}$$

will carry over directly.

Thus when the cross-sections are planes of elastic symmetry the problem is reduced to the solution of (1) subject to the condition (8) on the boundary (7) and this is precisely the problem of torsion of an isotropic beam.

Having solved this isotropic problem by methods already explained in Chapter IV, we obtain the solution of the corresponding anisotropic problem by expressing $\psi(\xi, \eta)$ in terms of x, y by 6.52 (19).

6.56. Solution of the equation satisfied by ψ

The equation is 6.52 (1)

$$D^{(2)}\psi \equiv k_{44}\frac{\partial^2\psi}{\partial x^2} - 2k_{45}\frac{\partial^2\psi}{\partial x\,\partial y} + k_{55}\frac{\partial^2\psi}{\partial y^2} = Px + Qy - 2\tau\,, \tag{1}$$

where P, Q are given by 6.52 (2).

The method of 2.6 shows that when the right-hand side of (1) is put equal to zero, the resulting homogeneous equation has the solution

$$'W(x + \lambda y) + {}'\overline{W}(x + \bar{\lambda} y)\,, \tag{2}$$

where λ, $\bar{\lambda}$ are the roots of the characteristic equation

$$k_{55}\lambda^2 - 2k_{45}\lambda + k_{44} = 0 \tag{3}$$

so that, incidentally,

$$k_{55}\lambda - k_{45} = \frac{k_{45}\lambda - k_{44}}{\lambda}\,. \tag{4}$$

The equation $D^{(2)}\psi = Px$ has the particular solution

$$\frac{Px^3}{6k_{44}} \tag{5}$$

A particular solution of $D^{(2)}\psi = -2\tau$ is given as in 6.41 (11). Thus for a particular solution of (1) we have

$$\psi_0 = \frac{Px^3}{6k_{44}} + \frac{Qy^3}{6k_{55}} - 2\tau \frac{k_{55}x^2 + 2k_{45}xy + k_{44}y^2}{4(k_{44}k_{55} - k_{45}^2)}. \tag{6}$$

Thus the general solution of (1) is

$$\psi = {}'W(x+\lambda y) + {}'\overline{W}(x+\bar{\lambda}y) + \psi_0. \tag{7}$$

6.6. The stress component $\widehat{zz}$

For the pure antiplane problem with an antiplane of elastic symmetry we have from 6.5

$$\nu_4 = \nu_5 = 0,\ \widehat{xx} = \widehat{xy} = \widehat{yy} = 0. \tag{1}$$

Therefore from 2.2 (8), 2.3 (14), 2.32 (2) we get

$$\widehat{zz} = \frac{1}{k_{33}}\{(A_1x + B_1y + C_1m)R + A_2x + B_2y + C_2m\} \tag{2}$$

It follows that of the equations of equilibrium 2.1 (1)—(3), the first two are satisfied identically, while the third becomes

$$\frac{\partial\widehat{xz}}{\partial x} + \frac{\partial\widehat{yz}}{\partial y} = -\frac{1}{k_{33}}(A_1x + B_1y + C_1m), \tag{3}$$

where

$$\widehat{xz} = \frac{\partial\psi}{\partial y} - \frac{1}{2k_{33}}(A_1x^2 + C_1mx), \tag{4}$$

$$\widehat{yz} = -\frac{\partial\psi}{\partial x} - \frac{1}{2k_{33}}(B_1y^2 + C_1my). \tag{5}$$

6.61. Determination of the constants $A_1, B_1, C_1; A_2, B_2, C_2$

Referring to fig. 3.22, we suppose that the stress distribution applied to the antiplane reduces, when the origin is taken as base point, to a force (P_1, P_2, P_3) and a moment (M_1, M_2, M_3). It is convenient to regard M_3 as negative and to write

$$-M_3 = M. \tag{1}$$

The moment M will be called the *twisting moment*. Since the lateral surface is free of applied force, 6.32 (9), (10) give

$$A_1 = -\frac{k_{33}P_1}{I_y},\ B_1 = -\frac{k_{33}P_2}{I_x},\ C_1 = 0. \tag{2}$$

From 6.32 (8) and 6.33 (4), (5) we have

$$A_2 = \frac{k_{33}M_2}{I_y},\ B_2 = \frac{-k_{33}M_1}{I_x},\ C_2m = \frac{-k_{33}P_3}{S}, \tag{3}$$

where S is the area of the cross-section.

Therefore from 6.6 (2), (3)

$$\widehat{zz} = -\left(\frac{P_1 x}{I_y} + \frac{P_2 y}{I_x}\right) R + \frac{M_2 x}{I_y} - \frac{M_1 y}{I_x} - \frac{P_3}{S}, \tag{4}$$

$$\frac{\partial \widehat{xz}}{\partial x} + \frac{\partial \widehat{yz}}{\partial y} = \frac{P_1 x}{I_y} + \frac{P_2 y}{I_x}, \tag{5}$$

while from 6.5 (7)

$$k_{44} \frac{\partial^2 \psi}{\partial x^2} - 2 k_{45} \frac{\partial^2 \psi}{\partial x \partial y} + k_{55} \frac{\partial^2 \psi}{\partial y^2}$$

$$= -2\tau + \left[\frac{(k_{63} + k_{45}) P_1}{I_y} - \frac{2 k_{13} P_2}{I_x}\right] x + \left[\frac{2 k_{23} P_1}{I_y} - \frac{(k_{63} + k_{45}) P_2}{I_x}\right] y \,. \tag{6}$$

When the material is orthotropic, so that additionally

$$k_{45} = k_{16} = k_{26} = k_{36} = 0 \tag{7}$$

equation (6) simplifies still further to

$$k_{44} \frac{\partial^2 \psi}{\partial x^2} + k_{55} \frac{\partial^2 \psi}{\partial y^2} = -2\tau - \frac{2 k_{13} P_2 x}{I_x} + \frac{2 k_{23} P_1 y}{I_y} \,. \tag{8}$$

6.62. Boundary conditions

From 6.6 (4), (5) and 6.61 (2) we find that

$$\widehat{xz} = \frac{\partial \psi}{\partial y} + \frac{P_1 x^2}{2 I_y}, \quad \widehat{yz} = -\frac{\partial \psi}{\partial x} + \frac{P_2 y^2}{2 I_x} \,. \tag{1}$$

Since the lateral surface is unloaded we have

$$l\,\widehat{xz} + m\,\widehat{yz} = 0 \text{ on the lateral surface}$$

and therefore since $l = dy/ds$, $m = -dx/ds$

$$\frac{d\psi}{ds} = -\frac{P_1 x^2}{2 I_y} \frac{dy}{ds} + \frac{P_2 y^2}{2 I_x} \frac{dx}{ds} \text{ on the lateral surface.} \tag{2}$$

When $P_1 = P_2 = 0$, (2) gives

$$\psi = \text{constant on the lateral surface} \,. \tag{3}$$

When the region occupied by a cross-section is simply connected we can take the constant to be zero. When P_1 and P_2 do not vanish we can proceed as follows.

Just as we introduced Timoshenko's stress function, 5.41, we can here apply the same principle by replacing ψ by

$$T - B_1 \frac{'s(x)}{2 k_{33}} + A_1 \frac{'t(y)}{2 k_{33}} \tag{4}$$

and then (1) gives

$$\begin{aligned} \widehat{xz} &= \frac{\partial T}{\partial y} - \frac{1}{2 k_{33}} A_1 [x^2 - t(y)] \\ \widehat{yz} &= -\frac{\partial T}{\partial x} - \frac{1}{2 k_{33}} B_1 [y^2 - s(x)] \,. \end{aligned} \tag{5}$$

If therefore we can choose $s(x)$, $t(y)$ so that the contents of the square brackets vanish on the boundary we shall have

$$\widehat{xz} = \frac{\partial T}{\partial y}, \quad \widehat{yz} = -\frac{\partial T}{\partial x} \text{ on the boundary}$$

and therefore the boundary condition $l\,\widehat{xz} + m\,\widehat{yz} = 0$ gives $dT/ds = 0$ so that T is constant on the boundary. For a simply connected cross-section we can take

$$T = 0 \quad \text{on the boundary}. \tag{6}$$

To obtain the equation satisfied by T, substitute the expression (4) for ψ in 6.52 (1). This gives

$$k_{44}\frac{\partial^2 T}{\partial x^2} - 2k_{45}\frac{\partial^2 T}{\partial x\,\partial y} + k_{55}\frac{\partial^2 T}{\partial y^2}$$

$$= A_1\left\{-\nu_6 x - \frac{k_{45}x}{k_{33}} - 2\nu_2 y - \frac{k_{55}}{2k_{33}}t'(y)\right\} \tag{7}$$

$$+ B_1\left\{2\nu_1 x + \nu_6 y + \frac{k_{45}y}{k_{33}} + \frac{k_{44}}{2k_{33}}s'(x)\right\} - 2\tau.$$

6.63. Flexure of an elliptic cylinder with an antiplane of elastic symmetry

We take the cross-section to be bounded by the ellipse

$$\frac{x^2}{a^2} + \frac{y^2}{b^2} = 1 \tag{1}$$

and the force to be $(-P, 0, 0)$ i.e., P directed along the negative x-axis and applied at the centroid. Then

$$A_1 = \frac{k_{33}P}{I_y}, \quad B_1 = 0, \quad C_1 = 0 \tag{2}$$

while by symmetry $\tau = 0$. Therefore from 6.52 (1), (2)

$$k_{44}\frac{\partial^2\psi}{\partial x^2} - 2k_{45}\frac{\partial^2\psi}{\partial x\,\partial y} + k_{55}\frac{\partial^2\psi}{\partial x^2} = -\left(\nu_6 + \frac{k_{45}}{k_{33}}\right)A_1 x - 2\nu_2 A_1 y. \tag{3}$$

Introduce from 6.62 (4) the stress function T given by

$$\psi = T + A_1\frac{t(y)}{2k_{33}} \tag{4}$$

$$\widehat{xz} = \frac{\partial T}{\partial y} - \frac{1}{2k_{33}}A_1[x^2 - t(y)], \quad \widehat{yz} = -\frac{\partial T}{\partial x}. \tag{5}$$

To make $\widehat{xz} = \frac{\partial T}{\partial y}$, $\widehat{yz} = -\frac{\partial T}{\partial x}$ on the boundary we take

$$t(y) = a^2 - \frac{a^2}{b^2}y^2 \tag{6}$$

and then on the boundary we can take

$$T = 0 \tag{7}$$

for dT/ds is zero there.

The equation satisfied by T is then

$$k_{44}\frac{\partial^2 T}{\partial x^2} - 2k_{45}\frac{\partial^2 T}{\partial x\,\partial y} + k_{55}\frac{\partial^2 T}{\partial y^2} = A_1\left\{\left(\frac{k_{55}a^2}{k_{33}b^2} - 2\nu_2\right)y - \left(\nu_6 + \frac{k_{45}}{k_{33}}\right)x\right\}. \quad (8)$$

To find a solution which vanishes on the boundary put

$$T = p\,x\left(\frac{x^2}{a^2} + \frac{y^2}{b^2} - 1\right) + q\,y\left(\frac{x^2}{a^2} + \frac{y^2}{b^2} - 1\right). \quad (9)$$

Substitution in (8) then gives

$$x\left\{\frac{6k_{44}p}{a^2} + \frac{2k_{55}p}{b^2} - \frac{4k_{45}q}{a^2}\right\} + y\left\{\frac{2k_{44}q}{a^2} + \frac{6k_{55}q}{b^2} - \frac{4k_{45}p}{b^2}\right\}$$
$$= A_1\left\{\left(\frac{k_{55}a^2}{k_{33}b^2} - 2\nu_2\right)y - \left(\nu_6 + \frac{k_{45}}{k_{33}}\right)x\right\}.$$

Therefore

$$\left(\frac{6k_{44}}{a^2} + \frac{2k_{55}}{b^2}\right)p - \frac{4k_{45}q}{a^2} = -\left(\nu_6 + \frac{k_{45}}{k_{33}}\right)A_1 \quad (10)$$

$$\frac{-4k_{45}p}{b^2} + \left(\frac{6k_{55}}{b^2} + \frac{2k_{44}}{a^2}\right)q = A_1\left(\frac{k_{55}}{k_{33}}\frac{a^2}{b^2} - 2\nu_2\right) \quad (11)$$

whose solution gives p and q and so the stress function T.

6.64. The twisting moment

The moment M is given by

$$M = \int_S (x\cdot\widehat{yz} - y\cdot\widehat{xz})\,dS = -\int_S\left(x\frac{\partial\psi}{\partial x} + y\frac{\partial\psi}{\partial y}\right)dS$$
$$+\int_S\left(\frac{P_2 x y^2}{2I_x} - \frac{P_1 x^2 y}{2I_y}\right)dS.$$

Now

$$\int_S\left(x\frac{\partial\psi}{\partial x} + y\frac{\partial\psi}{\partial y}\right)dS = \int_S\left(\frac{\partial}{\partial x}(x\psi) + \frac{\partial}{\partial y}(y\psi) - 2\psi\right)dS$$
$$= \oint_C \psi(lx + my)\,ds - \int_S 2\psi\,dS.$$

Therefore

$$M = -\oint_C \psi(lx + my)\,ds + 2\int_S \psi\,dS + \int_S\left(\frac{P_2 x y^2}{2I_x} - \frac{P_1 x^2 y}{2I_y}\right)dS. \quad (1)$$

In the particular case in which

$$P_1 = P_2 = 0$$

we have $\psi =$ constant on the boundary and therefore

$$M = 2\int_S \psi\,dS + \text{constant}. \quad (2)$$

For a simply connected cross-section we can make $\psi = 0$ on the boundary and in this case

$$M = 2\int_S \psi\,dS. \quad (3)$$

6.65. The displacement

The components of the displacement are given by **2.3** (23) in terms of certain displacements u_1, v_1, w_1 and expressions which in this case reduce to

$$L = A_1 x + B_1 y,\ T = A_2 x + B_2 y + C_2 m,\ P = -A_2,\ P' = -B_2 \quad (1)$$

$$Q_1 = \frac{1}{2}\left[\nu_1 A_1 x^2 + 2\nu_1 B_1\, xy + (\nu_6 B_1 - \nu_2 A_1)\, y^2\right] - \tau y \quad (2)$$

$$Q_1' = \frac{1}{2}\left[(\nu_6 A_1 - \nu_1 B_1)\, x^2 + 2\nu_2 A_1 xy + \nu_2 B_1 y^2\right] + \tau x\,. \quad (3)$$

The displacements u_1, v_1, w_1 satisfy **2.5** (6), (7) namely

$$\begin{aligned} \frac{\partial u_1}{\partial x} &= \nu_1 (A_2 x + B_2 y + C_2 m) \\ \frac{\partial v_1}{\partial y} &= \nu_2 (A_2 x + B_2 y + C_2 m) \\ \frac{\partial v_1}{\partial x} + \frac{\partial u_1}{\partial y} &= \nu_6 (A_2 x + B_2 y + C_2 m) \end{aligned} \quad (4)$$

$$\begin{aligned} \frac{\partial w_1}{\partial x} &= k_{45}\widehat{yz} + k_{55}\widehat{xz} - Q_1 \\ \frac{\partial w_1}{\partial y} &= k_{44}\widehat{yz} + k_{45}\widehat{xz} - Q_1'\,. \end{aligned} \quad (5)$$

The arbitrary elements are already contained in the rigid body movement in **2.3** (23) so that only particular solutions of (4) and (5) are required.

We readily prove or verify that

$$\begin{aligned} u_1 &= \frac{M_2}{2I_y}(k_{13}x^2 - k_{23}y^2) - \frac{M_1}{I_x}\left(k_{13}xy + \frac{1}{2}k_{63}y^2\right) - \frac{P_3}{S}\left(k_{13}x + \frac{1}{2}k_{63}y\right) \\ v_1 &= \frac{M_2}{I_y}\left(k_{23}xy + \frac{1}{2}k_{63}x^2\right) - \frac{M_1}{2I_x}(k_{23}y^2 - k_{13}x^2) - \frac{P_3}{S}\left(k_{23}y + \frac{1}{2}k_{63}x\right). \end{aligned} \quad (6)$$

From **6.56** (7) we get ψ and then $\widehat{xz}, \widehat{yz}$ from **6.6** (4). Substitution in (5), using **6.56** (4), then leads to

$$\begin{aligned} \frac{\partial w_1}{\partial x} &= \left(k_{45} - \frac{k_{44}}{\lambda}\right) W(x + \lambda y) + \left(k_{45} - \frac{k_{44}}{\bar\lambda}\right) \overline{W}(x + \bar\lambda y) \\ &+ \frac{1}{2} A_1\left\{\left(\frac{k_{45}^2 - k_{44}k_{55}}{k_{33}k_{44}} + \nu_6\frac{k_{45}}{k_{44}} - \nu_1\right) x^2 - \nu_2 y^2\right\} - \nu_1 B_1\left(\frac{k_{45}}{k_{44}}x^2 + xy\right)\Big\} \end{aligned} \quad (7)$$

$$\begin{aligned} \frac{\partial w_1}{\partial y} &= (\lambda k_{45} - k_{44})\, W(x + \lambda y) + (\bar\lambda k_{45} - k_{44})\, \overline{W}(x + \bar\lambda y) \\ &- \nu_2 A_1\left(\frac{k_{45}}{k_{55}}y^2 + xy\right) + \frac{1}{2} B_1\left\{-\nu_1 x^2 + \left(\frac{k_{45}^2 - k_{44}k_{55}}{k_{33}k_{55}} + \nu_6\frac{k_{45}}{k_{55}} - \nu_2\right) y^2\right\}. \end{aligned} \quad (8)$$

Multiply these equations respectively by dx, dy add and integrate. We then get

$$w_1 = \frac{1}{\lambda}\,{}'W(x+\lambda y) + \frac{1}{\bar\lambda}\,{}'\overline{W}(x+\bar\lambda y)$$

$$+\frac{1}{2}A_1\left\{\left(\frac{k_{45}^2-k_{44}k_{55}}{k_{33}k_{44}} + \nu_6\frac{k_{45}}{k_{44}} - \nu_1\right)\frac{x^3}{3} - \nu_2 x y^2 - 2\nu_2\frac{k_{45}}{k_{55}}\frac{y^3}{3}\right\} \tag{9}$$

$$+\frac{1}{2}B_1\left\{-2\nu_1\frac{k_{45}}{k_{44}}\frac{x^3}{3} - \nu_1 x^2 y + \left(\frac{k_{45}^2-k_{44}k_{55}}{k_{33}k_{55}} + \nu_6\frac{k_{45}}{k_{55}} - \nu_2\right)\frac{y^3}{3}\right\}.$$

6.7. Orthotropic material

Referring to the beam and axes of reference described in 6.41, let the coordinate planes be planes of elastic symmetry, so that the material is now orthotropic. Then

$$k_{45} = 0,\ \nu_6 = 0\,. \tag{1}$$

Therefore the equation satisfied by ψ, 6.52 (1), becomes

$$k_{44}\frac{\partial^2\psi}{\partial x^2} + k_{55}\frac{\partial^2\psi}{\partial y^2} = 2\nu_1 B_1 x - 2\nu_2 A_1 y - 2\tau\,. \tag{2}$$

The affine transformation of 6.52 is now

$$x = \xi\sqrt{(\mu k_{44})},\ y = \eta\sqrt{(\mu k_{55})},\ \mu = \frac{2}{k_{44}+k_{55}} \tag{3}$$

and (2) transforms into

$$\frac{1}{\mu}\left(\frac{\partial^2\psi}{\partial\xi^2} + \frac{\partial^2\psi}{\partial\eta^2}\right) = 2\nu_1 B_1\xi\sqrt{(\mu k_{44})} - 2\nu_2 A_1\mu\sqrt{(\mu k_{55})} - 2\tau \tag{4}$$

while the boundary

$$f(x, y) = 0 \tag{5}$$

for the problem posed by (2) becomes the boundary

$$f\left[\xi\sqrt{(\mu k_{44})},\ \eta\sqrt{(\mu k_{55})}\right] = F(\xi, \eta) = 0 \tag{6}$$

for the problem posed by (4) which is a corresponding problem for isotropic material.

Thus if the solution for the isotropic problem can be obtained we can write down the solution for the orthotropic problem.

The equation corresponding to (2) which is satisfied by Timoshenko's stress function T is got from 6.62 (7) by putting $k_{45} = \nu_6 = C_1 = 0$. Thus

$$k_{44}\frac{\partial^2 T}{\partial x^2} + k_{55}\frac{\partial^2 T}{\partial y^2}$$

$$= A_1\left\{-2\nu_2 y - \frac{k_{55}}{2k_{33}}t'(y)\right\} + B_1\left\{2\nu_1 x + \frac{k_{44}}{2k_{33}}s'(x)\right\} - 2\tau\,. \tag{7}$$

In going from the isotropic to the orthotropic solution Poisson's ratio η has to be replaced by the appropriate combination of inverse moduli.

Comparison of the matrix 1.98 (5) in the isotropic case with the general matrix 1.72 (2) shows that possible replacements would be

$$\frac{k_{12}}{k_{33}}, \frac{k_{23}}{k_{33}}, \frac{k_{31}}{k_{33}}$$

the two latter ratios being ν_2, ν_1 respectively. Comparison of (2) with 2.9 (4) shows that in the "A_1 term" η should be replaced by

$$-\nu_2, \tag{8}$$

while for the "B_1 term" the proper replacement is $-\nu_1$.

To find the proper replacement for $1+\eta$ in going from the isotropic to the orthotropic solution we begin by observing that each of the ratios

$$\frac{k_{44}}{2k_{33}}, \frac{k_{55}}{2k_{33}}, \frac{k_{66}}{2k_{33}}$$

reduces to $1+\eta$ for isotropic material. If in 5.4 (5) we temporarily replace Poisson's ratio by η_0 and write (ξ, η) for (x, y), the term on the right-hand side which contains ξ^2 is

$$-\mu(1+\eta_0)A_1\xi^2 = -\mu(1+\eta_0)A_1\frac{x^2}{\mu k_{44}}$$

on use of (3). Comparing this with the corresponding term in 6.62 (5) we see the proper replacement for $1+\eta$ in the "A_1 term" is

$$\frac{k_{44}}{2k_{33}} \tag{9}$$

Similarly in the "B_1 term" the proper replacement is $k_{55}/(2k_{33})$.

6.72. Torsion of an orthotropic elliptic cylinder

For an isotropic cylinder whose cross-section is bounded by the ellipse

$$\frac{\xi^2}{\alpha^2}+\frac{\eta^2}{\beta^2}=1 \tag{1}$$

the torsion stress function, from 4.52 (3), is

$$\psi = \mu\tau\Omega = \mu\tau\frac{\alpha^2\beta^2}{\alpha^2+\beta^2}\left(1-\frac{\xi^2}{\alpha^2}-\frac{\eta^2}{\beta^2}\right). \tag{2}$$

If the cross-section of the orthotropic cylinder is bounded by the ellipse

$$\frac{x^2}{a^2}+\frac{y^2}{b^2}=1, \tag{3}$$

the affine transformation

$$x=\xi\sqrt{(\mu k_{44})}, \quad y=\eta\sqrt{(\mu k_{55})}, \tag{4}$$

gives

$$a=\alpha\sqrt{(\mu k_{44})}, \quad b=\beta\sqrt{(\mu k_{55})},$$

and therefore for the orthotropic cylinder

$$\psi = \frac{\tau}{\dfrac{k_{44}}{a^2}+\dfrac{k_{55}}{b^2}}\left(1-\frac{x^2}{a^2}-\frac{y^2}{b^2}\right).$$

6.74. Torsion of an orthotropic rectangular beam

Let the beam be that described in 4.61 but of orthotropic material, and let the faces of the beam be planes of elastic symmetry. Making the affine transformation 6.7 (3) the corresponding isotropic problem is that for a beam whose cross-section is a rectangle whose sides are 2α, 2β where

$$a = \alpha\sqrt{(\mu k_{44})},\ b = \beta\sqrt{(\mu k_{55})},\ \mu = \frac{2}{k_{44}+k_{55}}\,. \tag{1}$$

Now from 4.1 (3), (8) the torsion stress function for the isotropic beam in terms of the coordinates (ξ, η) is

$$\Omega = \psi - \frac{1}{2}(\xi^2 + \eta^2), \tag{2}$$

where from 4.61 (9)

$$\psi = -\frac{1}{2}(\xi^2 - \eta^2) - 4\sum_{n=0}^{\infty} \frac{(-1)^n \cos(\mu_n \xi)\cosh(\mu_n \eta)}{\alpha\mu_n^3 \cosh(\mu_n \beta)} \tag{3}$$

and

$$\mu_n = \frac{2n+1}{2\alpha}\pi\,. \tag{4}$$

For our problem denote by Ψ the torsion function. Then in the present notation

$$\Psi = \mu\tau\Omega = \mu\tau\left\{\psi - \frac{1}{2}(\xi^2+\eta^2)\right\}$$

$$\frac{\Psi}{\mu\tau} = -\xi^2 - 4\sum_{n=0}^{\infty} \frac{(-1)^n \cos(\mu_n\xi)\cosh(\mu_n\eta)}{\alpha\mu_n^3\cosh(\mu_n\beta)}\,.$$

Therefore going back to the (x, y) system

$$\frac{\Psi}{\tau} = -\frac{x^2}{k_{44}} - 4a^2\sum_{n=0}^{\infty} \frac{(-1)^n \cos\left[(2n+1)\frac{\pi x}{2a}\right]\cosh\left[(2n+1)\frac{\pi y}{2a}\sqrt{\frac{k_{44}}{k_{55}}}\right]}{k_{44}\left[(2n+1)\frac{\pi}{2}\right]^3\cosh\left[(2n+1)\frac{\pi b}{2a}\sqrt{\frac{k_{44}}{k_{55}}}\right]}\,.$$

6.76. Flexure of an orthotropic elliptic cylinder

We consider the orthotropic cylinder whose cross-section is bounded by the ellipse, fig. 5.42,

$$\frac{x^2}{a^2} + \frac{y^2}{b^2} = 1 \tag{1}$$

under the action of a force P negatively along the x-axis.

We take first the isotropic problem when the load is W. In 5.42 replace (x, y), (a, b), T by (ξ, η), (α, β), T/μ and denote Poisson's ratio by η_0. Then the Timoshenko stress function which conforms with 6.62 (5), is

$$T = \frac{2W}{\pi(1+\eta_0)\alpha\beta}\cdot\frac{\beta^2\eta_0 + \alpha^2(1+\eta_0)}{\beta^2+3\alpha^2}\cdot\left(\frac{\xi^2}{\alpha^2}+\frac{\eta^2}{\beta^2}-1\right)\eta \tag{2}$$

and the cross-section is bounded by the ellipse

$$\frac{\xi^2}{\alpha^2} + \frac{\eta^2}{\beta^2} = 1. \tag{3}$$

Apply the affine transformation 6.7 (3). Then (3) becomes (1) where

$$a = \alpha \sqrt{(\mu k_{44})}, \qquad b = \beta \sqrt{(\mu k_{55})} \tag{4}$$

while, using 6.7 (8), (9), the stress function (2) becomes

$$T = \frac{2W\sqrt{(\mu k_{44})}}{\pi a b} \frac{(a^2 k_{55} - 2k_{23} b^2)}{3a^2 k_{55} + k_{44} b^2} \left(\frac{x^2}{a^2} + \frac{y^2}{b^2} - 1\right) y. \tag{5}$$

In order to make this correspond to a downward force P we substitute in 6.62 (7) wherein

$$A_1 = \frac{4k_{33} P}{\pi a^3 b}, \; B_1 = 0, \; \tau = 0$$

and we find without difficulty that $W\sqrt{(\mu k_{44})} = P$. Thus the required stress function is

$$T = \frac{2P}{\pi a b} \frac{a^2 k_{55} - 2k_{23} b^2}{3a^2 k_{55} + k_{44} b^2} \left(\frac{x^2}{a^2} + \frac{y^2}{b^2} - 1\right) y. \tag{6}$$

6.8. Methods of approximation

The principle of virtual work, 4.71, and the principle of virtual stresses, 4.8, are available to obtain approximate solutions. The application of these principles is exactly the same as in the case of isotropic material, of which illustrations have been given in Chapters IV and V.

EXAMPLES VI

1. Verify the correctness of the displacements given by 6.1 (5) for the bending of a beam by moments.
2. Obtain the transformations (1) and (2) of 6.32 for the stress integrals.
3. Fill out the detailed calculations which lead to 6.33 (3), (4),(5).
4. With the notation of 6.4 (9) show that a particular integral of this equation is

$$\psi_0 = \frac{\gamma_1}{2 l_{45}} x y$$

and explain why this simple form is, in general, unsuitable.

5. In the case of bending without twisting, 6.42, prove that

$$\frac{M_2}{M_1} = \frac{k_{43} k_{53}}{4 k_{33}} \frac{D^*}{I_x} \cdot \frac{1}{1 + \dfrac{k_{53}^2}{4k_{33}} \dfrac{D^*}{I_x}}$$

$$\frac{M}{M_1} = -\frac{k_{53}}{2} \frac{D^*}{I_x} \cdot \frac{1}{1 + \dfrac{k_{53}^2}{4k_{33}} \dfrac{D^*}{I_x}}.$$

6. Obtain the conditions for bending without twisting, 6.42, when the beam bends in a principal plane which is (i) $x = 0$; (ii) $y = 0$.

7. In the case of bending without twisting 6.42 show that the equation of the deformed axis is

$$y' = \frac{M_1 k_{33}}{2 I_x} \frac{1}{1 + \frac{k_{53}^2}{4 k_{33}} \frac{D^*}{I_x}} (L - R)^2 .$$

8. The beam of 6.42 is bent by the couple M_1 only i.e., we take $M_2 = 0$. Show that for the bent axis

$$y' = \frac{M_1 k_{33}}{2 I_x} (L - R)^2 .$$

Comparing this result with that of the preceding example show that the deflexion of the axis when a beam is bent by 2 couples M_1 and M_2 without torsion is less than the deflexion when the same beam is bent by the couple M_1 alone.

9. The elliptic cylinder of 6.44 is bent without torsion in the principal plane $x = 0$. Prove that the appropriate couples M_1, M_2 to achieve this are given by

$$\frac{M_2}{M_1} = \frac{k_{43} k_{53}}{k_{33}} \frac{1}{k_{55} + l_{44} \left(\frac{b}{a}\right)^2}$$

$$\frac{M_1}{M} = - \frac{k_{55} + l_{44} \left(\frac{b}{a}\right)^2}{2 k_{53}} .$$

10. For an elliptic cylinder bent without torsion as in Ex. 9 prove that

$$\widehat{zz} = \frac{4}{\pi a b} \left(\frac{M_1 y}{b^2} - \frac{M_2 x}{a^2} \right)$$

and evaluate this expression in terms of the twisting couple M.

11. A circular cylinder whose cross-section has radius a is bent in an axial plane without torsion. Prove that the rigidity is $\pi a^4 / (k_{44} + k_{55})$.

12. Write out in full the k-matrix and the l-matrix

(i) when the x, y plane is a plane of elastic symmetry

(ii) when the material is orthotropic and the coordinate planes are planes of elastic symmetry.

13. Solve the equations 6.63 (10), (11) and hence express the flexure stress function T for the elliptic cylinder with an antiplane of symmetry.

14. Derive equations 6.65 (7), (8), (9).

15. Find the displacement (u, v, w) of 6.65 in the case of orthotropic material.

16. Prove that for the rectangular beam of 6.74 the twisting couple is

$$M = \frac{\tau a b^3}{k_{65}} \left\{ \frac{16}{3} - \frac{b \sqrt{k_{44}}}{a \sqrt{k_{55}}} \left(\frac{4}{\pi}\right)^5 \sum_{n=0}^{\infty} \frac{1}{(2n+1)^5} \tanh \frac{(2n+1) \pi a \sqrt{k_{55}}}{2 b \sqrt{k_{44}}} \right\} .$$

17. The orthotropic beam of 6.74 is bent by a transverse load P along the negative x-axis. Show that the conditions of equilibrium are satisfied by the shear components

$$\widehat{xz} = \frac{\tau}{k_{55}}\left(\frac{\partial\phi}{\partial x} - y\right)$$
$$- \frac{P k_{33}}{I_y k_{55}}\left[\frac{\partial\chi}{\partial x} - \frac{1}{2}\frac{k_{13}}{k_{33}}x^2 + \frac{k_{55}k_{44} + k_{13}k_{44} + 2k_{23}k_{55}}{2k_{33}k_{55}}y^2\right],$$
$$\widehat{yz} = \frac{\tau}{k_{44}}\left(\frac{\partial\phi}{\partial y} + x\right) - \frac{P k_{33}}{I_y k_{44}}\left[\frac{\partial\chi}{\partial y} + \frac{k_{44}(k_{55} + k_{13})}{k_{33}k_{55}}xy\right]$$

where

$$\nabla^2\phi = 0, \ \nabla^2\chi = 0 .$$

18. In the preceding example show that the conditions at the (unloaded) lateral surface are

$$\frac{l}{k_{55}}\frac{\partial\phi}{\partial x} + \frac{m}{k_{44}}\frac{\partial\phi}{\partial y} = \frac{l y}{k_{55}} - \frac{m x}{k_{44}},$$
$$\frac{l}{k_{55}}\frac{\partial\chi}{\partial x} + \frac{m}{k_{44}}\frac{\partial\chi}{\partial y} = -\frac{l}{2k_{55}}\left(-\frac{k_{13}x^2}{k_{33}} + \frac{k_{44}k_{55} + k_{13}k_{44} + 2k_{23}k_{55}}{k_{33}k_{55}}y^2\right)$$
$$- \frac{m}{k_{33}}(k_{55} + k_{13})\,xy ,$$

where $(l, m, 0)$ are the direction cosines of the normal to the lateral surface.

19. The Timoshenko flexure function, 5.44,

$$T = \frac{1}{6}B_1(x - a)\ [(x + 2a)^2 - y]$$

determines the flexure of an isotropic equilateral triangle prism when Poisson's ratio is 1/2. Determine the shape of the orthotropic prism to which this solution corresponds by the affine transformation of 6.52.

20. Obtain the Timoshenko flexure function T for an orthotropic elliptic cylinder by direct substitution of a suitable form in the equation satisfied by T, when the force consists of components P along Ox and Q along Oy, O being the centroid and the axes being principal axes.

Chapter VII

General linear and cylindrical anisotropy

In this chapter we consider some aspects of linear anisotropy of general form and cylindrical anisotropy.

7.1. Generalized plane deformation

Consider material bounded by one or more cylindrical surfaces all of whose generators are parallel and extend to infinity in both senses.

Examples are an infinite tube, an infinite cylindrical hole in otherwise occupied space, a half-space, i.e., one in which the material occupies one side of an unbounded plane.

Choose as the antiplane any plane S perpendicular to the generators. The only state of antiplane stress of which the material is capable is one in which all stress and displacement components are independent of the applicate R. Otherwise such components would increase indefinitely with increase of R.

When the material is isotropic or when S is a plane of elastic symmetry, the displacement component w vanishes and cross-sections remain plane after deformation. This is plane deformation [MILNE-THOMSON (6)].

In a material with general anisotropy plane deformation becomes impossible in the sense that we can not satisfy the equations of equilibrium and the equations of Hooke's law 2.5 (6), (7) by postulating that $w \equiv 0$.

All that we can achieve is that stress and displacement are independent of R. We shall call such a state *generalized plane deformation*.

7.12. The complex stresses for generalized plane deformation

Since the displacement components are independent of R it follows from 2.3 (23)—(29) that L, T, P, P', Q_1, Q_1' are all zero and that

$$A_1 = B_1 = C_1 = A_2 = B_2 = C_2 = \tau = \omega_1 = \omega_2 = 0\,. \tag{1}$$

Therefore when body force is absent in 2.6 (1), (2) we can take the special solutions χ_0 and ψ_0 to be zero. Thus the stress components of 2.7 are

$$\left.\begin{aligned}
\widehat{xx} &= \frac{1}{4}\sum_{\nu=1}^{3}\left[\lambda_\nu^2 W_\nu(z_\nu) + \bar{\lambda}_\nu^2 \overline{W}_\nu(\bar{z}_\nu)\right]\\
\widehat{yy} &= \frac{1}{4}\sum_{\nu=1}^{3}\left[W_\nu(z_\nu) + \overline{W}_\nu(\bar{z}_\nu)\right]\\
\widehat{xy} &= -\frac{1}{4}\sum_{\nu=1}^{3}\left[\lambda_\nu W_\nu(z_\nu) + \bar{\lambda}_\nu \overline{W}_\nu(\bar{z}_\nu)\right]\\
\widehat{xz} &= \frac{1}{4}\sum_{\nu=1}^{3}\left[\lambda_\nu \mu_\nu W_\nu(z_\nu) + \bar{\lambda}_\nu \bar{\mu}_\nu \overline{W}_\nu(\bar{z}_\nu)\right]\\
\widehat{yz} &= -\frac{1}{4}\sum_{\nu=1}^{3}\left[\mu_\nu W_\nu(z_\nu) + \bar{\mu}_\nu \overline{W}_\nu(\bar{z}_\nu)\right]\\
\widehat{zz} &= -\frac{1}{k_{33}}\left(k_{13}\widehat{xx} + k_{23}\widehat{yy} + k_{43}\widehat{yz} + k_{53}\widehat{xz} + k_{63}\widehat{xy}\right).
\end{aligned}\right\} \tag{2}$$

From 2.72 (10)—(12) the displacements are

$$\left.\begin{aligned} u &= \sum_{\nu=1}^{3} [L_{1\nu}\,{}^{\backprime}W_\nu(z_\nu) + \bar{L}_{1\nu}\,{}^{\backprime}\overline{W}_\nu(\bar{z}_\nu)] + \alpha - \omega_3 y \\ v &= \sum_{\nu=1}^{3} \left[\frac{L_\nu}{\lambda_\nu}\,{}^{\backprime}W_\nu(z_\nu) + \frac{\bar{L}_\nu}{\bar{\lambda}_\nu}\,{}^{\backprime}\overline{W}_\nu(\bar{z}_\nu)\right] + \beta + \omega_3 x \\ w &= \sum_{\nu=1}^{3} [L_{5\nu}\,{}^{\backprime}W_\nu(z_\nu) + \bar{L}_{5\nu}\,{}^{\backprime}\overline{W}_\nu(\bar{z}_\nu)] + \gamma \end{aligned}\right\} \qquad (3)$$

where (α, β, γ) is a small translation and ω_3 is a small rotation about the R-axis.

As to boundary conditions we observe that Z_n must be zero so that the loading on the lateral surface is $(X_n, Y_n, 0)$ where X_n, Y_n are independent of R, and the effective loading coincides with the actual loading.

Therefore from 2.75 we have

$$\widehat{nn} + i\,\widehat{ns} = (X_n + i\,Y_n)\,e^{-i\alpha} \qquad (4)$$

with the stress boundary conditions

$$\left.\begin{aligned} &\sum_{\nu=1}^{3}\left[\delta_\nu W_\nu(z_\nu)\,m_\nu'(\sigma) - \frac{1}{\sigma^2}\bar{\gamma}_\nu \overline{W}_\nu(\bar{z}_\nu)\,\bar{m}_\nu'\left(\frac{1}{\sigma}\right)\right] = 2(\widehat{nn} + i\,\widehat{ns})\,m'(\sigma) \\ &\sum_{\nu=1}^{3}\left[\gamma_\nu W_\nu(z_\nu)\,m_\nu'(\sigma) - \frac{1}{\sigma^2}\bar{\delta}_\nu \overline{W}_\nu(\bar{z}_\nu)\bar{m}_\nu'\left(\frac{1}{\sigma}\right)\right] = -\frac{2}{\sigma^2}(\widehat{nn} - i\,\widehat{ns})\,\bar{m}'\left(\frac{1}{\sigma}\right) \\ &\sum_{\nu=1}^{3}\left[\mu_\nu W_\nu(z_\nu)\,m_\nu'(\sigma) - \frac{1}{\sigma^2}\bar{\mu}_\nu \overline{W}_\nu(\bar{z}_\nu)\,\bar{m}_\nu'\left(\frac{1}{\sigma}\right)\right] = 0\,. \end{aligned}\right\} \qquad (5)$$

The boundary condition for given displacements are obtained from (3) and 2.75 (18).

7.2. Line force applied to an elastic half-plane

Let the elastic material occupy the half-space $y \geqq 0$. Let the loading be a force $(S, -N, 0)$ per unit length applied along the R-axis, so that S is a shear and N is a normal tension

We shall imagine this force to be due to the limit of a distribution of stress as explained below.

In order to arrive at the above state of a concentrated line force let us first suppose that the whole surface $y = 0$ is loaded by a stress distribution

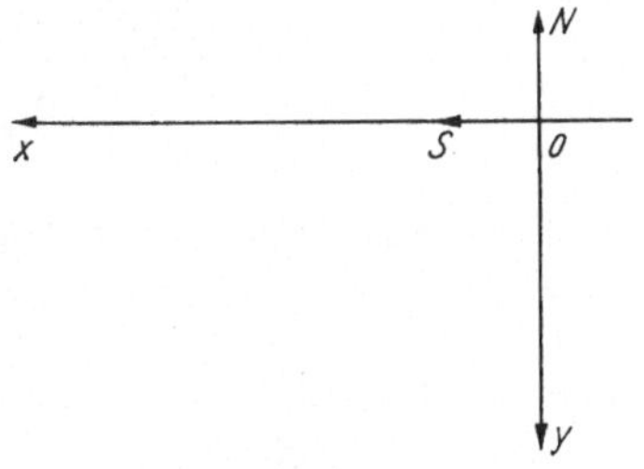

Fig. 7.2

$$(s(\xi), -n(\xi), 0)$$

where ξ is the abscissa. Then the boundary conditions are

$$\widehat{yy} = n(\xi), \quad \widehat{xy} = -s(\xi), \quad \widehat{yz} = 0 \text{ on } y = 0\,,$$

and therefore from 7.12

$$\sum_{\nu=1}^{3} [W_\nu(\xi) + \overline{W}_\nu(\xi)] = 4n(\xi)\,, \tag{1}$$

$$\sum_{\nu=1}^{3} [\lambda_\nu W_\nu(\xi) + \bar{\lambda}_\nu \overline{W}_\nu(\xi)] = 4s(\xi)\,, \tag{2}$$

$$\sum_{\nu=1}^{3} [\mu_\nu W_\nu(\xi) + \bar{\mu}_\nu \overline{W}_\nu(\xi)] = 0\,. \tag{3}$$

Let the function $f(z)$ be holomorphic when z is in the half-plane $y > 0$. Since

$$\bar{f}(z) = \overline{f(\bar{z})} \tag{4}$$

the function $\bar{f}(z)$ is defined and is holomorphic when z is in the half-plane $y < 0$, for then $\bar{z}$ is in the half-plane $y > 0$ and so the right-hand member of (4) is by hypothesis defined and holomorphic.

If in (4) we let $z \to \xi$, a point of the boundary $y = 0$, we get

$$\bar{f}(\xi) = \overline{f(\xi)}\,.$$

Therefore if t is in the half-plane $y > 0$ Cauchy's integral formula gives

$$\frac{1}{2\pi i}\int_{-\infty}^{\infty} \frac{f(\xi)\,d\xi}{\xi - t} = f(t), \quad \frac{1}{2\pi i}\int_{-\infty}^{\infty} \frac{\bar{f}(\xi)\,d\xi}{\xi - t} = 0\,, \tag{5}$$

subject to the limitation that the corresponding integral round the infinite semicircle vanishes.

Apply this result to (1), (2), (3). Then

$$\sum_{\nu=1}^{3} W_\nu(t) = \frac{1}{2\pi i}\int_{-\infty}^{\infty} \frac{4n(\xi)\,d\xi}{\xi - t} \tag{6}$$

$$\sum_{\nu=1}^{3} \lambda_\nu W_\nu(t) = \frac{1}{2\pi i}\int_{-\infty}^{\infty} \frac{4s(\xi)\,d\xi}{\xi - t} \tag{7}$$

$$\sum_{\nu=1}^{3} \mu_\nu W_\nu(t) = 0\,. \tag{8}$$

These equations can be solved to give expressions for

$$W_1(t),\ W_2(t),\ W_3(t)$$

each as a function of t. In these expressions we can write z_1, z_2, z_3 respectively for t and so obtain

$$W_1(z_1),\ W_2(z_2),\ W_3(z_3)\,.$$

Thus (6), (7), (8) solve the problem for a distribution of the kind described.

To adapt this to the case of a concentrated line force let

$$n(\xi) = \frac{N}{2\varepsilon} \text{ for } -\varepsilon < \xi < \varepsilon, \quad n(\xi) = 0 \text{ otherwise}$$

$$s(\xi) = \frac{S}{2\varepsilon} \text{ for } -\varepsilon < \xi < \varepsilon, \quad s(\xi) = 0 \text{ otherwise.}$$

We then take the limit of this case when $\varepsilon \to 0$. The limit of the integral in (6) is equal to

$$\lim_{\varepsilon \to 0} \frac{1}{2\pi i} \frac{2N}{\varepsilon} \int_{-\varepsilon}^{\varepsilon} \frac{d\xi}{\xi - t} = \frac{N}{\pi i} \lim_{\varepsilon \to 0} \frac{1}{\varepsilon} \left[\ln\left(1 - \frac{\varepsilon}{t}\right) - \ln\left(1 + \frac{\varepsilon}{t}\right)\right] = \frac{2Ni}{\pi t}$$

on using the logarithmic series.

Thus we get

$$W_1(z_1) = \frac{N(\mu_2\lambda_3 - \mu_3\lambda_2) - S(\mu_2 - \mu_3)}{(\mu_2\lambda_3 - \mu_3\lambda_2) + (\mu_3\lambda_1 - \mu_1\lambda_3) + (\mu_1\lambda_2 - \mu_2\lambda_1)} \cdot \frac{2i}{\pi z_1}$$

$$W_2(z_2) = \frac{N(\mu_3\lambda_1 - \mu_1\lambda_3) - S(\mu_3 - \mu_1)}{(\mu_2\lambda_3 - \mu_3\lambda_2) + (\mu_3\lambda_1 - \mu_1\lambda_3) + (\mu_1\lambda_2 - \mu_2\lambda_1)} \cdot \frac{2i}{\pi z_2}$$

$$W_3(z_3) = \frac{N(\mu_1\lambda_2 - \mu_2\lambda_1) - S(\mu_1 - \mu_2)}{(\mu_2\lambda_3 - \mu_3\lambda_2) + (\mu_3\lambda_1 - \mu_1\lambda_3) + (\mu_1\lambda_2 - \mu_2\lambda_1)} \cdot \frac{2i}{\pi z_3}.$$

7.3. Induced mappings for the region exterior to an ellipse

The ellipse

$$\frac{x^2}{a^2} + \frac{y^2}{b^2} = 1 \tag{1}$$

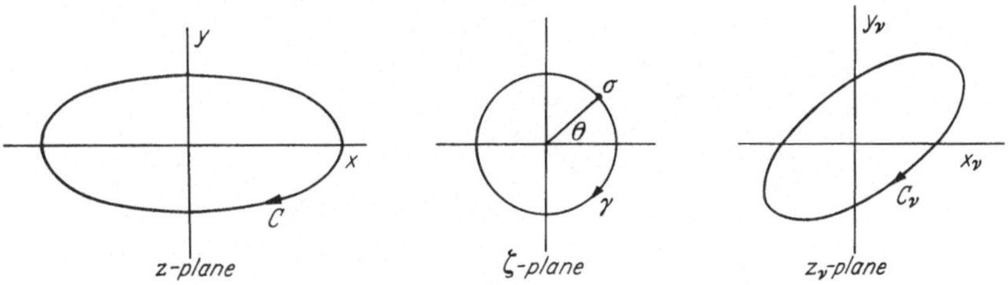

Fig. 7.3

is mapped on the unit circumference γ, $\sigma = e^{i\theta}$ by

$$z = \frac{1}{2}(a+b)\sigma + \frac{1}{2}(a-b)\frac{1}{\sigma} = m(\sigma) \tag{2}$$

and the region exterior to the ellipse is mapped on the region exterior to γ by

$$z = m(\zeta) \tag{3}$$

which gives

$$\zeta = \frac{z + (z^2 - a^2 + b^2)^{1/2}}{a+b} \tag{4}$$

the positive sign being taken before the square root since the exteriors correspond.

Corresponding to the root $\lambda_\nu = \alpha_\nu + i\,\beta_\nu$ of the characteristic equation 2.6 (6) we have

$$z_\nu = x_\nu + i\,y_\nu = x + \alpha_\nu y + i\,\beta_\nu y = x + \lambda_\nu y\,. \tag{5}$$

Thus with the definitions 2.7 (1) we have for points on the boundary of the induced circuit C_ν

$$\begin{aligned} z_\nu &= \gamma_\nu z + \delta_\nu \bar z = \gamma_\nu m(\sigma) + \delta_\nu m\left(\frac{1}{\sigma}\right) \\ &= \frac{1}{2}(a - i\lambda_\nu b)\,\sigma + \frac{1}{2}(a + i\lambda_\nu b)\,\frac{1}{\sigma} = m_\nu(\sigma) \end{aligned} \tag{6}$$

and for points exterior to C_ν.

$$z_\nu = m_\nu(\zeta) \tag{7}$$

whence

$$\zeta = \frac{z_\nu + \{z_\nu^2 - (a^2 + \lambda_\nu^2 b^2)\}^{1/2}}{a - i\lambda_\nu b} = \frac{a + i\lambda_\nu b}{z_\nu - \{z_\nu^2 - (a^2 + \lambda_\nu^2 b^2)\}^{1/2}}\,. \tag{8}$$

It follows that

$$\frac{d z_\nu}{d\zeta} = m_\nu'(\zeta) = \frac{\{z_\nu^2 - (a^2 + \lambda_\nu^2 b^2)\}^{1/2}}{\zeta}\,. \tag{9}$$

7.32. Elliptic cylindrical hole in an infinite elastic space

Let an infinite elastic space have a cylindrical hole whose cross-section is bounded by the ellipse C

$$\frac{x^2}{a^2} + \frac{y^2}{b^2} = 1 \tag{1}$$

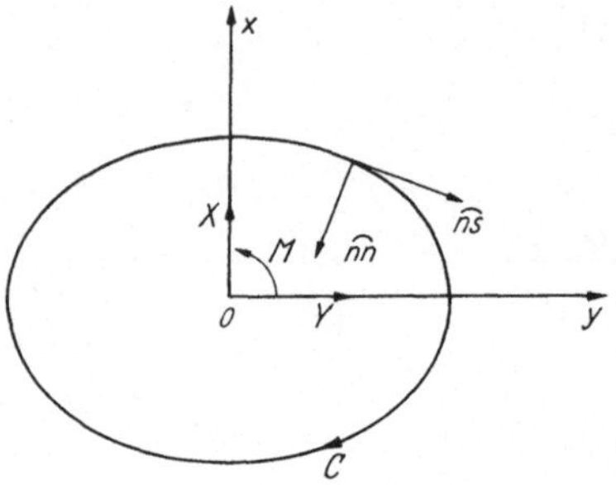

Fig. 7.32

which we map on the unit circumference γ by

$$\begin{aligned} z = m(\sigma) &= \frac{1}{2}(a + b)\,\sigma \\ &+ \frac{1}{2}(a - b)\,\frac{1}{\sigma} \end{aligned} \tag{2}$$

as described in 7.3.

Let the boundary of the hole be loaded with given normal tension $\widehat{nn}$ and shear $\widehat{ns}$, constant along a generator of the cylinder and let there be no body-field.

With the origin as base-point let the resultant loading be a force (X, Y) and a couple M. Then from 2.75 (7)

$$i(X + iY) = \oint_C (\widehat{nn} + i\,\widehat{ns})\,dz \tag{3}$$

while

$$\begin{aligned} M = \operatorname{Re}\oint_C i(x + iy)\,(X_n - i\,Y_n)\,ds &= -\frac{1}{2}\oint_C z(\widehat{nn} - i\,\widehat{ns})\,d\bar z \\ &- \frac{1}{2}\oint_C \bar z(\widehat{nn} + i\,\widehat{ns})\,dz\,. \end{aligned} \tag{4}$$

We shall regard X, Y, M as known quantities. We shall suppose the stress at infinity to be given. Let $W_\nu(z_\nu)$, $\nu = 1, 2, 3$, be the complex stresses. Then from 2.76 (13)

$$W_\nu(z_\nu) = a_{\nu 0} + \frac{a_{\nu 1}}{z_\nu} + W_\nu^*(z_\nu)\,, \tag{5}$$

where $a_{\nu 0}$, $a_{\nu 1}$ are obtained by solving the equations 2.76 (7)—(10), in which everything is known except $a_{\nu 0}$, $a_{\nu 1}$ and their complex conjugates. As explained in 2.76 $W_\nu^*(z_\nu)$ is holomorphic in the neighbourhood of infinity and has there an expansion of the form 2.76 (12).

We note that if $X = Y = 0$, then $a_{\nu 1} = 0$, while if the stress at infinity is zero, $a_{\nu 0} = 0$.

The boundary conditions 2.75 (13)—(15) then become

$$\sum_{\nu=1}^{3}\left[\delta_\nu W_\nu^*(z_\nu)\, m_\nu'(\sigma) - \frac{1}{\sigma^2}\,\bar\gamma_\nu\, \overline{W}_\nu^*(\bar z_\nu)\, \overline{m}_\nu'\left(\frac{1}{\sigma}\right)\right]$$
$$= 2\,(\widehat{nn} + i\,\widehat{ns})\, m'(\sigma) \tag{6}$$
$$-\sum_{\nu-1}^{3}\left[\delta_\nu\left(a_{\nu 0} + \frac{a_{\nu 1}}{m_\nu(\sigma)}\right) m_\nu'(\sigma) - \frac{1}{\sigma^2}\,\bar\gamma_\nu\left(\bar a_{\nu 0} + \frac{\bar a_{\nu 1}}{\overline{m}_\nu\left(\frac{1}{\sigma}\right)}\right)\overline{m}_\nu'\left(\frac{1}{\sigma}\right)\right] = A_1(\sigma)\,,$$

$$\sum_{\nu=1}^{3}\left[\gamma_\nu W_\nu^*(z_\nu)\, m_\nu'(\sigma) - \frac{1}{\sigma^2}\,\delta_\nu\, \overline{W}_\nu^*(\bar z_\nu)\, \overline{m}_\nu'\left(\frac{1}{\sigma}\right)\right]$$
$$= -\frac{2}{\sigma^2}\,(\widehat{nn} - i\,\widehat{ns})\, \overline{m}'\left(\frac{1}{\sigma}\right) \tag{7}$$
$$-\sum_{\nu=1}^{3}\left[\gamma_\nu\left(a_{\nu 0} + \frac{a_{\nu 1}}{m_\nu(\sigma)}\right) m_\nu'(\sigma) - \frac{1}{\sigma^2}\,\delta_\nu\left(\bar a_{\nu 0} + \frac{\bar a_{\nu 1}}{\overline{m}_\nu\left(\frac{1}{\sigma}\right)}\right)\overline{m}_\nu'\left(\frac{1}{\sigma}\right)\right] = A_2(\sigma)\,,$$

$$\sum_{\nu=1}^{3}\left[\mu_\nu W_\nu^*(z_\nu)\, m_\nu'(\sigma) - \frac{1}{\sigma^2}\,\bar\mu_\nu\, \overline{W}_\nu^*(\bar z_\nu)\, \overline{m}_\nu'\left(\frac{1}{\sigma}\right)\right] \tag{8}$$
$$= -\sum_{\nu=1}^{3}\left[\mu_\nu\left(a_{\nu 0} + \frac{a_{\nu 1}}{m_\nu(\sigma)}\right) m_\nu'(\sigma) - \frac{1}{\sigma^2}\,\bar\mu_\nu\left(\bar a_{\nu 0} + \frac{\bar a_{\nu 1}}{\overline{m}_\nu\left(\frac{1}{\sigma}\right)}\right)\overline{m}_\nu'\left(\frac{1}{\sigma}\right)\right] = A_3(\sigma).$$

7.33. Determination of the complex stresses

For the complex stresses of 7.32 we have

$$W_\nu^*(z_\nu) = W_\nu^*[m_\nu(\zeta)] = w_\nu^*(\zeta) \tag{1}$$

say, where from 7.3

$$z_\nu = m_\nu(\zeta) = \frac{1}{2}(a - i\lambda_\nu b)\,\zeta + \frac{1}{2}(a + i\lambda_\nu b)\,\frac{1}{\zeta}$$

and therefore as $|z_\nu| \to \infty$

$$\frac{z_\nu}{\zeta} \to \frac{1}{2}(a - i\lambda_\nu b).$$

Thus to the expansion 2.76 (12) there corresponds an expansion

$$W_\nu^*(z_\nu) = w_\nu^*(\zeta) = \frac{b_{\nu 2}}{\zeta^2} + \frac{b_{\nu 3}}{\zeta^3} + \cdots, \tag{2}$$

where $b_{\nu 2}, b_{\nu 3}, \ldots$ are expressible in terms of $a_{\nu 2}, a_{\nu 3}, \ldots$ Therefore on the boundary C

$$W_\nu^*(z_\nu) = w_\nu^*(\sigma) = \frac{b_{\nu 2}}{\sigma^2} + \frac{b_{\nu 3}}{\sigma^3} + \cdots. \tag{3}$$

We can now determine the $W_\nu(z_\nu)$ as follows:

$$A_1(\sigma),\, A_2(\sigma),\, A_3(\sigma) \tag{4}$$

are defined by 7.32 (6), (7), (8).

Multiply 7.32 (6) by $d\sigma/[2\pi i(\sigma - \zeta)]$, where ζ is a point in the region exterior to the unit circumference γ, and integrate round γ. Then

$$\sum_{\nu=1}^{3}\left[\frac{1}{2\pi i}\delta_\nu \oint_\gamma \frac{w_\nu^*(\sigma)\, m_\nu'(\sigma)}{\sigma - \zeta}\, d\sigma - \frac{1}{2\pi i}\gamma_\nu \oint_\gamma \frac{\overline{w}_\nu^*\left(\frac{1}{\sigma}\right) \overline{m}_\nu'\left(\frac{1}{\sigma}\right) d\sigma}{\sigma^2(\sigma - \zeta)}\right] = \frac{1}{2\pi i}\oint_\gamma \frac{A_1(\sigma)\, d\sigma}{\sigma - \zeta}. \tag{5}$$

By Cauchy's integral formula the first integral is equal to $w_\nu^*(\zeta)\, m_\nu'(\zeta)$ while the second integral vanishes, for on account of (3) the integrand is holomorphic inside γ when ζ lies outside γ.

Denote the value of the integral on the right hand side of (5) by $B_1(\zeta)$. Thus we get the first of the following results, the others being obtained by similar steps.

$$\left.\begin{aligned} \sum_{\nu=1}^{3} \delta_\nu w_\nu^*(\zeta)\, m_\nu'(\zeta) &= B_1(\zeta) \\ \sum_{\nu=1}^{3} \gamma_\nu w_\nu^*(\zeta)\, m_\nu'(\zeta) &= B_2(\zeta) \\ \sum_{\nu=1}^{3} \mu_\nu w_\nu^*(\zeta)\, m_\nu'(\zeta) &= B_3(\zeta) \end{aligned}\right\} \tag{6}$$

In the determinant

$$\Delta = \begin{vmatrix} \delta_1 & \delta_2 & \delta_3 \\ \gamma_1 & \gamma_2 & \gamma_3 \\ \mu_1 & \mu_2 & \mu_3 \end{vmatrix} \tag{7}$$

let δ_{11} denote the cofactor of δ_1 divided by Δ, with similar meanings* for $\delta_{22}, \delta_{33}, \gamma_{11}, \gamma_{22}, \gamma_{33}, \mu_{11}, \mu_{22}, \mu_{33}$. Then solving the system (6) we get

$$\left.\begin{aligned} w_1^*(\zeta)\, m_1'(\zeta) &= \delta_{11} B_1(\zeta) + \gamma_{11} B_2(\zeta) + \mu_{11} B_3(\zeta) \\ w_2^*(\zeta)\, m_2'(\zeta) &= \delta_{22} B_1(\zeta) + \gamma_{22} B_2(\zeta) + \mu_{22} B_3(\zeta) \\ w_3^*(\zeta)\, m_3'(\zeta) &= \delta_{33} B_1(\zeta) + \gamma_{33} B_2(\zeta) + \mu_{33} B_3(\zeta) \end{aligned}\right\} \tag{8}$$

Therefore using (1)

$$\begin{aligned} 2\pi i\, W_1^*(z_1)\, m_1'(\zeta) = \delta_{11} \oint_\gamma \frac{A_1(\sigma)\, d\sigma}{\sigma - \zeta} + \gamma_{11} \oint_\gamma \frac{A_2(\sigma)\, d\sigma}{\sigma - \zeta} \\ + \mu_{11} \oint_\gamma \frac{A_3(\sigma)\, d\sigma}{\sigma - \zeta}, \quad z_1 = m_1(\zeta) \end{aligned} \tag{9}$$

$$\begin{aligned} 2\pi i\, W_2^*(\zeta_2)\, m_2'(\zeta) = \delta_{22} \oint_\gamma \frac{A_1(\sigma)\, d\sigma}{\sigma - \zeta} + \gamma_{22} \oint_\gamma \frac{A_2(\sigma)\, d\sigma}{\sigma - \zeta} \\ + \mu_{22} \oint_\gamma \frac{A_3(\sigma)\, d\sigma}{\sigma - \zeta}, \quad z_2 = m_2(\zeta) \end{aligned} \tag{10}$$

$$\begin{aligned} 2\pi i\, W_3^*(z_3)\, m_3'(\zeta) = \delta_{33} \oint_\gamma \frac{A_1(\sigma)\, d\sigma}{\sigma - \zeta} + \gamma_{33} \oint_\gamma \frac{A_2(\sigma)\, d\sigma}{\sigma - \zeta} \\ + \mu_{33} \oint_\gamma \frac{A_3(\sigma)\, d\sigma}{\sigma - \zeta}, \quad z_3 = m_3(\zeta)\,. \end{aligned} \tag{11}$$

The appropriate values of ζ in terms of z_ν are given by 7.3 (8), but in numerical calculations it is often preferable to use the forms given by (8) wherein ζ is regarded as a parameter.

From 7.3 (9) we have

$$m_\nu'(\zeta) = \frac{\{z_\nu^2 - (a^2 + \lambda_\nu^2 b^2)\}^{1/2}}{\zeta}\,.$$

7.35. Elliptic hole under hydrostatic pressure

Let the elliptic hole of 7.32 be under hydrostatic pressure p. Then the boundary conditions are

$$\widehat{nn} = -p, \; \widehat{ns} = 0 \tag{1}$$

and therefore in 7.32 (5), $a_{\nu 1} = 0$.

If, in addition, the material is unstressed at infinity we have $a_{\nu 0} = 0$, so that

$$W_\nu(z_\nu) = W_\nu^*(z_\nu)\,,$$

where $W_\nu^*(z_\nu)$ is $O(1/z_\nu^2)$ for $|z_\nu|$ large.

* They are in fact elements of the reciprocal of the determinant Δ.

Therefore in 7.32 (6) — (8)

$$A_1(\sigma) = -2p\,m'(\sigma),\ A_2(\sigma) = \frac{2p}{\sigma^2}\,\overline{m}'\left(\frac{1}{\sigma}\right),\ A_3(\sigma) = 0\,, \tag{2}$$

where $m(\sigma) = \frac{1}{2}(a+b)\sigma + \frac{1}{2}(a-b)\frac{1}{\sigma}$, so that

$$A_1(\sigma) = -p(a+b) + \frac{p(a-b)}{\sigma^2},\ A_2(\sigma) = -p(a-b) + \frac{p(a+b)}{\sigma^2}$$

and thus

$$B_1(\zeta) = \frac{1}{2\pi i}\oint\limits_{\gamma} \frac{A_1(\sigma)\,d\sigma}{\sigma-\zeta} = \frac{p(a-b)}{\zeta^2},\ B_2(\zeta) = \frac{p(a+b)}{\zeta^2},\ B_3(\zeta) = 0\,.$$

Substitution in 7.33 (9)—(11) and use of 7.3 (8), (9) gives

$$W_1(z_1) = [\delta_{11}(a-b) + \gamma_{11}(a+b)]\frac{p}{\zeta^2 m_1'(\zeta)}\,,\quad \zeta = \frac{z_1 + \{z_1^2 - (a^2+\lambda_1^2 b^2)^{1/2}\}}{a - i\lambda_1 b} \tag{3}$$

$$W_2(z_2) = [\delta_{22}(a-b) + \gamma_{22}(a+b)]\frac{p}{\zeta^2 m_2'(\zeta)}\,,\quad \zeta = \frac{z_2 + \{z_2^2 - (a^2+\lambda_2^2 b^2)\}^{1/2}}{a - i\lambda_2 b} \tag{4}$$

$$W_3(z_3) = [\delta_{33}(a-b) + \gamma_{33}(a+b)]\frac{p}{\zeta^2 m_3'(\zeta)}\,,\quad \zeta = \frac{z_3 + \{z_3^2 - (a^2+\lambda_3^2 b^2)\}^{1/2}}{a - i\lambda_3 b}\,. \tag{5}$$

It appears from 7.3 (8) and (9) that

$$\zeta^2 m_\nu'(\zeta) = \frac{[z_\nu + \{z_\nu^2 - (a^2+\lambda_\nu^2 b^2)\}^{1/2}]\,\{z_\nu^2 - (a^2+\lambda_\nu^2 b^2)\}^{1/2}}{a - i\lambda_\nu b}\,. \tag{6}$$

7.37. Unloaded elliptic cylindrical hole in a space under stress

By an unloaded hole we mean that at its bounding surface

$$\widehat{nn} = 0,\ \widehat{ns} = 0\,.$$

The state of stress at infinity will be specified by the constants

$$a_{10},\ a_{20},\ a_{30} \tag{1}$$

given by 2.76 (10), (11).

The complex stresses are then of the form

$$W_\nu(z_\nu) = a_{\nu 0} + W_\nu^*(z_\nu)\,, \tag{2}$$

where $W_\nu^*(z_\nu)$ is $O(1/z_\nu^2)$ for $|z_\nu|$ large.

We then have from 7.32 (6)—(8)

$$A_1(\sigma) = -\sum_{\nu=1}^{3}\left[\delta_\nu a_{\nu 0} m'_\nu(\sigma) - \bar{\gamma}_\nu \bar{a}_{\nu 0}\frac{1}{\sigma^2}\bar{m}'_\nu\left(\frac{1}{\sigma}\right)\right]$$

$$= \frac{1}{2}\sum_{\nu=1}^{3}[a_{\nu 0}\delta_\nu(a + i\lambda_\nu b) + \bar{a}_{\nu 0}\bar{\gamma}_\nu(a + i\bar{\lambda}_\nu b)]\frac{1}{\sigma^2} + \text{constant.}$$

Therefore

$$B_1(\zeta) = \frac{1}{2\pi i}\oint_\gamma \frac{A_1(\sigma)\,d\sigma}{\sigma - \zeta} = \frac{C_1}{2\zeta^2} \tag{3}$$

where

$$C_1 = \sum_{\nu=1}^{3}[a_{\nu 0}\delta_\nu(a + i\lambda_\nu b) + \bar{a}_{\nu 0}\bar{\gamma}_\nu(a + i\bar{\lambda}_\nu b)]. \tag{4}$$

Similarly

$$B_2(\zeta) = \frac{C_2}{2\zeta^2},\ B_3(\zeta) = \frac{C_3}{2\zeta^2} \tag{5}$$

where

$$C_2 = \sum_{\nu=1}^{3}[a_{\nu 0}\gamma_\nu(a + i\lambda_\nu b) + \bar{a}_{\nu 0}\delta_\nu(a + i\bar{\lambda}_\nu b)] \tag{6}$$

$$C_3 = \sum_{\nu=1}^{3}[a_{\nu 0}\mu_\nu(a + i\lambda_\nu b) + \bar{a}_{\nu 0}\bar{\mu}_\nu(a + i\bar{\lambda}_\nu b)]. \tag{7}$$

Therefore the complex stresses are given by (2) in which from 7.33 (8)

$$W_1^*(z_1) = \frac{1}{2}(\delta_{11}C_1 + \gamma_{11}C_2 + \mu_{11}C_3)\frac{1}{m'_1(\zeta)\zeta^2},\ z_1 = m_1(\zeta)$$

$$W_2^*(z_2) = \frac{1}{2}(\delta_{22}C_1 + \gamma_{22}C_2 + \mu_{22}C_3)\frac{1}{m'_2(\zeta)\zeta^2},\ z_2 = m_2(\zeta)$$

$$W_3^*(z_3) = \frac{1}{2}(\delta_{33}C_1 + \gamma_{33}C_2 + \mu_{33}C_3)\frac{1}{m'_3(\zeta)\zeta^2},\ z_3 = m_3(\zeta),$$

where

$$\zeta^2 m'_\nu(\zeta) = \frac{[z_\nu + \{z_\nu^2 - (a^2 + \lambda_\nu^2 b^2)\}^{1/2}]\,\{z_\nu^2 - (a^2 + \lambda_\nu^2 b^2)\}^{1/2}}{a - i\lambda_\nu b}.$$

7.4. Bending of a cantilever by a transverse force at the free end

We consider the case of general linear anisotropy. Take for x- and y-axes principal axes of inertia at the centroid G of the free end.

To abbreviate the writing we shall suppose that body-field is absent and that the lateral surface is unloaded.

Let the bending force be P along the y-axis, and let

$$I = I_x = \int_S y^2\,dS.$$

Then from 6.61 (2), (3) and 6.33 (6)

$$\left.\begin{aligned} A_1 &= 0,\ B_1 = -\frac{k_{33}P}{I}, \quad C_1 = 0,\\ A_2 &= B_2 = 0, \quad C_2 m = \frac{-k_{34}P}{2S}, \end{aligned}\right\} \tag{1}$$

where S is the area of the cross-section.

Therefore from 2.3 (22)

$$e_{zz} = B_1 \left\{ yR + \frac{1}{2}(\nu_5 xy + \nu_4 y^2) + \frac{\nu_4 I}{2S} \right\}.$$

From 2.32 (2) it follows that

$$\widehat{zz} = \frac{e_{zz}}{k_{33}} - \nu_1 \widehat{xx} - \nu_2 \widehat{yy} - \nu_4 \widehat{yz} - \nu_5 \widehat{xz} - \nu_6 \widehat{xy}. \tag{2}$$

The remaining stresses are given (2.52) by

$$\begin{aligned} \widehat{xx} &= \frac{\partial^2 \chi}{\partial y^2}, \quad \widehat{xy} = -\frac{\partial^2 \chi}{\partial x \, \partial y}, \quad \widehat{yy} = \frac{\partial^2 \chi}{\partial x^2}, \\ \widehat{xz} &= \frac{\partial \psi}{\partial y}, \quad \widehat{yz} = -\frac{\partial \psi}{\partial x} + \frac{P}{2I} y^2, \end{aligned} \tag{3}$$

where in the notation of 2.52 the stress functions χ and ψ satisfy the equations

$$D^{(4)} \chi - D^{(3)} \psi = B_1 \left(-\nu_1 \nu_4 + \frac{1}{2} \nu_5 \nu_6 + \frac{l_{14}}{k_{33}} \right) \tag{4}$$

$$D^{(3)} \chi - D^{(2)} \psi = B_1 \left\{ \left(\frac{1}{2} \nu_5^2 - 2\nu_1 \right) x + \left(\frac{1}{2} \nu_4 \nu_5 - \nu_6 - \frac{l_{45}}{k_{33}} \right) y \right\} + 2\tau. \tag{5}$$

The twist τ per unit length is obtained by equating to zero the twisting couple which, from 6.33 (7), gives

$$\int_S \psi \, dS = 0. \tag{6}$$

At this point we can, if we wish, introduce the complex stress $W_\nu(z_\nu)$ of 2.6.

From 2.3 (23) the displacements are

$$\begin{aligned} u &= u_1 + \frac{1}{4} \nu_5 B_1 R^2 y + \left[\frac{1}{2} B_1 (2\nu_1 xy + \nu_6 y^2) - \tau y \right] R + \alpha + \omega_2 R - \omega_3 y \\ v &= v_1 - \frac{1}{6} B_1 R^3 - \frac{1}{4} \nu_5 B_1 R^2 x \\ &\quad + \left[\frac{1}{2} B_1 (-\nu_1 x^2 + \nu_2 y^2) + \tau x \right] R + \beta + \omega_3 x - \omega_1 R \\ w &= w_1 + \frac{1}{2} B_1 R^2 y + \frac{1}{2} B_1 \left(\nu_5 xy + \nu_4 y^2 + \frac{\nu_4 I}{2S} \right) R + \gamma + \omega_1 y - \omega_2 x, \end{aligned} \tag{7}$$

where u_1, v_1, w_1 satisfy 2.32 (6), (7).

Let us fix the end $(0, 0, L)$ of the axis by the conditions

$$\begin{gathered} u = v = w = 0 \\ \frac{\partial u}{\partial R} = 0, \quad \frac{\partial v}{\partial R} = 0 \end{gathered} \tag{8}$$

and let u_{10}, v_{10}, w_{10} be the values of u_1, v_1, w_1 at the point $(0, 0, R)$. These values are independent of R.

We then find that

$$u = (u_1 - u_{10}) + \frac{1}{4} \nu_5 B_1 y R^2 + \left[\frac{1}{2} B_1 (2\nu_1 x y + \nu_6 y^2) - \tau y\right] R - \omega_3 y$$

$$v = (v_1 - v_{10}) - \frac{1}{6} B_1 (R^3 - 3L^2 R + 2L^3) - \frac{1}{4} \nu_5 B_1 x R^2$$
$$+ \left[\frac{1}{2} B_1 (-\nu_1 x^2 + \nu_2 y^2) + \tau x\right] R + \omega_3 x$$

$$w = (w_1 - w_{10}) + \frac{1}{2} B_1 y R^2 + \frac{1}{2} B_1 (\nu_5 x y + \nu_4 y^2) R$$
$$+ B_1 \frac{\nu_4 I}{2S} (R - L) .$$

Thus at any point $(0, 0, R)$ of the axis

$$u = 0, \tag{9}$$

$$v = \frac{k_{33} P}{6I} (R^3 - 3L^2 R + 2L^3) , \tag{10}$$

$$w = \frac{-k_{43} P}{2S} (R - L) . \tag{11}$$

Thus the bent axis of the beam lies in the axial plane which contains the force P.

7.41. Centre of flexure

Stevenson's definition of the centre of flexure must fail in the case of general anisotropy, for as is easily seen from 5.16, the local twist, calculated as in 5.2 is a function of the applicate R.

Take principal axes of inertia at the centroid G as x, y-axes. In the case of general anisotropy bending by a transverse load applied along one of these axes, unless it is an axis of symmetry, is accompanied by torsion characterized by a constant τ.

This torsion can be avoided if we choose the load point properly.

We define the *centre of flexure* C_f as that point, which when taken as load point for an arbitrary transverse force makes the characteristic torsion constant zero.

Let C_f be the point (x_f, y_f). A force

$$P = P_1 + i P_2$$

applied at C_f is equivalent to a force (P_1, P_2) applied at the centroid G together with a twisting moment

$$M_0 = x_f P_2 - y_f P_1 .$$

To find the position of C_f we therefore solve the flexure problem for the loading (P_1, P_2) at the centroid G on the supposition that $\tau = 0$.

We then calculate the twisting moment

$$M_0 = \int_S (x \cdot \widehat{yz} - y \cdot \widehat{xz}) \, dS .$$

The centre of flexure is then from 3.27

$$x_f = \frac{\partial M_0}{\partial P_2}, \; y_f = -\frac{\partial M_0}{\partial P_1} .$$

or alternatively

$$\bar{z}_f = 2i \frac{\partial M_0}{\partial P} .$$

7.42. Timoshenko's stress function

We can always replace the stress function ψ by a stress function T, like the stress function introduced by Timoshenko in the isotropic case (5.41), such that

$$T = \text{constant on } C, \tag{1}$$

where C is the boundary of a cross-section. When the area of the cross-section is simply connected we can take the constant to be zero

$$T = 0 \text{ on } C . \tag{2}$$

To see this we have from 2.52

$$\begin{aligned} \widehat{xz} &= \frac{\partial \psi}{\partial y} - \frac{1}{2k_{33}} [A_1 x^2 + (C_1 - \varepsilon) m x] \\ \widehat{yz} &= -\frac{\partial \psi}{\partial x} - \frac{1}{2k_{33}} [B_1 y^2 + (C_1 - \varepsilon) m y] . \end{aligned} \tag{3}$$

In this write

$$\psi = T + \frac{A_1}{2k_{33}} \, 't(y) - \frac{B_1}{2k_{33}} \, 's(x) \tag{4}$$

where $'s(x)$, $'t(y)$ are functions to be chosen subsequently. We then get

$$\begin{aligned} \widehat{xz} &= \frac{\partial T}{\partial y} - \frac{1}{2k_{33}} [A_1 \{x^2 - t(y)\} + (C_1 - \varepsilon) m x] \\ \widehat{yz} &= -\frac{\partial T}{\partial x} - \frac{1}{2k_{33}} [B_1 \{y^2 - s(x)\} + (C_1 - \varepsilon) m y] . \end{aligned} \tag{5}$$

If we now choose $s(x)$ and $t(y)$ such that

$$\begin{aligned} A_1 \{x^2 - t(y)\} + (C_1 - \varepsilon) m x &= 0 \text{ on } C \\ B_1 \{y^2 - s(x)\} + (C_1 - \varepsilon) m y &= 0 \text{ on } C , \end{aligned} \tag{6}$$

we shall have

$$\widehat{xz} = \frac{\partial T}{\partial y}, \; \widehat{yz} = -\frac{\partial T}{\partial x} \text{ on } C .$$

But the boundary condition is $l\,\widehat{xz} + m\,\widehat{yz} = 0$ on C where $l = \frac{dy}{ds}$, $m = -\frac{dx}{ds}$. Therefore

$$\frac{dT}{ds} = 0 \text{ on } C$$

and therefore

$$T = \text{constant on } C\,.$$

When $s(x)$ and $t(y)$ have been chosen, the shear is determined from (5). Again

$$\begin{aligned} D^{(3)}\,\psi &= D^{(3)}\,T + \frac{A_1}{2k_{33}}\,D^{(3)}\,`t(y) - \frac{B_1}{2k_{33}}\,D^{(3)}\,`s(x) \\ &= D^{(3)}\,T - \frac{A_1 l_{15}}{2k_{33}}\,t''(y) - \frac{B_1 l_{24}}{2k_{33}}\,s''(x) \end{aligned} \tag{7}$$

$$D^{(2)}\,\psi = D^{(2)}\,T + \frac{A_1 l_{55}}{2k_{33}}\,t'(y) - \frac{B l_{44}}{2k_{33}}\,s'(x)\,. \tag{8}$$

Thus, if in equations 2.52 (8) and (10) we replace ψ by T, the zeros on the right hand sides must be replaced by

$$-\frac{A_1 l_{15}}{2k_{33}}\,t''(y) - \frac{B_1 l_{24}}{2k_{33}}\,s''(x), \text{ and } \frac{A_1 l_{55}}{2k_{33}}\,t'(y) - \frac{B l_{44}}{2k_{33}}\,s'(x) \tag{9}$$

respectively.

7.44. Flexure of a cylinder with elliptic or circular cross-section

Let the contour of the cross-section be the ellipse C

$$\frac{x^2}{a^2} + \frac{y^2}{b^2} = 1 \tag{1}$$

and let the bending force be P along the y-axis applied at the centroid of the free end. Let there be no lateral loading or body-field. Let $I = I_x$ be the second moment of area of the cross-section about the x-axis. Then from 6.32 (9), (10)

$$A_1 = 0,\; B = \frac{-k_{33}P}{I}, \quad C_1 = 0\,. \tag{2}$$

Therefore to obtain a Timoshenko stress function T we put in 7.42 (5)

$$s(x) = -b^2\left(\frac{x^2}{a^2} - 1\right), \quad t(y) = 0$$

so that

$$\psi = T + \frac{B_1 b^2}{2k_{33}}\left(\frac{x^3}{3a^2} - x\right),$$

$$\widehat{xz} = \frac{\partial T}{\partial y},\; \widehat{yz} = -\frac{\partial T}{\partial x} - \frac{B_1 b^2}{2k_{33}}\left(\frac{x^2}{a^2} + \frac{y^2}{b^2} - 1\right) \tag{3}$$

and therefore $T =$ constant on the boundary C.

The boundary conditions 6.2 (12), (13) can therefore be taken in the form

$$\frac{\partial \chi}{\partial x} = \frac{\partial \chi}{\partial y} = T = 0 \text{ on } C\,. \tag{4}$$

From the symmetry the torsion $\tau = 0$, and therefore we can satisfy the boundary conditions (4) by taking

$$\chi = N_1\left(\frac{x^2}{a^2} + \frac{y^2}{b^2} - 1\right)^2, \; T = (N_2 x + N_3 y)\left(\frac{x^2}{a^2} + \frac{y^2}{b^2} - 1\right). \tag{5}$$

Substitute for χ and ψ in 7.4 (4), (5) and express that the resulting equations are identities. We then get

$$\begin{aligned} &8N_1\left(\frac{3l_{22}}{a^4} + \frac{2l_{12} + l_{66}}{a^2 b^2} + \frac{3l_{11}}{b^4}\right) - 2N_2\left(\frac{3l_{24}}{a^2} + \frac{l_{14} + l_{56}}{b^2}\right) \\ &+ 2N_3\left(\frac{l_{25} + l_{46}}{a^2} + \frac{3l_{15}}{b^2}\right) = -B_1\left(\nu_1\nu_4 - \frac{1}{2}\nu_5\nu_6 - \frac{l_{14}}{k_{33}} - \frac{b^2}{a^2}\frac{l_{24}}{k_{33}}\right) \end{aligned} \tag{6}$$

$$\begin{aligned} &8N_1\left(\frac{3l_{24}}{a^4} + \frac{l_{14} + l_{56}}{a^2 b^2}\right) - 2N_2\left(\frac{3l_{44}}{a^2} + \frac{l_{55}}{b^2}\right) + \frac{4N_3 l_{45}}{a^2} \\ &= B_1\left(\frac{1}{2}\nu_5^2 - 2\nu_1 + \frac{b^2}{a^2}\frac{l_{44}}{k_{33}}\right) \end{aligned} \tag{7}$$

$$\begin{aligned} &8N_1\left(\frac{l_{25} + l_{46}}{a^2 b^2} + \frac{3l_{15}}{b^4}\right) - \frac{4N_2 l_{45}}{b^2} + 2N_3\left(\frac{l_{44}}{a^2} + \frac{3l_{55}}{b^2}\right) \\ &= -B_1\left(\frac{1}{2}\nu_4\nu_5 - \nu_6 - \frac{l_{45}}{k_{33}}\right). \end{aligned} \tag{8}$$

These 3 linear equations determine N_1, N_2, N_3 and therefore χ and T from (5).

7.5. Cylindrical anisotropy

The definition of curvilinear anisotropy has been given in 1.8. When the set of triply orthogonal surfaces of the definition reduce to coaxial circular cylinders, planes through the common axis of these cylinders and planes perpendicular to this axis we have (circular) cylindrical isotropy. The common axis of the cylinders will be called the *axis of anisotropy*.

Consider now a cylindrical or prismatic body of material whose axis of anisotropy is parallel to the generators. This axis may lie inside or outside the body and will meet a chosen cross-section, which we take for antiplane, at the point O. This point will be taken as origin, and the axis of anisotropy will be taken as R-axis.

It will be found more natural and convenient to use cylindrical coordinates (r, θ, R) rather than cartesians (x, y, R) so that

$$x = r\cos\theta,\; y = r\sin\theta\,. \tag{1}$$

We shall suppose the body-field to derive from a potential V independent of R so that the radial, transverse and axial components will be $(b_r, b_\theta, 0,)$ where

$$b_r = \frac{\partial V}{\partial r}, \quad b_\theta = \frac{\partial V}{r\,\partial\theta}. \tag{2}$$

The external loading on the lateral faces will be denoted by $(r_N, \theta_N, 0)$, where r_N is the radial and θ_N the transverse component.

Let $\boldsymbol{i}_r, \boldsymbol{i}_\theta, \boldsymbol{i}_R$ be unit vectors in the radial, transverse and axial directions. Then the stress tensor can be written in the form

$$\begin{aligned} \mathsf{S} = {} & (\boldsymbol{i}_r; \boldsymbol{i}_r)\widehat{rr} + (\boldsymbol{i}_r; \boldsymbol{i}_\theta)\widehat{r\theta} + (\boldsymbol{i}_r; \boldsymbol{i}_R)\widehat{rz} \\ & + (\boldsymbol{i}_\theta; \boldsymbol{i}_r)\widehat{r\theta} + (\boldsymbol{i}_\theta; \boldsymbol{i}_\theta)\widehat{\theta\theta} + (\boldsymbol{i}_\theta; \boldsymbol{i}_R)\widehat{\theta z} \\ & + (\boldsymbol{i}_R; \boldsymbol{i}_r)\widehat{rz} + (\boldsymbol{i}_R; \boldsymbol{i}_\theta)\widehat{\theta z} + (\boldsymbol{i}_R; \boldsymbol{i}_R)\widehat{zz} \end{aligned} \tag{3}$$

so that the normal stresses are $\widehat{rr}$, $\widehat{\theta\theta}$, $\widehat{zz}$ while the shears are $\widehat{r\theta}$, $\widehat{\theta z}$, $\widehat{rz}$.

Of the unit vectors $\boldsymbol{i}_R$ is independent of position, whereas $\boldsymbol{i}_r$, $\boldsymbol{i}_\theta$ depend on θ. For their rates of change we have [MILNE-THOMSON (5)].

$$\begin{aligned} & \frac{\partial \boldsymbol{i}_r}{\partial r} = \frac{\partial \boldsymbol{i}_\theta}{\partial r} = 0 \\ & \frac{\partial \boldsymbol{i}_r}{\partial \theta} = \boldsymbol{i}_\theta, \quad \frac{\partial \boldsymbol{i}_\theta}{\partial \theta} = -\boldsymbol{i}_r. \end{aligned} \tag{4}$$

Hamilton's operator nabla is given by

$$\nabla = \boldsymbol{i}_r \frac{\partial}{\partial r} + \frac{\boldsymbol{i}_\theta}{r}\frac{\partial}{\partial\theta} + \boldsymbol{i}_R \frac{\partial}{\partial R}. \tag{5}$$

The displacement $\boldsymbol{q}$ will be written

$$\boldsymbol{q} = \boldsymbol{i}_r u_r + \boldsymbol{i}_\theta u_\theta + \boldsymbol{i}_R u_z. \tag{6}$$

The component here denoted by u_z is the same as the component w of 2.2.

With these notations the equation of equilibrium is

$$\nabla \cdot \mathsf{S} = \boldsymbol{i}_r b_r + \boldsymbol{i}_\theta b_\theta \tag{7}$$

and the deformation tensor is

$$\mathsf{D} = \frac{1}{2}(\nabla; \boldsymbol{q} + \boldsymbol{q}; \nabla). \tag{8}$$

Since, by hypothesis, there is no axial loading on the lateral surface, and no body-field in the axial direction, it follows that

$$\widehat{zz} \text{ is independent of } R. \tag{9}$$

Therefore, since we are dealing with antiplane stress, the stress tensor S is independent of R.

From (3), (4), (5), (7), (9) we therefore obtain the equations of equilibrium in the form

$$\left.\begin{aligned}
&\frac{\partial \widehat{rr}}{\partial r}+\frac{1}{r}\frac{\partial \widehat{r\theta}}{\partial \theta}+\frac{\widehat{rr}-\widehat{\theta\theta}}{r}=b_r\\
&\frac{\partial \widehat{r\theta}}{\partial r}+\frac{1}{r}\frac{\partial \widehat{\theta\theta}}{\partial \theta}+\frac{2\widehat{r\theta}}{r}=b_\theta\\
&\frac{\partial \widehat{rz}}{\partial r}+\frac{1}{r}\frac{\partial \widehat{\theta z}}{\partial \theta}+\frac{\widehat{rz}}{r}=0\,.
\end{aligned}\right\}\tag{10}$$

From (4), (5), (6), (8) we get the strain coefficients

$$\left.\begin{aligned}
&e_{rr}=\frac{\partial u_r}{\partial r},\ e_{\theta\theta}=\frac{1}{r}\frac{\partial u_\theta}{\partial \theta}+\frac{u_r}{r},\ e_{zz}=\frac{\partial u_z}{\partial R}\\
&2e_{\theta z}=\frac{1}{r}\frac{\partial u_z}{\partial \theta}+\frac{\partial u_\theta}{\partial R},\ 2e_{zr}=\frac{\partial u_r}{\partial R}+\frac{\partial u_z}{\partial r},\ 2e_{r\theta}=\frac{\partial u_\theta}{\partial r}-\frac{u_\theta}{r}+\frac{1}{r}\frac{\partial u_r}{\partial \theta}
\end{aligned}\right\}\tag{11}$$

7.52. The displacement in cylindrical anisotropy

With the hypotheses of 7.5 we saw the stress tensor S is independent of R. Also in terms of the inverse Hooke's tensor 1.7 (5) we have

$$\mathsf{D}=\mathsf{K}_{(4)}\cdot\cdot\,\mathsf{S}\,.\tag{1}$$

It follows from (1) that the deformation tensor D is likewise independent of R.

Therefore modification of the column matrices of 1.72 (2) to

$$\begin{bmatrix} e_{rr}\\ e_{\theta\theta}\\ e_{zz}\\ 2e_{\theta z}\\ 2e_{rz}\\ 2e_{r\theta}\end{bmatrix}\qquad\begin{bmatrix} \widehat{rr}\\ \widehat{\theta\theta}\\ \widehat{zz}\\ \widehat{\theta z}\\ \widehat{rz}\\ \widehat{r\theta}\end{bmatrix}$$

and use of 7.5 (11) give the following equations

$$\frac{\partial u_r}{\partial r}=e_{rr}=k_{11}\widehat{rr}+k_{12}\widehat{\theta\theta}+k_{13}\widehat{zz}+k_{14}\widehat{\theta z}+k_{15}\widehat{rz}+k_{16}\widehat{r\theta}\tag{2}$$

$$\frac{1}{r}\frac{\partial u_\theta}{\partial \theta}+\frac{u_r}{r}=e_{\theta\theta}=k_{21}\widehat{rr}+k_{22}\widehat{\theta\theta}+k_{23}\widehat{zz}+k_{24}\widehat{\theta z}+k_{25}\widehat{rz}+k_{26}\widehat{r\theta}\tag{3}$$

$$\frac{\partial u_z}{\partial R}=e_{zz}=k_{31}\widehat{rr}+k_{32}\widehat{\theta\theta}+k_{33}\widehat{zz}+k_{34}\widehat{\theta z}+k_{35}\widehat{rz}+k_{36}\widehat{r\theta}\tag{4}$$

$$\frac{1}{r}\frac{\partial u_z}{\partial \theta}+\frac{\partial u_\theta}{\partial R}=2e_{\theta z}=k_{41}\widehat{rr}+k_{42}\widehat{\theta\theta}+k_{43}\widehat{zz}+k_{44}\widehat{\theta z}+k_{45}\widehat{rz}+k_{46}\widehat{r\theta}\tag{5}$$

$$\frac{\partial u_r}{\partial R}+\frac{\partial u_z}{\partial r}=2e_{rz}=k_{51}\widehat{rr}+k_{52}\widehat{\theta\theta}+k_{53}\widehat{zz}+k_{54}\widehat{\theta z}+k_{55}\widehat{rz}+k_{56}\widehat{r\theta}\tag{6}$$

$$\begin{aligned}\frac{\partial u_\theta}{\partial r}+\frac{1}{r}\frac{\partial u_r}{\partial \theta}-\frac{u_\theta}{r}=2e_{r\theta}=k_{61}\widehat{rr}+k_{62}\widehat{\theta\theta}+k_{63}\widehat{zz}+k_{64}\widehat{\theta z}&\\+k_{65}\widehat{rz}+k_{66}\widehat{r\theta}&\end{aligned}\tag{7}$$

where all the strain coefficients $e_{\alpha\beta}$ are independent of R.

Integration of (4) with respect to R gives

$$u_z = R e_{zz} + U_z^\circ\,, \tag{8}$$

where $U_z^\circ = U_z^\circ(r, \theta)$ is an arbitrary function. Substitute this in (5) and (6). Then

$$\frac{\partial u_\theta}{\partial R} = -R\frac{1}{r}\frac{\partial e_{zz}}{\partial \theta} + 2e_{\theta z} - \frac{1}{r}\frac{\partial U_z^\circ}{\partial \theta}$$

$$\frac{\partial u_r}{\partial R} = -R\frac{\partial e_{zz}}{\partial r} + 2e_{rz} - \frac{\partial U_z^\circ}{\partial r}$$

and therefore by integration with respect to R

$$u_r = -\frac{1}{2}R^2\frac{\partial e_{zz}}{\partial r} + R\left(2e_{rz} - \frac{\partial U_z^\circ}{\partial r}\right) + U_r^\circ\,, \tag{9}$$

$$u_\theta = -\frac{1}{2}R^2\frac{1}{r}\frac{\partial e_{zz}}{\partial \theta} + R\left(2e_{\theta z} - \frac{1}{r}\frac{\partial U_z^\circ}{\partial \theta}\right) + U_\theta^\circ\,, \tag{10}$$

where U_r°, U_θ° are arbitrary functions independent of R.

Substituting (9), (10) in (2), (3), (7) we get

$$-\frac{1}{2}R^2\frac{\partial^2 e_{zz}}{\partial r^2} + R\frac{\partial}{\partial r}\left(2e_{rz} - \frac{\partial U_z^\circ}{\partial r}\right) + \frac{\partial U_r^\circ}{\partial r} - e_{rr} = 0, \tag{11}$$

$$\begin{aligned} -\frac{1}{2}R^2\left(\frac{1}{r^2}\frac{\partial^2 e_{zz}}{\partial \theta^2} + \frac{1}{r}\frac{\partial e_{zz}}{\partial r}\right) + \frac{1}{r}R\left\{\frac{\partial}{\partial \theta}\left(2e_{\theta z} - \frac{1}{r}\frac{\partial U_z^\circ}{\partial \theta}\right) + 2e_{rz} - \frac{\partial U_z^\circ}{\partial r}\right\} \\ + \frac{1}{r}\frac{\partial U_\theta^\circ}{\partial \theta} + \frac{U_r^\circ}{r} - e_{\theta\theta} = 0, \end{aligned} \tag{12}$$

$$\begin{aligned} &-R^2\left(\frac{1}{r}\frac{\partial^2 e_{zz}}{\partial r\,\partial \theta} - \frac{1}{r^2}\frac{\partial e_{zz}}{\partial \theta}\right) \\ &+ R\left\{\frac{\partial}{\partial r}\left(2e_{\theta z} - \frac{1}{r}\frac{\partial U_z^\circ}{\partial \theta}\right) + \frac{1\partial}{r\partial \theta}\left(2e_{rz} - \frac{\partial U_z^\circ}{\partial r}\right) - \frac{1}{r}\left(2e_{\theta z} - \frac{1}{r}\frac{\partial U_z^\circ}{\partial \theta}\right)\right\} \\ &+ \frac{\partial U_\theta^\circ}{\partial r} - \frac{U_\theta^\circ}{r} + \frac{1}{r}\frac{\partial U_r^\circ}{\partial \theta} - 2e_{r\theta} = 0\,. \end{aligned} \tag{13}$$

In each of these three identities in R equate to zero the coefficient of R^2. Then

$$\frac{\partial^2 e_{zz}}{\partial r^2} = 0, \frac{\partial e_{zz}}{\partial r} + \frac{1}{r}\frac{\partial^2 e_{zz}}{\partial \theta^2} = 0, \frac{\partial^2 e_{zz}}{\partial r \partial \theta} - \frac{1}{r}\frac{\partial e_{zz}}{\partial \theta} = 0$$

whence it follows that

$$e_{zz} = A_2 r\cos\theta + B_2 r\sin\theta + C_2 m \tag{14}$$

where A_2, B_2, C_2 are constants and m is the length introduced in 2.2 (7) to keep A_2, B_2, C_2 of the same dimensions.

It follows from (14) and (4) that

$$\begin{aligned} \widehat{zz} = \frac{1}{k_{33}}\{A_2 r\cos\theta + B_2 r\sin\theta + C_2 m - (k_{13}\widehat{rr} + k_{23}\widehat{\theta\theta} + k_{34}\widehat{\theta z} \\ + k_{35}\widehat{rz} + k_{36}\widehat{r\theta})\}\,. \end{aligned} \tag{15}$$

Write

$$l_{\alpha\beta} = k_{\alpha\beta} - \frac{k_{\alpha 3} k_{\beta 3}}{k_{33}}, \qquad \nu_\alpha = \frac{k_{\alpha 3}}{k_{33}}, \tag{16}$$

where the $l_{\alpha\beta}$ are modified inverse moduli as in 2.32. Then with the aid of (15) we can eliminate $\widehat{zz}$ from (2)—(7) to give

$$\left.\begin{aligned}
e_{rr} &= l_{11}\widehat{rr} + l_{12}\widehat{\theta\theta} + l_{14}\widehat{\theta z} + l_{15}\widehat{rz} + l_{16}\widehat{r\theta} \\
&\qquad + \nu_1(A_2 r\cos\theta + B_2 r\sin\theta + C_2 m) \\
e_{\theta\theta} &= l_{21}\widehat{rr} + l_{22}\widehat{\theta\theta} + l_{24}\widehat{\theta z} + l_{25}\widehat{rz} + l_{26}\widehat{r\theta} \\
&\qquad + \nu_2(A_2 r\cos\theta + B_2 r\sin\theta + C_2 m) \\
2e_{\theta z} &= l_{41}\widehat{rr} + l_{42}\widehat{\theta\theta} + l_{44}\widehat{\theta z} + l_{45}\widehat{rz} + l_{46}\widehat{r\theta} \\
&\qquad + \nu_4(A_2 r\cos\theta + B_2 r\sin\theta + C_2 m) \\
2e_{rz} &= l_{51}\widehat{rr} + l_{52}\widehat{\theta\theta} + l_{54}\widehat{\theta z} + l_{55}\widehat{rz} + l_{56}\widehat{r\theta} \\
&\qquad + \nu_5(A_2 r\cos\theta + B_2 r\sin\theta + C_2 m) \\
2e_{r\theta} &= l_{61}\widehat{rr} + l_{62}\widehat{\theta\theta} + l_{64}\widehat{\theta z} + l_{65}\widehat{rz} + l_{66}\widehat{r\theta} \\
&\qquad + \nu_6(A_2 r\cos\theta + B_2 r\sin\theta + C_2 m)
\end{aligned}\right\} \tag{17}$$

By equating to zero the coefficients of R in (11)—(13) we get

$$\left.\begin{aligned}
&\frac{\partial}{\partial r}\left(2e_{rz} - \frac{\partial U_z^\circ}{\partial r}\right) = 0 \\
&\frac{\partial}{\partial \theta}\left(2e_{\theta z} - \frac{1}{r}\frac{\partial U_z^\circ}{\partial \theta}\right) + 2e_{rz} - \frac{\partial U_z^\circ}{\partial r} = 0 \\
&r\frac{\partial}{\partial r}\left(2e_{\theta z} - \frac{1}{r}\frac{\partial U_z^\circ}{\partial \theta}\right) + \frac{\partial}{\partial \theta}\left(2e_{rz} - \frac{\partial U_z^\circ}{\partial r}\right) - \left(2e_{\theta z} - \frac{1}{r}\frac{\partial U_z^\circ}{\partial \theta}\right) = 0
\end{aligned}\right\} \tag{18}$$

Again from the terms independent of R in (11)—(13)

$$\left.\begin{aligned}
&\frac{\partial U_r^\circ}{\partial r} = e_{rr} \\
&\frac{1}{r}\frac{\partial U_\theta^\circ}{\partial \theta} + \frac{U_r^\circ}{r} = e_{\theta\theta} \\
&\frac{1}{r}\frac{\partial U_r^\circ}{\partial \theta} + \frac{\partial U_\theta^\circ}{\partial r} - \frac{U_\theta^\circ}{r} = 2e_{r\theta}
\end{aligned}\right\} \tag{19}$$

In (18) and (19) we can suppose the strain coefficients e_{rs} to be expressed in terms of the stresses by means of (17).

Introduce an arbitrary rigid body movement consisting of a small rotation $(\omega_1, \omega_2, \omega_3)$ and a small translation (α, β, γ) with respect to axes Ox, Oy, OR. Then in cylindrical coordinates we can write

$$\left.\begin{aligned}
U_r^\circ &= U_r + \alpha\cos\theta + \beta\sin\theta \\
U_\theta^\circ &= U_\theta - \alpha\sin\theta + \beta\cos\theta + \omega_3 r \\
U_z^\circ &= U_z + \omega_1 r\sin\theta - \omega_2 r\cos\theta + \gamma
\end{aligned}\right\} \tag{20}$$

Then from (17) and (19)

$$\left.\begin{aligned}\frac{\partial U_r}{\partial r} &= l_{11}\widehat{rr} + l_{12}\widehat{\theta\theta} + l_{14}\widehat{\theta z} + l_{15}\widehat{rz} + l_{16}\widehat{r\theta}\\ &\quad + \nu_1(A_2 r\cos\theta + B_2 r\sin\theta + C_2 m)\\ \frac{1}{r}\frac{\partial U_\theta}{\partial\theta} + \frac{U_r}{r} &= l_{21}\widehat{rr} + l_{22}\widehat{\theta\theta} + l_{24}\widehat{\theta z} + l_{25}\widehat{rz} + l_{26}\widehat{r\theta}\\ &\quad + \nu_2(A_2 r\cos\theta + B_2 r\sin\theta + C_2 m)\\ \frac{1}{r}\frac{\partial U_r}{\partial\theta} + \frac{\partial U_\theta}{\partial r} - \frac{U_\theta}{r} &= l_{61}\widehat{rr} + l_{62}\widehat{\theta\theta} + l_{64}\widehat{\theta z} + l_{65}\widehat{rz} + l_{66}\widehat{r\theta}\\ &\quad + \nu_6(A_2 r\cos\theta + B_2 r\sin\theta + C_2 m)\end{aligned}\right\} \tag{21}$$

Observe that only particular solutions of these equations are required. The arbitrary rigid body movement has already been allowed for in (20).

Integrating $(18)_{1,\,2}$ we get, using $(20)_3$

$$2e_{rz} - \frac{\partial U_z}{\partial r} - \omega_1\sin\theta + \omega_2\cos\theta = f_1'(\theta)$$

$$2e_{\theta z} \quad \frac{1}{r}\frac{\partial U_z}{\partial\theta} - \omega_1\cos\theta - \omega_2\sin\theta = -f_1(\theta) + f_2(r)$$

$$r\frac{\partial}{\partial r}\{-f_1(\theta) + f_2(r)\} + f_1''(\theta) + f_1(\theta) - f_2(r) = 0$$

so that if $f_1''(\theta) + f_1(\theta) = 0$, then $rf_2'(r) - f_2(r) = 0$.

Therefore we take

$$f_1(\theta) = \omega_1\cos\theta + \omega_2\sin\theta, \quad f_2(r) = \tau r,$$

where τ is a constant. Thus, using (17),

$$\left.\begin{aligned}\frac{\partial U_z}{\partial r} &= l_{51}\widehat{rr} + l_{52}\widehat{\theta\theta} + l_{54}\widehat{\theta z} + l_{55}\widehat{rz} + l_{56}\widehat{r\theta}\\ &\quad + \nu_5(A_2 r\cos\theta + B_2 r\sin\theta + C_2 m)\\ \frac{\partial U_z}{r\partial\theta} &= l_{41}\widehat{rr} + l_{42}\widehat{\theta\theta} + l_{44}\widehat{\theta z} + l_{45}\widehat{rz} + l_{46}\widehat{r\theta}\\ &\quad + \nu_4(A_2 r\cos\theta + B_2 r\sin\theta + C_2 m) - \tau r\end{aligned}\right\} \tag{22}$$

Therefore from (8)—(10) the displacements are

$$\begin{aligned}u_r = U_r - \frac{1}{2}R^2(A_2\cos\theta + B_2\sin\theta) + R(-\omega_1\sin\theta + \omega_2\cos\theta)&\\ + \alpha\cos\theta + \beta\sin\theta&\end{aligned} \tag{23}$$

$$\begin{aligned}u_\theta = U_\theta - \frac{1}{2}R^2(-A_2\sin\theta + B_2\cos\theta) + R(\tau r - \omega_1\cos\theta - \omega_2\sin\theta)&\\ - \alpha\sin\theta + \beta\cos\theta + \omega_3 r&\end{aligned} \tag{24}$$

$$\begin{aligned}u_z = U_z + R(A_2 r\cos\theta + B_2 r\sin\theta + C_2 m) + \omega_1 r\sin\theta&\\ - \omega_2 r\cos\theta + \gamma,&\end{aligned} \tag{25}$$

where U_r, U_θ, U_z are particular solutions of (21), (22).

It appears from these expressions that if the material extends to infinity in the R direction, the displacements can be bounded only if

$$A_2 = B_2 = C_2 = 0\,, \quad \tau = 0\,, \quad \omega_1 = \omega_2 = 0\,. \tag{26}$$

The condition $\omega_1 = \omega_2 = 0$ can always be achieved by proper choice of the arbitrary rigid body movement.

7.54. Lateral and end conditions

Let $\boldsymbol{n}$ be the unit normal vector drawn outwards at a point of the lateral surface, that is to say pointing away from the material. Then in terms of the unit vectors of 7.5 we can write

$$\boldsymbol{n} = l\boldsymbol{i}_r + m\boldsymbol{i}_\theta\,. \tag{1}$$

The stress boundary condition is then

$$\mathsf{S}\cdot\boldsymbol{n} = r_N\boldsymbol{i}_r + \theta_N\boldsymbol{i}_\theta$$

or from 7.5 (3)

$$\left.\begin{aligned} l\widehat{rr} + m\widehat{r\theta} &= r_N \\ l\widehat{r\theta} + m\widehat{\theta\theta} &= \theta_N \\ l\widehat{rz} + m\widehat{\theta z} &= 0 \end{aligned}\right\} \text{ at the lateral surface} \tag{2}$$

In the antiplane let there be applied an axial force P_3 at the origin and couples M_1 about the x-axis and M_2 about the y-axis. Then from 3.1 (3), (4)

$$M_1 = -\int_S y\cdot\widehat{zz}_0\,dS,\ M_2 = \int_S x\cdot\widehat{zz}_0\,dS,\ P_3 = -\int_S \widehat{zz}_0\,dS\,. \tag{3}$$

Let $(x_G, y_G, 0)$ be the centroid of the antiplane. We put

$$\int_S y^2\,dS = I_x,\ \int_S x^2\,dS = I_y,\ \int_S xy\,dS = H\,. \tag{4}$$

Then A_2, B_2, C_2 are determined by the following equations obtained by substituting in (3) the expression 7.52 (15) for $\widehat{zz}$.

$$\begin{aligned}(A_2x_G + B_2y_G + C_2m)\,S \\ = \int_S (k_{31}\widehat{rr} + k_{32}\widehat{\theta\theta} + k_{34}\widehat{\theta z} + k_{35}\widehat{rz} + k_{36}\widehat{r\theta})\,dS - k_{33}P_3\,,\end{aligned} \tag{5}$$

$$\begin{aligned}A_2I_y + B_2H + C_2mSx_G \\ = \int_S (k_{31}\widehat{rr} + k_{32}\widehat{\theta\theta} + k_{34}\widehat{\theta z} + k_{35}\widehat{rz} + k_{36}\widehat{r\theta})\,r\cos\theta\,dS + k_{33}M_2,\end{aligned} \tag{6}$$

$$\begin{aligned}A_2H + B_2I_x + C_2mSy_G \\ = \int_S (k_{31}\widehat{rr} + k_{32}\widehat{\theta\theta} + k_{34}\widehat{\theta z} + k_{35}\widehat{rz} + k_{36}\widehat{r\theta})\,r\sin\theta\,dS - k_{33}M_1\,.\end{aligned} \tag{7}$$

7.6. Equations satisfied by the stress functions

The equations of equilibrium 7.5 (10) are satisfied identically, when the body-force derives from a potential V, by

$$\widehat{rr} = \frac{1}{r}\frac{\partial \chi}{\partial r} + \frac{1}{r^2}\frac{\partial^2 \chi}{\partial \theta^2} + V, \quad \widehat{r\theta} = -\frac{\partial^2}{\partial r\, \partial \theta}\left(\frac{\chi}{r}\right), \quad \widehat{\theta\theta} = \frac{\partial^2 \chi}{\partial r^2} + V, \tag{1}$$

$$\widehat{rz} = \frac{1}{r}\frac{\partial \psi}{\partial \theta}, \quad \widehat{\theta z} = -\frac{\partial \psi}{\partial r}. \tag{2}$$

To obtain the equations satisfied by the stress functions χ and ψ we can proceed as follows. By means of (1) and (2) we can express the right-hand sides of 7.52 (21), (22) in terms of the derivates of χ and ψ. We then substitute these results in the identities

$$\begin{aligned} &\left(\frac{\partial^2}{\partial r\, \partial \theta} + \frac{1}{r}\frac{\partial}{\partial \theta}\right)\left(\frac{1}{r}\frac{\partial U_r}{\partial \theta} + \frac{\partial U_\theta}{\partial r} - \frac{U_\theta}{r}\right) \\ &= \frac{\partial^2}{\partial r^2}\left(\frac{\partial U_\theta}{\partial \theta} + U_r\right) + \left(\frac{1}{r}\frac{\partial^2}{\partial \theta^2} - \frac{\partial}{\partial r}\right)\frac{\partial U_r}{\partial r}, \end{aligned} \tag{3}$$

$$\frac{\partial}{\partial \theta}\left(\frac{\partial U_z}{\partial r}\right) = \frac{\partial}{\partial r}\left(\frac{\partial U_z}{\partial \theta}\right) \tag{4}$$

of which the first can be derived from the compatibility equation 1.32 (7) when expressed in cylindrical coordinates by means of 7.5 (4) and (5).

After a straightforward reduction we obtain, in the absence of body-force, the equations

$$\begin{aligned} D^{(4)}\chi - D_1^{(3)}\psi + 2\cos\theta\,[A_2(\nu_2 - \nu_1) - B_2\nu_6] \\ + 2\sin\theta\,[A_2\nu_6 + B_2(\nu_2 - \nu_1)] = 0 \end{aligned} \tag{5}$$

$$\begin{aligned} D_2^{(3)}\chi - D^{(2)}\psi + r\cos\theta\,[2A_2\nu_4 - B_2\nu_5) \\ + r\sin\theta\,(A_2\nu_5 + B_2\nu_4) + C_2 m\nu_4 - 2\tau r = 0 \end{aligned} \tag{6}$$

where the operators are

$$\begin{aligned} D^{(4)} = {} & l_{22} r\frac{\partial^4}{\partial r^4} - 2l_{26}\frac{\partial^4}{\partial r^3 \partial \theta} + (2l_{12} + l_{66})\frac{1}{r}\frac{\partial^4}{\partial r^2 \partial \theta^2} \\ & - 2l_{16}\frac{1}{r^2}\frac{\partial^4}{\partial r\, \partial \theta^3} + l_{11}\frac{1}{r^3}\frac{\partial^4}{\partial \theta^4} \\ & + 2l_{22}\frac{\partial^3}{\partial r^3} - (2l_{12} + l_{66})\frac{1}{r^2}\frac{\partial^3}{\partial r\, \partial \theta^2} + 2l_{16}\frac{1}{r^3}\frac{\partial^3}{\partial \theta^3} \\ & - l_{11}\frac{1}{r}\frac{\partial^2}{\partial r^2} - 2(l_{16} + l_{26})\frac{1}{r^2}\frac{\partial^2}{\partial r\, \partial \theta} + (2l_{11} + 2l_{12} + l_{66})\frac{1}{r^3}\frac{\partial^2}{\partial \theta^2} \\ & + l_{11}\frac{1}{r^2}\frac{\partial}{\partial r} + 2(l_{16} + l_{26})\frac{1}{r^3}\frac{\partial}{\partial \theta} \end{aligned} \tag{7}$$

$$\begin{aligned} D_1^{(3)} = {} & l_{24} r\frac{\partial^3}{\partial r^3} - (l_{25} + l_{46})\frac{\partial^3}{\partial r^2 \partial \theta} + (l_{14} + l_{56})\frac{1}{r}\frac{\partial^3}{\partial r\, \partial \theta^2} - l_{15}\frac{1}{r^2}\frac{\partial^3}{\partial \theta^3} \\ & - (l_{14} - 2l_{24})\frac{\partial^2}{\partial r^2} - (l_{46} - l_{15})\frac{1}{r}\frac{\partial^2}{\partial r\, \partial \theta} - l_{15}\frac{1}{r^2}\frac{\partial}{\partial \theta} \end{aligned} \tag{8}$$

$$D_2^{(3)} = l_{42} r \frac{\partial^3}{\partial r^3} - (l_{46} + l_{25}) \frac{\partial^3}{\partial r^2 \partial \theta} + (l_{14} + l_{56}) \frac{1}{r} \frac{\partial^3}{\partial r \partial \theta^2} - l_{15} \frac{1}{r^2} \frac{\partial^3}{\partial \theta^3}$$

$$+ (l_{14} + l_{24}) \frac{\partial^2}{\partial r^2} + (l_{46} - l_{15}) \frac{1}{r} \frac{\partial^2}{\partial r \partial \theta} \tag{9}$$

$$- (l_{14} + l_{56}) \frac{1}{r^2} \frac{\partial^2}{\partial \theta^2} - l_{46} \frac{1}{r^2} \frac{\partial}{\partial \theta}$$

$$D^{(2)} = l_{44} r \frac{\partial^2}{\partial r^2} - 2 l_{45} \frac{\partial^2}{\partial r \partial \theta} + l_{55} \frac{1}{r} \frac{\partial^2}{\partial \theta^2} + l_{44} \frac{\partial}{\partial r}. \tag{10}$$

7.7. Circular tube under pressure

Consider a tube of length L bounded by coaxial cylinders the outer of radius a the inner of radius b, subjected to external hydrostatic pressure p_a and internal hydrostatic pressure p_b, so that the stress boundary conditions are

$$\widehat{bb} = - p_b, \; \widehat{aa} = - p_a \tag{1}$$

$$\widehat{a\theta} = \widehat{b\theta} = 0 .$$

We suppose further that at the free end there is an axial force $P = - P_3$, and a twisting moment $M = - M_3$, where

$$P = 2\pi \int_a^b \widehat{zz} r \, dr, \; M = 2\pi \int_a^b \widehat{\theta z} r^2 \, dr. \tag{2}$$

There is no body-force or lateral force.

Let the material have cylindrical anisotropy the axis of anisotropy coinciding with the axis of the cylinder. In these circumstances everything is a function of r only, and therefore $A_2 = B_2 = 0$, while the stresses are derived from stress functions

$$\chi = \chi(r), \; \psi = \psi(r) \tag{3}$$

which give the stress components

$$\widehat{rr} = \frac{\chi'}{r}, \quad \widehat{\theta\theta} = \chi'', \quad \widehat{r\theta} = 0, \quad \widehat{zr} = 0, \quad \widehat{\theta z} = - \psi' . \tag{4}$$

Therefore

$$\widehat{zz} = \frac{1}{k_{33}} [C_2 m - (k_{13} \widehat{rr} + k_{23} \widehat{\theta\theta} + k_{34} \widehat{\theta z})] . \tag{5}$$

The equations satisfied by the stress functions are

$$l_{22} r \chi'''' + 2 l_{22} \chi''' - l_{11} \frac{1}{r} \chi'' + l_{11} \frac{1}{r^2} \chi' - l_{24} r \psi''' + (l_{14} - 2 l_{24}) \psi'' = 0 \tag{6}$$

$$l_{24} r \chi''' + (l_{14} + l_{24}) \chi'' - l_{44} r \psi'' - l_{44} \psi' + \nu_4 C_2 m - 2 \tau r = 0. \tag{7}$$

Consider first the homogeneous equations got by putting $C_2 = 0$, $\tau = 0$. Then the substitutions

$$\chi_1' = Q r^n, \; \psi_1' = S r^k \tag{8}$$

lead to

$$n = k + 1 \tag{9}$$

and

$$(n-1)\,[(l_{22}n^2 - l_{11})\,Q - (l_{24}n - l_{14})\,S] = 0$$
$$n\,[(l_{24}n + l_{14})\,Q - l_{44}S] = 0\,.$$

Thus

$$n = 0,\ 1,\ c,\ -c$$
$$k = -1,\ 0,\ c-1,\ -c-1$$

where

$$c^2 = \frac{l_{11}l_{44} - l_{14}^2}{l_{22}l_{44} - l_{24}^2} \tag{10}$$

$$\frac{Q}{S} = \frac{l_{44}}{l_{24}n + l_{14}},\ n \neq 0;\ \frac{Q}{S} = \frac{l_{14}}{l_{11}},\ n = 0\,.$$

Thus for the complementary function we have

$$\chi_1' = K_0 l_{14} + K_1 l_{44} r + K_c l_{44} r^c + K_{-c} l_{44} r^{-c} \tag{11}$$

$$\begin{aligned}\psi_1' = \frac{K_0 l_{11}}{r} + K_1(l_{14} + l_{24}) + K_c(l_{14} + l_{24}c)\,r^{c-1} \\ + K_{-c}(l_{14} - l_{24}c)\,r^{-c-1}\end{aligned} \tag{12}$$

where the K_s $(s = 0, 1, c, -c)$ are arbitrary constants.

It is easy to prove or verify that particular integrals are

$$\chi_0' = \alpha\tau r^2,\ \psi_0' = \beta\tau r + \frac{\nu_4 C_2 m}{l_{44}}, \tag{13}$$

where

$$\frac{\alpha}{l_{14} - 2l_{24}} = \frac{\beta}{l_{11} - 4l_{22}} = \frac{1}{4(l_{22}l_{44} - l_{24}^2) - (l_{11}l_{44} - l_{14}^2)} \tag{14}$$

and the complete solutions are given by

$$\begin{aligned}\chi' &= \chi_1' + \chi_0' \\ \psi' &= \psi_1' + \psi_0'\end{aligned} \tag{15}$$

and the stresses and displacements can now be determined.

7.71. Determination of the stresses

We have from 7.7 (4)

$$\widehat{rr} = \frac{K_0 l_{14}}{r} + K_1 l_{44} + K_c l_{44} r^{c-1} + K_{-c} l_{44} r^{-c-1} + \alpha\tau r \tag{1}$$

$$\widehat{\theta\theta} = K_1 l_{44} + cK_c l_{44} r^{c-1} - cK_{-c} l_{44} r^{-c-1} + 2\alpha\tau r \tag{2}$$

$$\begin{aligned}\widehat{\theta z} = -\frac{K_0 l_{11}}{r} - K_1(l_{14} + l_{24}) - K_c(l_{14} + c\,l_{24})\,r^{c-1} \\ - K_{-c}(l_{14} - c\,l_{24})\,r^{-c-1} - \beta\tau r - \frac{\nu_4 C_2 m}{l_{44}}\end{aligned} \tag{3}$$

$$\widehat{r\theta} = \widehat{zr} = 0\,. \tag{4}$$

Since the displacement component U_z is independent of θ we have from 7.52 (22)

$$0 = \frac{\partial U_z}{r\partial\theta} = l_{41}\widehat{rr} + l_{42}\widehat{\theta\theta} + l_{44}\widehat{\theta z} + \nu_4 C_2 m - \tau r\,.$$

Substituting the above stresses and noting that,

$$\alpha l_{14} + 2\alpha l_{24} - \beta l_{44} - 1 = 0$$

we find that $K_0 = 0$.

Since $\partial U_\theta/\partial\theta = 0$ we have from 7.52 (21) expressions for U_r and $\partial U_r/\partial r$. These must agree and by equating the two values of $\partial U_r/\partial r$ we find that

$$\frac{l_{44}K_1}{C_2 m} = \frac{\nu_4(l_{24} - l_{14}) + (\nu_1 - \nu_2)l_{44}}{(l_{22}l_{44} - l_{14}^2) - (l_{11}l_{44} - l_{24}^2)} \tag{5}$$

Introduce the abbreviations

$$d_s = a^{c+s}b^{c+s},\ f_s = \frac{a^{c+s} - b^{c+s}}{a^{2c} - b^{2c}},\ h_s = \frac{l_{14} + s l_{24}}{l_{44}}\,.$$

Then using the boundary conditions (1) we find that

$$\widehat{rr} = \frac{p_b b^{c+1} - p_a a^{c+1}}{a^{2c} - b^{2c}} r^{c-1} + \frac{p_a b^{c-1} - p_b a^{c-1}}{a^{2c} - b^{2c}} d_1 r^{-c-1}$$
$$+ K_1 l_{44}(1 - f_1 r^{c-1} - f_{-1} d_1 r^{-c-1}) + \alpha\tau(r - f_2 r^{c-1} - f_{-2} d_2 r^{-c-1}) \tag{7}$$

$$\widehat{\theta\theta} = \frac{p_b b^{c+1} - p_a a^{c+1}}{a^{2c} - b^{2c}} c r^{c-1} - \frac{p_a b^{c-1} - p_b a^{c-1}}{a^{2c} - b^{2c}} c d_1 r^{-c-1}$$
$$+ K_1 l_{44}(1 - f_1 c r^{c-1} + f_{-1} c d_1 r^{-c-1}) + \alpha\tau(2r - f_2 c r^{c-1} + f_{-2} c d_2 r^{-c-1}) \tag{8}$$

$$\widehat{\theta z} = -\frac{p_b b^{c+1} - p_a a^{c+1}}{a^{2c} - b^{2c}} h_c r^{c-1} - \frac{p_a b^{c-1} - p_b a^{c-1}}{a^{2c} - b^{2c}} h_{-c} d_1 r^{-c-1}$$
$$+ K_1 l_{44}\left(-h_1 - \frac{\nu C_2 m}{K_1 l_{44}^2} + f_1 h_c r^{c-1} + f_{-1} d_1 h_c r^{-c-1}\right) \tag{9}$$
$$+ \tau(-\beta r + \alpha f_2 h_c r^{c-1} + \alpha f_{-2} h_{-c} d_2 r^{-c-1})\,.$$

It remains to determine K_1 (or C_2) and τ. Equations for this, which we omit, are obtained from the relations 7.7 (2).

7.72. Lamé's problem of the tube under pressure

Lamé studied the problem of plane deformation of a tube subjected to internal and external hydrostatic pressure. If in 7.7 we put $C_2 = 0$, $\tau = 0$ we have the corresponding problem of generalized plane deformation for the case of cylindrical anisotropy about the axis of the tube.

We then get from 7.71 (7)—(9)

$$\widehat{rr} = Q_1 r^{c-1} + Q_2 d_1 r^{-c-1},\quad \widehat{\theta\theta} = c Q_1 r^{c-1} - c Q_2 d_1 r^{-c-1},$$
$$\widehat{\theta z} = -Q_1 h_c r^{c-1} - Q_2 h_{-c} d_1 r^{-c-1} \tag{1}$$

where
$$Q_1 = \frac{p_b b^{c+1} - p_a a^{c+1}}{a^{2c} - b^{2c}}, \quad Q_2 = \frac{p_a b^{c-1} - p_b a^{c-1}}{a^{2c} - b^{2c}}.$$
From 7.52 (21), (22)
$$\frac{\partial U_z}{\partial r} = l_{51}\widehat{rr} + l_{52}\widehat{\theta\theta} + l_{54}\widehat{\theta z} \tag{3}$$
$$\frac{\partial U_\theta}{\partial r} - \frac{U_\theta}{r} = l_{61}\widehat{rr} + l_{62}\widehat{\theta\theta} + l_{64}\widehat{\theta z} \tag{4}$$
Thus from (1) and (3)
$$c\,U_z = Q_1(l_{51} + c l_{52} - h_c l_{54}) r^c - Q_2 d_1 (l_{51} - c l_{52} - h_{-c} l_{54}) d_1 r^{-c} \tag{5}$$
and from this we infer that the cross-sections are warped, for U_z is not a linear function of r.

Again from (4) we get
$$\frac{\partial}{\partial r}\left(\frac{U_\theta}{r}\right) = Q_1 (l_{61} + c l_{62} - h_c l_{64}) r^{c-2} + Q_2 (l_{61} - c l_{62} - h_{-c} l_{64}) d_1 r^{-c-2}.$$
Therefore
$$\frac{U_\theta}{r} = \frac{Q_1}{c-1}(l_{61} + c l_{62} - h_c l_{64}) r^{c-1}$$
$$- \frac{Q_2}{c+1}(l_{61} - c l_{62} - h_{-c} l_{64}) d_1 r^{-c-1} + \text{constant}$$
Thus U_θ is not a linear function of r. Therefore the radii are also warped as well as the cross-sections.

EXAMPLES VII

1. Find the complex stresses which correspond
   (i) to a state of simple tension T_1 parallel to the x-axis,
   (ii) to a state of simple tension T_2 parallel to the y-axis,
   (iii) to a state of all-round tension T.

2. An elastic space has an unloaded circular cylindrical hole and the space is stressed by simple tension T perpendicular to the axis of the hole. Investigate the distribution of hoop stress round the hole and find its maximum value.

3. Investigate the form of the equations 7.4 (4), (5) when body-force and surface loading are included.

4. For the beam loaded and bent as in 7.4 show that Euler's law of bending applies, namely that the bending moment is proportional to the curvature of the bent axis.

5. For the anisotropic beam loaded and bent as in 7.4 show that the rotation of an element of the R-axis is
$$-\frac{P k_{35}}{4 I_x}(L - R)(L + R) - \tau (L - R)$$
and that the rate of change of bending along the axis is
$$\frac{P k_{35}}{2 I_x} R + \tau.$$

Prove further that this rate of change is constant when the beam has an antiplane of elastic symmetry.

6. In the flexure of an elliptic cylinder (7.44) determine the constants N_1, N_2, N_3.

7. Determine the Timoshenko stress function for a circular cylinder loaded as in 7.44 and discuss the values of the stresses on the diameters $x = 0$, and $y = 0$.

8. Determine the flexure of an elliptic cylinder whose cross-section is

$$\frac{x^2}{a^2} + \frac{y^2}{b^2} = 1$$

when bent by a force Q along the x-axis.

9. Determine the flexure of an elliptic cylinder when bent by a transverse force applied at the centroid of the free end.

10. In the flexure of the elliptic cylinder in 7.44 there is an antiplane of elastic symmetry. Prove that $\widehat{xx} = \widehat{yy} = \widehat{xy} = 0$ and find the shears.

11. If the cantilever of 7.44 is orthotropic, the coordinate planes being planes of elastic symmetry, prove that the maximum shear occurs on the axis of the cylinder and has the value

$$\frac{2P}{\pi ab} \; \frac{2k_{44}b^2 + (k_{55} + 2k_{13})\,a^2}{3k_{44}b^2 + k_{55}a^2}$$

Deduce that for an isotropic rod with elliptic cross-section the maximum shear is

$$\frac{2P}{\pi ab} \cdot \frac{\dfrac{1}{1+\eta} + \dfrac{2b^2}{a^2}}{1 + \dfrac{3b^2}{a^2}} \,.$$

12. Obtain the stress equilibrium equations in cylindrical coordinates, 7.5 (10).

13. Obtain the strain coefficients in cylindrical coordinates, 7.5 (11).

14. Express the equations of compatibility of strain in cylindrical coordinates, and hence deduce the identity 7.6 (3).

15. Verify that the strain coefficients derived from the expressions 7.52 (23)—(25) for the displacements are in fact independent of R.

16. Prove that if there is a body-field derived from a potential V, the right-hand side of the equation 7.6 (5) must be increased by

$$-\,(l_{12} + l_{22})\,r\,\frac{\partial^2 V}{\partial r^2} + (l_{16} + l_{26})\,\frac{\partial^2 V}{\partial r\,\partial\theta} - (l_{11} + l_{12})\,\frac{1}{r}\,\frac{\partial^2 V}{\partial\theta^2}$$
$$+\,(l_{11} - 2l_{22} + l_{12})\,\frac{\partial V}{\partial r} + (l_{16} + l_{26})\,\frac{\partial V}{\partial\theta}$$

and that the right-hand side of the equation 7.6 (6) must be increased by

$$(l_{14} + l_{24})\,r\left(\frac{\partial V}{\partial r} - \frac{V}{r}\right) - (l_{15} + l_{25})\,r\,\frac{\partial V}{\partial\theta}\,.$$

17. Determine the displacements in the case of the tube of **7.7** subjected to internal and external pressure.

18. In the notation of **7.71** show that if the antiplane is one of elastic symmetry, then $h_c = h_{-c} = 0$, and prove that if $C_2 = \tau = 0$, the cross-sections are not warped.

19. For a tube in a state of generalized plane deformation (**7.72**) prove that the cross-sections will not warp if they are planes of elastic symmetry.

Prove further that if every axial plane is one of elastic symmetry, the radii will not warp.

20. In the case of the tube of **7.72** in a state of generalized plane deformation, if the cross-sections are planes of elastic symmetry and if

$$l_{11} l_{44} - l_{14}^2 = l_{22} l_{44} - l_{24}^2 ,$$

prove that the stresses are the same as for a tube of isotropic material.

21. In Lamé's problem, **7.72**, show that $\widehat{zz}$ is given by

$$- k_{33} \widehat{zz} = \frac{p_b b^{c+1} - p_a a^{c+1}}{a^{2c} - b^{2c}} (k_{13} + c k_{23} - h_c k_{34}) r^{c-1}$$
$$+ \frac{p_a b^{c-1} - p_b a^{c-1}}{a^{2c} - b^{2c}} d_1 (k_{13} - c k_{23} - h_{-c} k_{34}) r^{-c-1} .$$

22. In Lamé's problem, **7.72**, show that there is an axial force P where

$$- \frac{P k_{33} (a^{2c} - b^{2c})}{2\pi} = (p_a a^{c+1} - p_b b^{c+1}) (a^{c+1} - b^{c+1}) \frac{k_{13} + c k_{23} - h_c k_{34}}{1 + c}$$
$$- (p_a b^{c-1} - p_b a^{c-1}) (a^{c-1} - b^{c-1}) a^2 b^2 \frac{k_{13} - c k_{23} - h_{-c} k_{34}}{1 - c} .$$

23. In Lamé's problem, **7.72**, show that there is a twisting moment M given by

$$\frac{M (a^{2c} - b^{2c})}{2\pi} = - (p_a a^{c+1} - p_b b^{c+1}) (a^{c+2} - b^{c+2}) \frac{h_c}{2 + c}$$
$$+ (p_a b^{c-1} - p_b a^{c-1}) (a^{c-2} - b^{c-2}) \frac{h_{-c}}{2 - c} .$$

24. In Lamé's problem, **7.72**, let $t = (a - b)/a$ be the thickness ratio. Prove that the hoop stress at the inner surface is

$$\widehat{\theta\theta} = c p_b \frac{1 + (1 - t)^{2c}}{1 - (1 - t)^{2c}} - c p_a \frac{2 (1 - t)^{c-1}}{1 - (1 - t)^{2c}} .$$

Find also the hoop stress at the outer surface.

25. Obtain the stresses $\widehat{rr}$, $\widehat{\theta\theta}$ in Lamé's problem when the material is isotropic by inserting the appropriate values of the elastic constants in the corresponding stresses given in **7.72**.

References

CASTIGLIANO, A.: Nuova teoria intorno all'equilibrio dei sistemi elastici. Atti accad. sci. Torino **10**, 380—423 (1875).

DEUTSCH, E.: Sur le problème de la torsion de certains cylindres élastiques isotropes. C. R. Acad. Sci. (Paris) **251**, 2281—2283 (1960).

FILON, L. N. G.: (1) On the approximate solution for the bending of a beam of rectangular cross-section. Phil. Trans. (**A**) **201**, 63—155 (1903); (2) On antiplane stress in an elastic solid. Proc. Roy. Soc. (**A**) **160**, 137—154 (1937).

HILBERT, D.: Über eine Anwendung der Integralgleichungen auf ein Problem der Funktionentheorie. Verhandlungen des III. internationalen Mathematiker-Kongresses, Heidelberg 1904.

HOOKE, ROBERT: Lectures de Potentia Restitutiva, or of Spring Explaining the Power of Springing Bodies. London 1678. Facsimile reproduction in R. T. Gunther: Early Science in Oxford **8**, 331—356 (1931).

LAMÉ, G.: Leçons sur la théorie mathématique de l'élasticité des corps solides Paris (1852).

LEIBENZON, L. S.: Variational methods of solution of problems in the theory of elasticity (in Russian). Moscow-Leningrad 1943.

LEKHNITSKII, S. G.: Teoriya uprugosti anizotropnogo tela. Moscow-Leningrad 1950.

LOVE, A. E. H.: Mathematical theory of elasticity, 4th edition. Cambridge 1934.

MCCONNELL, A. J.: Absolute differential calculus. London 1931.

MILNE-THOMSON, L. M.: (1) Consistency equations for the stresses in isotropic elastic and plastic materials. J. Lond. Math. Soc. **17**, 115—128 (1942); (2) Cálculo tensorial por métodos directos. Rev. Mat. Hispano-Americana, 4th series **10**, 1—27 (1950); (3) Deformaciones finitas y elasticidad. Rev. Mat. Hispano-Americana, 4th series **12**, 1—27 (1952); (4) The Calculus of finite differences. London: Macmillan 1960; (5) Theoretical hydrodynamics, 4th ed. London: Macmillan 1960; (6) Plane elastic systems. Berlin-Göttingen-Heidelberg: Springer 1960; (7) Bounds for the torsional rigidity of isotropic beams. Proc. Camb. Phil. Soc. **58**, 216—219 (1962).

MURNAGHAN, F. D.: Finite deformations of an elastic solid. Amer. J. Math. **59**, 235—260 (1937).

MUSKHELISHVILI, N. I.: Some basic problems of the mathematical theory of elasticity (translated into English from the Russian). Groningen (Holland): Noordhoff 1953.

SOKOLNIKOFF, I. S.: Mathematical theory of elasticity, 2nd edition. London-New York-Toronto: McGraw-Hill 1956.

SOLOMON, L.: Some remarks on Saint-Venant's problem (to appear).

STEVENSON, A. C.: Flexure with shear and associated torsion in prisms of uniaxial and asymmetric cross-sections. Phil. Trans. Roy. Soc. (**A**) **237**, 161—229 (1938).

TEODORESCU, P. P.: Probleme plane in Teoria elasticitâtii, Vol. 1; Bucharest, 1961.

TODHUNTER, I., and K. PEARSON: A history of the theory of elasticity and of the strength of materials. Cambridge 1886—1893.

TRUESDELL, C.: The mechanical foundations of elasticity and fluid dynamics. J. Rat. Mechan. Anal. **1**, 125—300 (1952); **2**, 593—616 (1953).

VOLTERRA, V.: Sur l'équilibre des corps élastiques multiplement connexes. Ann. École norm. (3) **24**, 401—517 (1907).

WEINGARTEN, G.: Sulle superficie di discontinuità nella teoria della elasticità dei corpi solidi, Roma. Accad. Lincei Rend. (5) **10**, 57 (1901).

Index